Springer
Tokyo
Berlin
Heidelberg
New York
Barcelona
Hong Kong
London
Milan
Paris
New Delhi
Singapore

M. Kato (Ed.)

The Biology of Biodiversity

With 82 Figures, Including 6 in Color

Masahiro Kato
Department of Biological Sciences
Graduate School of Science
University of Tokyo
7-3-1 Hongo, Tokyo 113-8654, Japan
e-mail: sorang@biol.s.u-tokyo.ac.jp

ISBN 4-431-70262-8 Springer-Verlag Tokyo Berlin Heidelberg New York

Library of Congress Cataloging-in-Publication Data

The biology of biodiversity / M. Kato (ed.)
p. cm.
Includes bibliographical references and index.
ISBN 4-431-70262-8 (hardcover: alk. paper)
1. Biological diversity Congresses. 2. Biological diversity conservation Congresses.
I. Kato, Masahiro, 1946–
II. Symposium of the International Prize for Biology (14th: 1998: Hayama-machi, Japan)
QH541.15.B56B635 1999
333.95'11—dc21 99-40935

Printed on acid-free paper

Typesetting: from the authors' electronic files
Printing and Binding: Best-set Typesetter Ltd., Hong Kong
SPIN: 10728579

Preface

Biological diversity, or biodiversity, refers to the universal attribute of all living organisms that each individual being is unique — that is, no two organisms are identical. The biology of biodiversity must include all the aspects of evolutionary and ecological sciences analyzing the origin, changes, and maintenance of the diversity of living organisms. Today biodiversity, which benefits human life in various ways, is threatened by the expansion of human activities. Biological research in biodiversity contributes not only to understanding biodiversity itself but also to its conservation and utilization.

The Biology of Biodiversity was the specialty area of the 1998 International Prize for Biology. The International Prize for Biology was established in 1985 in commemoration of the sixty-year reign of the Emperor Showa and his longtime devotion to biological research. The 1998 Prize was awarded to Professor Otto Thomas Solbrig, Harvard University, one of the authors of this book.

In conjunction with the awarding of the International Prize for Biology, the 14th International Symposium with the theme of The Biology of Biodiversity was held in Hayama on the 9th and 10th of December 1998, with financial support by an international symposium grant from the Ministry of Education, Science, Sports and Culture of Japan. The invited speakers were chosen so as to cover four basic aspects of biodiversity: species diversity and phylogeny, ecological biodiversity, development and evolution, and genetic diversity of living organisms including human beings.

This book is one of the outcomes of the symposium, presenting a broad review of current biological research on biodiversity. It synthesizes modern approaches, the current status of research, and future perspectives of biodiversity science for a better understanding of the origin, changes, and maintenance of biodiversity. The book's 20 chapters in four parts focus on recent multidisciplinary approaches to major problems in biodiversity, with an emphasis on the increasing use of molecular approaches to the problems of speciation, phylogeny, developmental morphology, and genetic biodiversity. The first part, Species Diversity and Phylogeny, reviews molecular phylogenetic and genetic approaches, in order to better understand the origin of reproductive isolation, phylogeography and conservation, phylogenetic relationships, species taxonomy, and adaptive radiation. The second part, Ecological Biodiversity, discusses the significance of biodiversity science, ecosystem processes, relationships between reproduction and landscape fragmentation, population persistence and community diversity, biodiversity maintenance, and ecological factors of biodiversity enhancement. Included here is Professor Solbrig's paper on theory and practice in the science of biodiversity. Development and Evolution, the book's third part, reviews the current explosion of evolutionary and molecular developmental research on the body-plan diversity of animals and plants owing to the evolution of transcription factor-encoding genes. The final part, Genetic Biodiversity, looks at recent progress in studies on the genetic diversity and evolution of advanced animals.

The symposium was organized by Dr. N. Takahata, the Graduate University for Advanced Studies; Dr. N. Satoh, Kyoto University; Dr. Y. Iwasa, Kyushu University; Dr. R. Ueshima, the University of Tokyo; and myself. I am indebted to the organizers and participants for the success of the symposium, without which this book would not have appeared. I also gratefully acknowledge Dr. K. Endo of the University of Tokyo, and Dr. N. Satoh and Dr. R. Ueshima for reviewing some of the papers and providing useful suggestions. Publication of this book was supported by a 1999 Grant-in-Aid (No. 206002) for Publication of Scientific Research Results from the Japan Society for the Promotion of Science.

M. Kato

Contents

Contributors

Shin-ichiro Aiba	Faculty of Science, Kagoshima University, Kagoshima 890-0065, Japan
Michael Akam	University Museum of Zoology, Department of Zoology, Downing Street, Cambridge, CB2 3 EJ, England
Peter S. Ashton	Arnold Arboretum and Organismic and Evolutionary Biology, 22 Divinity Avenue, Harvard University, Cambridge, MA 02138, USA
Stephane Boissinot	Laboratory of Molecular and Cellular Biology, NIDDK, NIH Bethesda, MD 20892-0830, USA
Dedy Darnaedi	Botanical Gardens, Bogor, Jl. Ir. H. Juanda 13, Bogor 16122, Indonesia
Elise Eller	Department of Anthropology, University of Utah, Salt Lake City, UT 84112, USA
Thomas J. Givnish	Department of Botany, University of Wisconsin, Madison, WI 53706, USA
Stefan Gleissberg	Institute of Systematic Botany, University of Mainz, 55099 Mainz, Germany
Yoshito Harada	Department of Zoology, Graduate School of Science, Kyoto University, Sakyo-ku, Kyoto 606-8502, Japan
Henry C. Harpending	Department of Anthropology, University of Utah, Salt Lake City, UT 84112, USA
Susan Harrison	Department of Environmental Science and Policy, University of California, Davis, One Shields Avenue, Davis, CA 95616, USA
Mitsuyasu Hasebe	National Institute for Basic Biology, 38 Nishigonaka, Myodaiji-cho, Okazaki 444-8585, Japan
David Hewett-Emmett	Human Genetics Center, University of Texas, 6901 Bertner Ave., P.O. Box 20334, Houston, TX 77225, USA

Kohji Hotta	Department of Zoology, Graduate School of Science, Kyoto University, Sakyo-ku, Kyoto 606-8502, Japan
Motomi Ito	Department of Biology, Faculty of Science, Chiba University, 1-33 Yayoicho, Inage-ku, Chiba 263-0022, Japan
Hisako Iwasaki	Department of Botany, Graduate School of Science, Kyoto University, Kitashirakawa-Oiwake-cho, Sakyo-ku, Kyoto 606-8502, Japan
Kunio Iwatsuki	Faculty of Science, Rikkyo University, 3-34-1 Nishi-Ikebukuro, Toshima-ku, Tokyo 171-0021, Japan
Judy Jernstedt	Department of Agronomy and Range Science, University of California, Davis, CA 95616, USA
Minsung Kim	Section of Plant Biology, University of California, Davis, CA 95616, USA
John Klicka	J. F. Bell Museum of Natural History, University of Minnesota, St. Paul, Minnesota 55108, USA
Takashi Kohyama	Graduate School of Environmental Earth Science, Hokkaido University, Sapporo, Hokkaido 060-0810, Japan; Harvard University Herbaria, Cambridge, MA 02138, USA
Yoshinori Kumazawa	Department of Earth and Planetary Sciences, Nagoya University, Furo-cho, Chikusa-ku, Nagoya 464-8602, Japan
J.V. LaFrankie	Center for Tropical Forest Science, Nanyang Technical University, National Institute of Education, 469 Bukit Timah Road, Singapore 1025, Republic of Singapore
John H. Lawton	NERC Centre for Population Biology, Imperial College, Silwood Park, Ascot, SL5 7PY, UK
Wen-Hsiung Li	Human Genetics Center, University of Texas, 6901 Bertner Ave., P.O. Box 20334, Houston, TX 77225, USA; Current address: Department of Ecology and Evolution, University of Chicago, 1101 East 57th Street, Chicago, IL 60637, USA
Craig Moritz	Department of Zoology and the Cooperative Research Centre for Tropical Rainforest Ecology and Management, The University of Queensland, Qld, 4072, Australia

Noriaki Murakami	Department of Botany, Graduate School of Science, Kyoto University, Kitashirakawa-Oiwake-cho, Sakyo-ku, Kyoto 606-8502, Japan
Mutsumi Nishida	Department of Marine Bioscience, Fukui Prefectural University, 1-1 Gakuen-machi, Obama 917-0003, Japan; Current address: Ocean Research Institute, University of Tokyo, 1-15-1 Minamidai, Nakano-ku, Tokyo 164-8639, Japan
Keiichi Omoto	International Research Center for Japanese Studies, 3-2 Goryo Oeyama-cho, Nishikyo-ku, Kyoto 610-1192, Japan
Gouki Satoh	Department of Zoology, Graduate School of Science, Kyoto University, Sakyo-ku, Kyoto 606-8502, Japan
Nori Satoh	Department of Zoology, Graduate School of Science, Kyoto University, Sakyo-ku, Kyoto 606-8502, Japan
Kyoichi Sawamura	Department of Applied Biology, Kyoto Institute of Technology, Matsugasaki, Sakyo-ku, Kyoto 606-8585, Japan
Tatsuyuki Seino	Graduate School of Environmental Earth Science, Hokkaido University, Sapporo, Hokkaido 060-0810, Japan
Song-Kun Shyue	Institute of Biomedical Sciences, Academia Sinica, Taipei 11529, Taiwan
Neelima Sinha	Section of Plant Biology, University of California, Davis, CA 95616, USA
Otto T. Solbrig	Department of Organismic and Evolutionary Biology, Harvard University, 22 Divinity Ave., Cambridge, MA 02138, USA
Douglas E. Soltis	Department of Botany, Washington State University, Pullman, WA 99164-4238, USA
Pamela S. Soltis	Department of Botany, Washington State University, Pullman, WA 99164-4238, USA
Eizi Suzuki	Faculty of Science, Kagoshima University, Kagoshima 890-0065, Japan
Kuni Tagawa	Department of Zoology, Graduate School of Science, Kyoto University, Sakyo-ku, Kyoto 606-8502, Japan
Shunsuke Taguchi	Department of Zoology, Graduate School of Science, Kyoto University, Sakyo-ku, Kyoto 606-8502, Japan

Hiroki Takahashi	Department of Zoology, Graduate School of Science, Kyoto University, Sakyo-ku, Kyoto 606-8502, Japan
Ying Tan	Human Genetics Center, University of Texas, 6901 Bertner Ave., P.O. Box 20334, Houston, TX 77225, USA; Current address: Department of Ecology and Evolution, University of Chicago, 1101 East 57th Street, Chicago, IL 60637, USA
Izumi Washitani	Institute of Biological Sciences, University of Tsukuba, Tsukuba 305-8572, Japan
Motoomi Yamaguchi	Department of Marine Bioscience, Fukui Prefectural University, 1-1 Gakuen-machi, Obama 917-0003, Japan; Current address: Ocean Research Institute, University of Tokyo, 1-15-1 Minamidai, Nakano-ku, Tokyo 164-8639, Japan
Yoko Yatabe	Department of Botany, Graduate School of Science, Kyoto University, Kitashirakawa-Oiwake-cho, Sakyo-ku, Kyoto 606-8502, Japan
Robert M. Zink	J. F. Bell Museum of Natural History, University of Minnesota, St. Paul, Minnesota 55108, USA

Part 1
Species Diversity and Phylogeny

1
The Origin of Reproductive Isolation: Biological Mechanisms of Genetic Incompatibility

KYOICHI SAWAMURA

Drosophila Genetic Resource Center, Kyoto Institute of Technology, Matsugasaki, Sakyo-ku, Kyoto 606-8585, Japan

Abstract

The origin of species has been the mystery of mysteries since Charles Darwin's era. According to the biological species concept advocated by the founders of the Modern Synthesis, the question 'how new species evolve' can be substituted by a more answerable question 'how reproductive isolation is established between populations'. Mechanisms preventing gene exchange between species are various, and most of them are thought to be the result, not the cause, of genetic diversity. Some disturb species recognition systems (prezygotic isolation), and others terminate generations of mixed genomes (postzygotic isolation). Examples of the latter whose biological mechanisms have recently been documented well (tumorigenesis, sexual reversion, irregular dosage compensation, nucleolar dominance, homeotic transformation, spermatogenic defects, mitotic defects, and maternal/zygotic transition failure in interspecific hybrids) are reviewed here. Gene interactions between loci from different species play important roles. As has been shown in developmental biology, transcriptional regulation may be a key factor. Genetic incompatibility in hybrids may be the manifestation of improper transcriptional regulation resulting from the coevolution of DNA-binding proteins and their binding sites.

Key words. biological species concept, dosage compensation, *Drosophila*, genetic incompatibility, homeotic transformation, hybrid inviability, hybrid sterility, maternal/zygotic transition, mitosis, nucleolar dominance, postmating reproductive isolation, sex determination, speciation, spermatogenesis, tumorigenesis

1 Introduction

Speciation has two aspects, horizontal and vertical. The nodes of a phylogenetic tree represent the former, which increases the number of species, while internodes represent the latter, which elaborate the form of species. Biodiversity has been shaped as the sum of speciation, and it is important to shed light on both these aspects. From the viewpoint of the biological species concept (Dobzhansky 1937; Mayr 1942), it is a fundamental question as to how reproductive isolation is acquired in diverged populations. Although all mechanisms preventing genetic exchange between populations are sometimes regarded as reproductive isolation, they are not homogeneous. Some mechanisms perturb inter-population matings (prezygotic isolation), and others produce abnormal hybrids of mixed genomes (postzygotic isolation). Although we cannot rule out the possibility that some mechanisms developed in order to reduce the cost of producing unadapted hybrids (reinforcement), most of the mechanisms are supposed to be the result, not the cause, of genetic diversity between populations. Regardless of the mode of speciation, whether sympatric or allopatric, genetic distance between populations increases, resulting in reproductive isolation as byproducts (Zouros 1989; Coyne and Orr 1989, 1998; Coyne 1992; Wu and Davis 1993; Wu and Palopoli 1994; Forejt 1996; Wu et al. 1996; Hutter 1997; Laurie 1997; Orr 1997).

In the present review, I will concentrate on mechanisms of postzygotic isolation which is the consequence of genetic incompatibility between species. This is not because postzygotic isolation is more important than prezygotic isolation in speciation, but merely because the former has been better elucidated in this century. Because each species has its own history and may have accumulated different modifications of genetic systems of development, the coexistence of genomes in interspecific hybrids may result in genetic incompatibility. The only force to conserve the conspecific genetic systems is the shared history bounded by reproduction. Any kinds of deviation from perfect development of individuals with mixed genomes should be regarded as incompatibility, although only inviability and sterility in F_1 hybrids and in descended generations (hybrid breakdown) have been seriously treated so far. Even defects seen in interspecific gene transformation, somatic cell hybrids, and organ transplantation can be included in this category.

2 Case Studies

2.1 Tumorigenesis

An example of genetic incompatibility whose molecular mechanism is well documented is melanoma formation in the poeciliid fish hybrids. This phenomenon has been well-known since classical descriptions of the hybrids [for reviews see Schwab (1987) and Schartl (1995)]. The platifish (*Xiphophorus maculatus*) and the swordtail (*X. helleri*) have a uniform grey body colouration due to small black pigment cells (micromelanophores). Some geographical races of platifish, however, exhibit spot patterns of melanophores that are much larger than normal (macromelanophores). These patterns are determined by a sex-linked locus which is closely linked to an oncogene, *Tumor* (*Tu*) (Weis and Schartl 1998). *Tu*-bearing platifish, by nature, do not suffer from melanoma, because they are assumed to have an autosomal dominant suppressor gene, *Regulator* (*R*). The platifish with spots are genotypically described as *Tu/Tu*; *R/R*. On the contrary, the swordtails have neither of these genes: in this respect, described as -/-; -/-. F_1 hybrids are *Tu*/-; R/-, and show enhanced expression of the pigmentation pattern (but still benign melanoma). When the F_1 hybrids are backcrossed to swordtails, four genotypic classes are produced. A quarter of the BC_1 hybrids are *Tu*/-; -/-, and suffer from malignant melanoma.

The *Tu* gene was cloned by Wittbrodt et al. (1989) and was found to be a novel putative receptor tyrosine kinase, now named *Xiphophorus melanoma receptor kinase* (*Xmrk*). The sequence analysis of the oncogene (*ONC-Xmrk*) and the proto-oncogene (*INV-Xmrk*) (Adam et al. 1993) indicated that *Tu* was an *Xmrk* duplication produced by a non-homologous recombination. The second copy has a 5' region derived from an anonymous locus, designated *D*, rather than the normal 5' end of the original copy. The consequence of this event is that *INV-Xmrk* and *ONC-Xmrk* are subject to different transcriptional regulation. Specific overexpression of *ONC-Xmrk* seems to be responsible for melanoma formation. It is assumed that the *R*-locus-dependent transcriptional control of the oncogene promoter allows high levels of expression only in pigment cells of certain hybrid genotypes. Then, what is the product of the *R* locus? Is it a transcription factor? The gene has been located to linkage group V and a candidate, *CDKN2-like*, is already cloned (Nairn et al. 1996). Anyway, the control of melanomas by the *R* locus can thus be viewed as “an accidental side effect of the regulation *R* exerts on *D*” (Friend 1993).

2.2 Sexual Reversion

Sexual reversion of hybrids is well-known in several crosses. A classical example is inter-racial hybrids of the gypsy moth (*Lymantria dispar*) [for a review see Goldschmidt (1940)]. This phenomenon was attributed to the geographical variation of the “strength” of sex determinants. In crosses where the strength of the sex determinants in the parental races differ, intersexual hybrids are produced. When the difference is slight, intersexes are so much like normal females or males that they are fertile. Where the difference is greater, however, the intersexes are sterile. When the difference is much greater, the intersexes transform into individuals of the sex opposite to the chromosomal sex. Similar cases are known in several lepidopteran interspecific crosses [cited in Haldane (1922)]. Although more detailed genetic bases have not been elucidated, it may be explicable by more elaborate terms of sex determination. In *Drosophila*, for example, sex is determined by the X-chromosome/autosome (X/A) ratio. The X chromosome has numerator genes (e.g., *sisterless-a*, *sisterless-b*) and autosomes have putative denominator genes. It is well-known that a maternal product (a transcription factor) of a second-linked gene, *daughterless*, calculates the X/A ratio and regulates the ON/OFF state of a key gene of sex determination, *Sex-lethal* (Cline 1993). If new genes are recruited to the numerator/denominator system or each numerator/denominator element evolves rapidly in independent populations, interspecific hybrids will show an imbalance of the system resulting in sex reversion. Because sex determination is related to dosage compensation of the X chromosome in some animals (see next section), the failure of numerator/denominator count may also cause sex-specific inviability of interspecific hybrids. Intersexuality accompanied with low viability of flies produced from mothers of certain genotypes of the *D. repleta*/*D. neorepleta* combination (Sturtevant 1946) may be an example of this.

Developmental bases of sexual reversion in mouse hybrids is well documented. When the Y chromosome is introduced from wild mice of certain *Mus domesticus* populations (e.g., the *poschiavinus* chromosomal race) into a particular strain of laboratory mouse, the chromosomal (XY) males develop as females or as hermaphrodites (Eicher et al. 1982; Nagamine et al. 1987). [Note that the laboratory mouse strain (a genetic mixture of *M. domesticus* and *M. musculus*) carries a Y chromosome of *M. musculus* origin (Bishop et al. 1985).] A *testis-determining Y-linked* gene (*Tdy* = *Sry*) is believed to be involved in this phenomenon: the *domesticus* (or *poschiavinus*) allele of the *Sry* gene does not function normally in the genetic background, homozygous for the laboratory mouse allele of *testis-determining autosomal* genes (*tda1*, *tda2*, and *tda3*) (Eicher and Washburn 1983; Coward et al. 1994; Eicher et al. 1995, 1996). It has been shown that the sex-reversed fetuses have a later onset of testicular development than the control fetuses (Taketo-Hosotani et al. 1989; Taketo et al. 1991; Palmer and Burgoyne 1991). The Sry product, a transcription factor, is normally produced but its inactivation is delayed in the gonads of sex-reversed fetuses. This delay seems to cause improper expression of genes downstream of it, which results in the sex reversion (Lee and Taketo 1994). Recent finding of high interspecific variability of the *Sry* gene sequences surprised molecular biologists (Whitfield et al. 1993; Tucker and Lundrigan 1993). Mammalian geneticists tend to assume that this gene may play a key role in causing sex specific defects in hybrids (Short 1997; Graves and O'Neill 1997). The incompatibility between the *Sry* gene and some X-linked or autosomal genes may explain the paucity and/or sterility of heterogametic (XY) sex in interspecific mammalian hybrids, the so-called Haldane's rule (Haldane1922).

2.3 Irregular Dosage Compensation

The X chromosome inactivation in female mammals, which compensates for the dose difference of X-linked genes between sexes, is cell-autonomous: whether maternally derived X is inactivated or paternally derived X is inactivated is usually determined at random (Lyon 1993). But this is not the case in some interspecific hybrids: e.g., paternal X (of donkey origin) is preferentially inactivated in a female mule (Hamerton et al. 1969). Such an irregularity is also seen in gray vole (*Microtus*) hybrids: the X chromosome with a heterochromatin block is preferentially inactivated in this case (Zakian et al. 1987, 1991). Improper dosage compensation, if any, is potentially responsible for abnormal development of hybrids, e.g., hybrid inviability and sterility. This should not be restricted to the X-linked genes. For example, if the maternal allele is active in one species but the paternal allele is active in the other at the early stages of embryogenesis (genomic imprinting) (Cattanach and Beechey 1990), hybrids may show improper expression of the loci resulting in hybrid inferiority.

Dosage compensation in insects is somewhat different from that in mammals, but it can also be involved in postzygotic reproductive isolation. The equalization of X-linked gene expression in male and female *Drosophila* is brought about by doubling the activity of the single X loci in males (Lucchesi 1998). If species-specific systems (e.g., cis-acting sequence differences) operate in the dosage compensation, hybrids may show improper gene expression resulting in abnormal development. Although proper dosage compensation has been shown in many cases of interspecific *Drosophila* (Dobzhansky 1957; Lakhotia et al. 1981; Mutsuddi et al. 1984; see also Orr 1989), there is at least one case of improper dosage compensation in *D. azteca*/*D. athabasca* hybrids. Males carrying the *D. azteca* X chromosome are larger ('hybrid giant males'), while those carrying the *D. athabasca* X are smaller and have low viability ('hybrid dwarf males'), if compared with their parents and sisters (Sturtevant and Dobzhansky 1936). The replication of the hemizygous X chromosome is asynchronized from autosomes in the hybrid giant males, and as a result their polyten chromosomes undergo one more endoreplication (Meer 1976, 1980).

2.4 Nucleolar Dominance

Nucleolar dominance is a phenomenon usually seen in interspecific hybrids in which an allele of nucleolar organizer (rRNA coding gene) derived from one of the parental species is transcriptionally inactivated. This is observed in various crosses of plants and animals [for reviews see Wallace and Langridge (1971), Rieger et al. (1979), Reeder (1984)]. Its molecular basis is well documented in amphibian (*Xenopus*) hybrids. The *X. laevis*/*X. borealis* hybrid embryos express only ribosomal genes of the former species (Honjo and Reeder 1973; Cassidy and Blacklar 1974). This is the consequence of competition between enhancers of ribosomal genes from two species (Reeder and Roan 1984): *X. laevis* has more copies of an enhancer element for the ribosomal gene promoter in the spacer region (La Volpe et al. 1983). It is, however, not always true that nucleolar dominance is a manifestation of a simple "allelic repression". It is assumed in mouse-human somatic hybrids that nucleolar dominance is due to the loss or inactivation of the gene for a specificity factor required to recognize the species-specific ribosomal gene promoter, since extracts from cell lines in which some of the human chromosomes are lost cannot initiate transcription of human ribosomal genes (Onishi et al. 1984; Miesfeld et al.

1984). In fact, no obvious shared consensus sequence has been recognized in the region of transcription initiation of ribosomal gene from different species of eukaryotes. If the change in a single promotor sequence of this transcription unit is complemented by a similar modification of polymerase I (or the transcription factor) DNA binding site, the factor from different species won't be able to induce transcription of the ribosomal gene. This has been shown in in vitro transcription systems by Grummt et al. (1982).

Another model system of nucleolar dominance is *Drosophila melanogaster/D. simulans* hybrids (Durica and Krider 1977). Regulatory elements of nucleolar dominance (ND) have been identified outside the nucleolar organizer region (NOR = *bobbed* gene) itself. By using chromosome aberrations Durica and Krider (1978) could map the NDs very precisely. Several elements which regulate ribosomal gene activity transcriptionally or replicationally have been characterized in *D. melanogaster* (Procunier and Tartof 1978; Goodrich-Young and Krider 1989), and it is intriguing to know whether any element(s) represent ND. Because low production of rRNA causes the bobbed phenotype or inviability in severe cases, nucleolar dominance-mediated postzygotic isolation is plausible. In the cross between *D. melanogaster* females and *D. simulans* males, hybrid males are inviable at the late larval stage (Sturtevant 1920), and can be rescued by a mutation of *D. simulans*, *Lethal hybrid rescue* (*Lhr*) (Watanabe 1979). If a *D. melanogaster* strain carrying an X-linked defect of ribosomal gene regulation is employed, the hybrid males cannot be rescued by *Lhr* because the Y chromosome of *D. simulans* origin lacks rDNA (Granadino et al. 1996). But it is not clear at the moment whether nucleolar dominance is directly involved in the hybrid male inviability. Apparent examples of nucleolar dominance-mediated hybrid inferiority are *D. mulleri/D. arizonae* hybrids (Bicudo and Richardson 1977, 1978), *D. hydei/D. neohydei* hybrids (Schäfer 1979), and inter-subspecific European meadow grasshopper (*Chorthippus parallelus*) hybrids (Bella et al. 1990; Virdee and Hewitt 1992).

2.5 Homeotic Transformation

Homeotic transformation in interspecific hybrids is somewhat anecdotal,e.g., flour beetle (*Tribolium*) hybrids (Wade and Johnson 1994; Wade et al. 1994) and *Drosophila virilis/D. lummei* hybrids (Heikkinen 1991). A well-documented case is the character of extra sex combs seen in hybrids between *D. madeirensis* females and *D. subobscura* males. Males of the pure species have sex combs only on the first pair of legs, but the hybrid males have extra sex combs on the second and the third pairs of legs. The character is similar to several homeotic mutations of *D. melanogaster*, e. g., *extra sex combs* (Slifer 1942). This character is caused by multi-genes, not by a single major homeotic gene (Papaceit et al. 1991; Khadem and Krimbas 1991, 1993, 1997). The genes might epistatically interact with some homeotic genes. It is well-known in a homeotic gene, *Ultrabithorax*, that such "modifiers" exist all over the genome of *D. melanogaster* (Gibson and Hogness 1996; Gibson and van Helden 1997).

2.6 Spermatogenic Defects

Evolutionary geneticists now have a consensus that spermatogenesis is one of the most sensitive processes of development in interspecific hybrids (Wu and Davis 1993; Wu and Palopoli 1994; Carvajal et al. 1996; Forejt 1996; Hollocher and Wu 1996; True et al. 1996; Wu et al. 1996; Laurie 1997; Presgraves and Orr 1998; Coyne et al. 1998; K. Sawamura, A. W. Davis and C.-I. Wu, submitted), and publication of papers dealing with this subject has burgeoned this decade. Here, I shall cite a case where the molecular mechanism is characterized well. Recently, a gene named *Odysseus* which causes male sterility when introduced from *Drosophila mauritiana* into *D. simulans* was cloned (Ting et al. 1998; see also Perez et al. 1993; Perez and Wu 1995). This is the first sequenced gene of postzygotic reproductive isolation *sensu stricto*. Surprisingly, the gene encodes a paired-like homeodomain protein which is highly conserved among nematodes, mice, and insects. Sequence analysis indicated that the homeobox domain of the gene evolved rapidly between closely related species of *Drosophila*. It may be, speculatively, that one of the redundant (duplicated) homeotic genes was recruited to the functional regulation of spermatogenesis in this species group. Such an accelerated evolution, presumably accompanied by an accessory function, in a particular species group of *Drosophila* is also known in another homeotic gene, *spalt* (Reuter et al. 1989). It has been assumed that genes involved in male sexual traits may evolve rapidly because of sexual selection (Wu and Davis 1993; Wu et al. 1996; Tsaur and Wu 1997; Civetta and Singh 1998a, b; Tsaur et al. 1998).

2.7 Mitotic Defects

Mitotic chromosome loss in the early development of interspecific hybrids, which sometimes results in inviability, is well-known since classical experiments. The extensive literature on this subject, which deals particularly with echinoderms and fishes, has been reviewed by Hertwig (1936) [for recent analyses see Bennet et al. 1976; Finch 1983; Fujiwara et al. 1997; references therein]. Endosymbiotic rickettsia, *Wolbachia*, are involved in some cases, e.g., in gynogenetic parasitic wasps (*Nasonia*) produced by interspecific fertilization (Breeuwer and Werren 1990). *Wolbachia* is well-known as an agent of cytoplasmic incompatibility responsible for intraspecific hybrid inviability in the crosses between uninfected females and infected males or between females and males infected by different strains (O'Neill and Karr 1990). *Nasonia* may be a special case in which haploid zygotes that result from the loss of the paternal genome can develop as males. A well-documented mitotic defect is the loss of the paternal dot chromosome and resulting eye wrinkling in the hybrids between *Drosophila virilis* females and *D. lummei* males (Orr 1990; Braverman et al. 1992) [but see Heikkinen (1991, 1992) and references therein]. Chromosome loss is also seen in mammalian somatic cell hybrids (Weiss and Green 1967). It is interesting that paternal chromosomes are preferentially lost in most of the cases. This may be caused by the incompatibility between maternally supplied proteins and paternal chromosomes (see also next section).

A mitotic defect is also seen in the larval lethal hybrid males from the crosses between *Drosophila melanogaster* females and males of its sibling species (*D. simulans, D. mauritiana,* or *D. sechellia*). The male larvae show a typical pathology of mitotic mutants and show a failure of chromatin condensation (Orr et al. 1997). Normal mitotic figures can be seen in hybrids which are rescued from the inviability by a mutation, *Hybrid male rescue* (Hutter and Ashburner 1990). The loss of the X chromosome of sibling species origin is frequently detected as mosaics in hybrid females (XX//XO), which may result from a milder mitotic defect than in the brothers. And, importantly, the XO clones (produced with a trick) are mostly seen in the abdomen not in the head nor the thorax. As Orr et al. (1997) speculate, maternally supplied products necessary for mitosis in the hybrids might be used up gradually and imaginal disc cells may be less tolerant than larval histoblast cells (the latter mainly forms abdominal cuticle).

2.8 Maternal/Zygotic Transition Failure

Proteins and mRNAs necessary for the very early stage of development are generally supplied from mothers. Interspecific hybrids have one set of genome derived from paternal species, so the maternally supplied transcription factors may not be able to regulate zygotic gene actions of the incompatible species. In such cases, hybrid embryos will die before the onset of zygotic transcriptions. In species groups of male heterogamety (i.e., XX females, XY males), an X chromosome of paternal origin exists in hybrid females but not in hybrid males. So, only hybrid females should suffer from the incompatibility, if the paternal X chromosome has a dominant incompatible gene(s). This could result in the exceptions to Haldane's rule (Sawamura et al. 1993a, c; Wu and Davis 1993; Wu et al. 1996; Sawamura 1996; Hutter 1997; Laurie 1997).

A well-known example is the embryonic inviability of hybrid females from the crosses between *Drosophila melanogaster* males and females of its sibling species (*D. simulans, D. mauritiana,* or *D. sechellia*) [for a review see Sawamura et al. (1993b); other examples listed in Sawamura (1996)]. The lethal embryos show a typical pathology of maternal/zygotic transition failure (K. Sawamura, C.-I Wu and T. L. Karr, unpublished observation). Two genetic components involved in the inviability have been detected as rescue mutations: one is a maternally acting gene located on the second chromosome of *D. simulans* (*maternal hybrid rescue*: *mhr*), and the other is a zygotic acting gene located on the X chromosome of *D. melanogaster* (*zygotic hybrid rescue*: *zhr*) (Sawamura et al. 1993a, c). Orr (1996) has described a dominant rescue gene(s) segregating in natural populations of *D. simulans,* which is potentially allelic to *mhr*. Its dominance may depend on the *D. melanogaster* strains employed, although further examination is necessary. Apparently, natural populations of *D. melanogaster* have a variation in the strength of the

X-linked factor involved (Sawamura et al. 1993b; K. Sawamura, unpublished observation).

The mutation, *zhr*, was found to be a deficiency of a specific region of the X heterochromatin, and has been precisely mapped by using deficiencies and duplications of the X heterochromatin (Sawamura and Yamamoto 1993). The viability/inviability of hybrids is determined by the absence/existence of the locus, so the incompatible component of *D. melanogaster* is neomorphic. The region where *zhr* was localized is rich in a 359 bp repetitive sequence of satellite DNA whose buoyant density is 1.688 g/cm^3 (Lohe et al. 1993). Further, it is assumed that the gene consists of a kind of repetitive sequence based on its quantitative effect (Sawamura et al. 1995; Sawamura and Yamamoto 1997). The 1.688 satellite DNA itself or some repetitive sequences embodied in it (e.g., transposable elements) seem to be the factor responsible for the embryonic hybrid inviability. If the sequences have some extent of homology to the recognition sites of a DNA binding protein [in fact, 1.688 satellite DNA-related sequences exist on several locations of the X euchromotin (di Bartolomeis et al. 1992)], they will titrate out the maternally-supplied binding factor(s) necessary to regulate some essential genes. Coevolution of the transcription factor and its binding sites including the titration region will cause this incompatibility between species. Besides transcriptional regulation, the DNA binding protein may play an important role in chromosome condensation and/or chromosome segregation. It is interesting in this context that some heterochromatic satellite DNAs bound by well-known DNA-binding proteins (e.g., GAGA factor, Proliferation Disruptor protein) are species-specific (Platero et al. 1998).

3 Conclusion

It is apparent that postzygotic reproductive isolation is the manifestation of interspecific genetic incompatibility. A gene that functions normally in a pure species may cause an anomaly in the background of different species. Most of the cases are not the consequences of allelic interaction (underdominance) but of gene interaction (epistasis): genetic incompatibility usually involves at least a pair of loci (Dobzhansky 1937; Muller 1940; Hutter et al. 1990; Sawamura et al. 1993c; Wu and Davis 1993; Wu and Palopoli 1994; Orr 1997; Coyne and Orr 1998). This is because populations cannot cross the adaptive valley if a single locus is responsible for the unadapted genotype (Fig. 1). Assume one locus, *A*, whose allelic status is indicated by subscripts (1 and 2). The missing link between A_1A_1 and A_2A_2 species should be A_1A_2, an inferior genotype. On the other hand, if two loci, *A* and *B*, are involved, $A_1A_1B_1B_1$ proto-species can evolve to $A_2A_2B_1B_1$ and $A_1A_1B_2B_2$ species through $A_1A_2B_1B_1$ and $A_1A_1B_1B_2$ genotypes, respectively, avoiding the inferior genotype equivalent to the interspecific hybrids, $A_1A_2B_1B_2$. Theoretically, any genetic interactions in pure species can cause genetic incompatibility in interspecific hybrids, and evolutionary geneticists haven't seriously cared about its biological mechanisms so far. Recent advances in developmental biology have shown that gene regu-

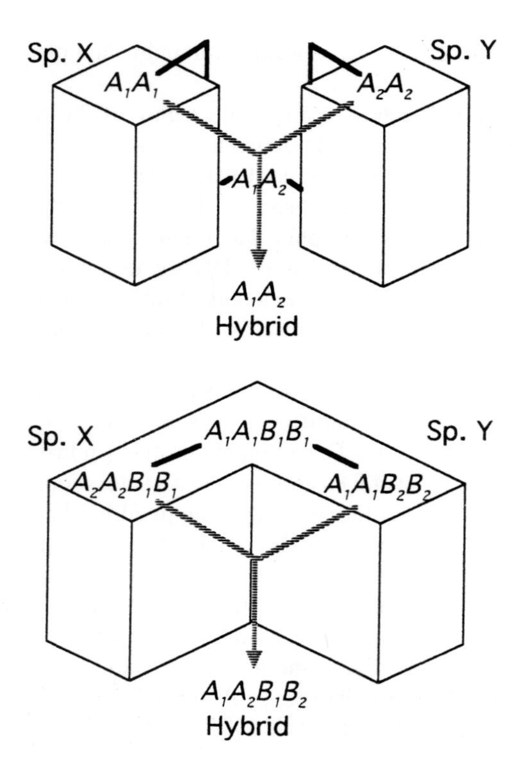

Fig. 1. Adaptive landscape of extant species (X and Y), the common ancestor (or missing link), and the hybrid. The fitness of each genotype is represented by the height. *Upper*: the allelic interaction (underdominance) model; one locus is involved in the incompatibility (A_1 <-> A_2). *Lower*: the gene interaction (epistasis) model; two loci are involved in the incompatibility (A_2 <-> B_2). In the former model, two species are isolated by the adaptive valley (genotype equivalent to the hybrid). In the latter model, two species are bridged by the common ancestor (genotype not equivalent to the hybrid). The latter model is more plausible

lation mediated by transcription factors plays a key role. Gene regulation may also be important in evolutionary biology, if such systems evolve rapidly (Wilson et al. 1977; Rose and Doolittle 1983; Dickinson 1988; Carroll 1995). It is predictable that a majority of genetic incompatibility may be the result of failure of gene regulation. In the context of transcriptional gene regulation, it is also predictable that developmental stages where transcription pattern is dramatically changing are sensitive in hybrids. Maternal/zygotic transition in the early embryogenesis and transcriptional silencing in post-meiotic spermatogenesis, both accompanied by dramatic change of chromosome structure, could be the two largest targets of this.

The evolution of cis-regulatory elements of gene expression has recently been characterized well in some sibling species of *Drosophila* and mice (Georgel et al. 1992; Kreitman and Ludwig 1996; Singh et al. 1998), and has been suggested to be the cause of species-specific tempo and mode of gene expression (Wang et al. 1996; Tamarina et al. 1997). The co-evolution of specificity of transcription factors and the sequences of its binding sites will provide a system causing genetic incompatibility in interspecific hybrids (Fig. 2). Comparative study of such elements will be a promising future research.

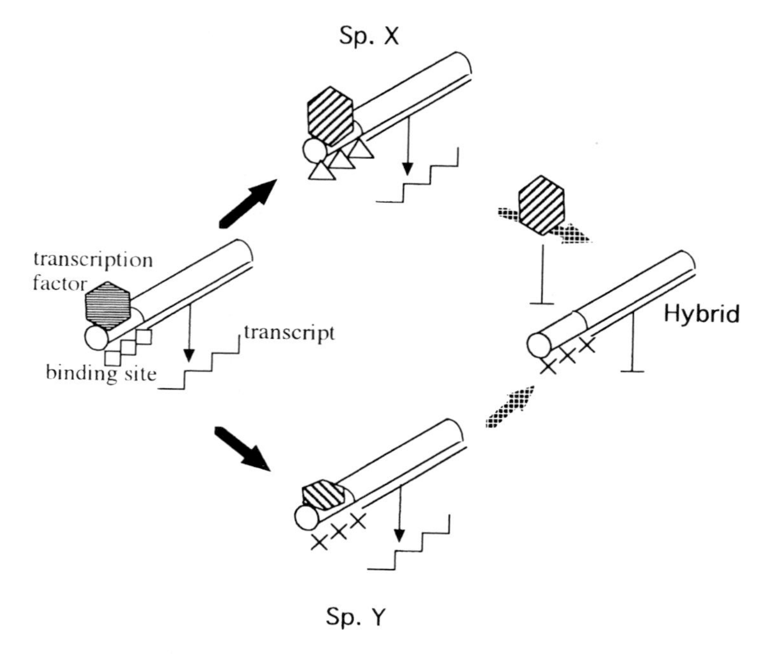

Fig. 2. Hybrid incompatibility caused by the interaction between a transcription factor and its binding sites. A transcription factor and its binding sites co-evolve independently in each isolated population. The transcription factor of species X may not be able to bind to the binding site of species Y, which results in the failure of transcription regulation in the hybrid

Acknowledgements

The author is currently a Research Fellow of the Japan Society for the Promotion of Science (JSPS). The present research was partly supported by a Grant-in-Aid for Scientific Research from the Ministry of Education, Science, Sports, and Culture of Japan.

References

Adam D, Dimitrijevic N, Schartl M (1993) Tumor suppression in *Xiphophorus* by an accidentally acquired promoter. Science 259:816-819

Bella JL, Hewitt GM, Gosálvez J (1990) Meiotic imbalance in laboratory-produced hybrid males of *Chorthippus parallelus parallelus* and *Chorthippus parallelus erythropus*. Genet Res 56:43-48

Bennet MD, Finch RA, Barclay IR (1976) The time rate and mechanism of chromosome elimination in *Hordeum* hybrids. Chromosoma 54:175-200

Bicudo HEMC, Richardson RH (1977) Gene regulation in *Drosophila mulleri, D. arizonensis*, and their hybrids: the nucleolar organizer. Proc Natl Acad Sci USA 74:3498-3502
Bicudo HEMC, Richardson RH (1978) Morphological and developmental studies of *Drosophila mulleri, D. arizonensis* and their hybrids. Biol Zentbl 97:195-203
Bishop CE, Boursot P, Baron B, Bonhomme F, Hatat D (1985) Most classical *Mus musculus* domesticus laboratory mouse strains carry a *Mus musculus musculus* Y chromosome. Nature 315:70-72
Braverman JM, Goñi B, Orr HA (1992) Loss of a paternal chromosome causes developmental anomalies among *Drosophila* hybrids. Heredity 69:416-422
Breeuwer JA, Werren JH (1990) Microorganisms associated with chromosome destruction and reproductive isolation between two insect species. Nature 346:558-560
Carroll SB (1995) Homeotic genes and the evolution of arthropods and chordates. Nature 376:479-485
Carvajal AR, Gandarela MR, Naveira HF (1996) A three-locus system of interspecific incompatibility underlies male inviability in hybrids between *Drosophila buzzattii* and *D. koepferae*. Genetica 98:1-19
Cassidy DM, Blacklar AW (1974) Repression of nucleolar organizer activity in an interspecific hybrid of the genus *Xenopus*. Dev Biol 41:84-96
Cattanach BM, Beechey CV (1990) Autosomal and X-chromosome imprinting. Dev 1990 Suppl:63-72
Civetta A, Singh RS (1998a) Sex-related genes, direct selection, and speciation. Mol Biol Evol 15:901-909
Civetta A, Singh RS (1998b) Sex and speciation: genetic architecture and evolutionary potential of sexual versus nonsexual traits in the sibling species of the *Drosophila melanogaster* complex. Evolution 52:1080-1092
Cline TW (1993) The *Drosophila* sex determination signal: how do flies count to two? Trends Genet 9:385-390
Coward P, Nagai K, Chen D, Thomas HD, Nagamine CM, Lau YFC (1994) Polymorphism of a CAG trinucleotide repeat within *Sry* correlates with B6.Y^{DOM} sex reversal. Nature Genet 6:245-250
Coyne JA (1992) Genetics and speciation. Nature 355:511-515
Coyne JA, Orr HA (1989) Two rules of speciation. In: Otte D, Endler J (Eds) Speciation and its consequences. Sinauer Associates, Sunderland, pp 180-207
Coyne JA, Orr HA (1998) The evolutionary genetics of speciation. Phil Trans Roy Soc Lond B 353:287-305
Coyne JA, Simeonidis S, Rooney P (1998) Relative paucity of genes causing inviability in hybrids between *Drosophila melanogaster* and *D. simulans*. Genetics 150:1091-1103
di Bartolomeis SM, Tartof KD, Jackson FR (1992) A superfamily of *Drosophila* satellite related (SR) DNA repeats restricted to the X chromosome euchromatin. Nucl Acids Res 20:1113-1116
Dickinson WJ (1988) On the architecture of regulatory systems: evolutionary insights and implications. BioEssays 8:204-208
Dobzhansky Th (1937) Genetics and the origin of species. Columbia Univ Press, New York
Dobzhansky Th (1957) The X-chromosome in the larval salivary glands of hybrids *Drosophila insularis* × *Drosophila tropicalis*. Chromosoma 8:691-698
Durica DS, Krider HM (1977) Studies on the ribosomal RNA cistrons in interspecific *Drosophila* hybrids. I. Nucleolar dominance. Dev Biol 59:62-74

Durica DS, Krider HM (1978) Studies on the ribosomal RNA cistrons in interspecific *Drosophila* hybrids. II. Heterochromatic regions mediating nucleolar dominance. Genetics 89:37-64

Eicher EM, Washburn LL (1983) Inherited sex reversal in mice: identification of a new primary sex-determining gene. J Exp Zool 228:297-304

Eicher EM, Washburn LL, Whiteny JBIII, Morrow KE (1982) *Mus poschiavinus* Y chromosome in the C57BL/6J murine genome causes sex reversal. Science 217:535-537

Eicher EM, Shown EP, Washburn LL (1995) Sex reversal in C57BL/6J-Y^{POS} mice corrected by a *Sry* transgene. Phil Trans Roy Soc Lond B 350:263-269

Eicher EM, Washburn LL, Schork NJ, Lee BK, Shown EP, Xu X, Dredge RD, Pringle MJ, Page DC (1996) Sex-determining genes on mouse autosomes identified by linkage analysis of C57BL/6J-Y^{POS} sex reversal. Nature Genet 14:206-209

Finch RA (1983) Tissue-specific elimination of alternative whole parental genomes in one barley hybrid. Chromosoma 88:386-393

Forejt J (1996) Hybrid sterility in the mouse. Trends Genet 12:412-417

Friend SH (1993) Genetic models for studying cancer susceptibility. Science 259:774-775

Fujiwara A, Abe S, Yamaha E, Yamazaki F, Yoshida MC (1997) Uniparental chromosome elimination in the early embryogenesis of the inviable salmonid hybrids between masu salmon female and rainbow trout male. Chromosoma 106:44-52

Georgel P, Bellard F, Dretzen G, Jagla K, Richards G (1992) GEBF-I in *Drosophila* species and hybrids: the co-evolution of an enhancer and its cognate factor. Mol Gen Genet 235:104-112

Gibson G, Hogness DS (1996) Effect of polymorphism in the *Drosophila* regulatory gene *Ultrabithorax* on homeotic stability. Science 271:200-203

Gibson G, van Helden S (1997) Is function of the *Drosophila* homeotic gene *Ultrabithorax* canalized? Genetics 147:1155-1168

Goldschmidt R (1940) The material basis of evolution. Yale Univ Press, New Haven

Goodrich-Young C, Krider HM (1989) Nucleolar dominance and replicative dominance in *Drosophila* interspecific hybrids. Genetics 123:349-358

Granadino B, Penalva LOF, Sanchéz L (1996) Indirect evidence of alteration in the expression of the rDNA genes in interspecific hybrids between *Drosophila melanogaster* and *Drosophila simulans*. Mol Gen Genet 250:89-96

Graves JAM, O'Neill RJW (1997) Sex chromosome evolution and Haldane's rule. J Hered 88:358-360

Grummt I, Roth E, Puale MR (1982) Ribosomal RNA transcription in vitro is species specific. Nature 296:173-174

Haldane JBS (1922) Sex ratio and unisexual sterility in hybrid animals. J Genet 12:101-109

Hamerton JL, Giannelli F, Collins F, Hallett J, Fryer A, McGuire VM, Short RV (1969) Non-random X-inactivation in the female mule. Nature 222:1277-1278

Heikkinen E (1991) Wrinkling of the eye in hybrids between *Drosophila virilis* and *Drosophila lummei* is caused by interaction of maternal and zygotic genes. Heredity 66:357-365

Heikkinen E (1992) Genetic basis of reduced eyes in the hybrids of *Drosophila virilis* phylad species. Hereditas 117:275-285

Hertwig P (1936) Vol. IIB Part 21 In: Baur E, Hartmann M (Eds) Handbuch der Vererbungswissenschaft. Borntraeger, Berlin, pp 1-140

Hollocher H, Wu CI (1996) The genetics of reproductive isolation in the *Drosophila simulans* clade: X vs. autosomal effects and male vs. female effects. Genetics 143:1243-1255

Honjo T, Reeder RH (1973) Preferential transcription of *Xenopus laevis* ribosomal RNA in interspecies hybrids between *Xenopus laevis* and *Xenopus mulleri*. J Mol Biol 80:217-228

Hutter P (1997) Genetics of hybrid inviability in *Drosophila*. Adv Genet 36:157-185

Hutter P, Ashburner M (1990) Genetic rescue of inviable hybrids between *Drosophila melanogaster* and its sibling species. Nature 327:331-333

Hutter P, Roote J, Ashburner M (1990) A genetic basis for the inviability of hybrids between sibling species of *Drosophila*. Genetics 124:909-920

Khadem M, Krimbas CB (1991) Studies of the species barrier between *Drosophila madeirensis* and *Drosophila subobscura*. II. Genetic analysis of developmental incompatibilities in hybrids. Hereditas 114:189-195

Khadem M, Krimbas CB (1993) Studies of the species barrier between *Drosophila subobscura* and *D. madeirensis*. III. How universal are the rules of speciation? Heredity 70:353-361

Khadem M, Krimbas CB (1997) Studies of the species barrier between *Drosophila subobscura* and *D. madeirensis*. IV. A genetic dissection of the X chromosome for speciation genes. J Evol Biol 10:909-920

Kreitman M, Ludwig M (1996) Tempo and mode of *even-skipped* stripe 2 enhancer evolution in *Drosophila*. Semin Cell Dev Biol 7:583-592

Lakhotia SC, Mishra A, Sinha P (1981) Dosage compensation of X-chromosome activity in interspecific hybrids of *Drosophila melanogaster* and *D. simulans*. Chromosoma 82:229-236

Laurie CC (1997) The weaker sex is heterogametic: 75 years of Haldane's rule. Genetics 147:937-951

La Volpe A, Taggart M, Macleod D, Bird A (1983) Coupled demethylation of sites in a conserved sequence of *Xenopus* ribosomal DNA. Cold Spring Harbor Symp Quant Biol 47:585-592

Lee CH, Taketo T (1994) Normal onset, but prolonged expression, of *Sry* gene in the B6.Y^{DOM} sex-reversed mouse gonad. Dev Biol 165:442-452

Lohe AR, Hilliker AJ, Roberts PA (1993) Mapping simple repeated DNA sequences in heterochromatin of *Drosophila melanogaster*. Genetics 134:1149-1174

Lucchesi JC (1998) Dosage compensation in flies and worms: the ups and downs of X-chromosome regulation. Curr Opin Genet Dev 8:179-184

Lyon MF (1993) Epigenetic inheritance in mammals. Trends Genet 9:123-128

Mayr E (1942) Systematics and the origin of species. Columbia Univ Press, New York

Meer B (1976) Anomalous development and differential DNA replication in the X-chromosome of a *Drosophila* hybrid. Chromosoma 57:235-260

Meer B (1980) Extra DNA replication and cell proliferation in wing imaginal discs of a *Drosophila* species hybrid. Roux's Arch Dev Biol 198:83-89

Miesfeld R, Sollner-Webb B, Croce C, Arnheim N (1984) The absence of a human-specific ribosomal DNA transcription factor leads to nucleolar dominance in mouse>human hybrid cells. Mol Cell Biol 4:1306-1312

Muller HJ (1940) Bearings of the *Drosophila* work on systematics. In: Huxley JS (Ed) The new systematics, Claredon Press, Oxford, pp 185-268

Mutsuddi (née Das) M, Mutsuddi D, Mukherjee AS, Duttagupta AK (1984) Conserved autonomy of replication of the X-chromosomes in hybrids of *Drosophila miranda* and *Drosophila persimilis*. Chromosoma 89:55-62

Nagamine CM, Taketo T, Koo GC (1987) Studies on the genetics of *tda-1* sex reversal in the mouse. Differentiation 33:223-231

Nairn RS, Kazianis S, McEntire BB, Della Coletta L, Walter RB, Morizot DC (1996) A *CDKN2-like* polymorphism in *Xiphophorus* LG V is associated with UV-B-induced melanoma formation in platifish-swordtail hybrids. Proc Natl Acad Sci USA 93:13042-13047

O'Neill SL, Karr TL (1990) Bidirectional incompatibility between conspecific populations of *Drosophila simulans*. Nature 348:178-180

Onishi T, Berglund C, Reeder RH (1984) On the mechanism of nucleolar dominance in mouse-human somatic cell hybrids. Proc Natl Acad Sci USA 81:484-487

Orr HA (1989) Does postzygotic isolation result from improper dosage compensation? Genetics 122: 891-894

Orr HA (1990) Developmental anomalies in *Drosophila* hybrids are apparently caused by loss of microchromosome. Heredity 64:255-262

Orr HA (1996) The unexpected recovery of hybrids in a *Drosophila* species cross: a genetic analysis. Genet Res 67:11-18

Orr HA (1997) Haldane's rule. Annu Rev Ecol Syst 28:195-218

Orr HA, Madden LD, Coyne JA, Goodwin R, Hawley RS (1997) The developmental genetics of hybrid inviability: a mitotic defect in *Drosophila* hybrids. Genetics 145:1031-1040

Palmer SJ, Burgoyne PS (1991) The *Mus musculus domesticus Tdy* allele acts later than the *Mus musculus musculus Tdy* allele: a basis for XY sex-reversal in C57BL/6-Y^{POS} mice. Development 113:709-714

Papaceit M, San Antonio J, Prevosti A (1991) Genetic analysis of extra sex combs in the hybrids between *Drosophila subobscura* and *D. madeirensis*. Genetica 84:107-114

Perez DE, Wu CI (1995) Further characterization of the *Odysseus* locus of hybrid sterility in *Drosophila*: one gene is not enough. Genetics 140:201-206

Perez DE, Wu CI, Johnson, NA, Wu ML (1993) Genetics of reproductive isolation in the *Drosophila simulans* clade: DNA marker-assisted mapping and characterization of a hybrid-male sterility gene, *Odysseus* (*Ods*). Genetics 133:261-275

Platero JS, Csink AK, Quintanilla A, Henikoff S (1998) Changes in chromosomal localization of heterochromatin-binding proteins during the cell cycle in *Drosophila*. J Cell Biol 140:1297-1306

Presgraves DC, Orr HA (1998) Haldane's rule in taxa lacking a hemizygous X. Science 282:952-954

Procunier JD, Tartof KD (1978) A genetic locus having trans and contiguous cis functions that control the disproportionate replication of ribosomal RNA genes in *Drosophila melanogaster*. Genetics 88:67-79

Reeder RH (1984) Enhancers and ribosomal gene spacers. Cell 38:349-351

Reeder RH, Roan JG (1984) The mechanism of nucleolar dominance in *Xenopus* hybrids. Cell 38:39-44

Reuter D, Schuh R, Jäckle H (1989) The homeotic gene *spalt* (*sal*) evolved during *Drosophila* speciation. Proc Natl Acad Sci USA 86:5483-5486

Rieger R, Nicoloff H, Anastassova-Kristeva M (1979) "Nucleolar dominance" in interspecific hybrids and translocation lines — a review. Biol Zentbl 98:385-398

Rose MR, Doolittle WF (1983) Molecular biological mechanisms of speciation. Science 220:157-162

Sawamura K (1996) Maternal effect as a cause of exceptions for Haldane's rule. Genetics 143:609-611

Sawamura K, Yamamoto MT (1993) Cytogenetical localization of *Zygotic hybrid rescue* (*Zhr*), a *Drosophila melanogaster* gene that rescues interspecific hybrids from embryonic lethality. Mol Gen Genet 239:441-449

Sawamura K, Yamamoto MT (1997) Characterization of a reproductive isolation gene, *zygotic hybrid rescue*, of *Drosophila melanogaster* by using minichromosomes. Heredity 79:97-103

Sawamura K, Taira T, Watanabe TK (1993a) Hybrid lethal systems in the *Drosophila melanogaster* species complex. I. The *maternal hybrid rescue* (*mhr*) gene of *Drosophila simulans*. Genetics 133:299-305

Sawamura K, Watanabe TK, Yamamoto MT (1993b) Hybrid lethal systems in the *Drosophila melanogaster* species complex. Genetica 88:175-185

Sawamura K, Yamamoto MT, Watanabe TK (1993c) Hybrid lethal systems in the *Drosophila melanogaster* species complex. II. The *Zygotic hybrid rescue* (*Zhr*) gene of *Drosophila melanogaster*. Genetics 133:307-313

Sawamura K, Fujita A, Yokoyama R, Taira T, Inoue YH, Park HS, Yamamoto MT (1995) Molecular and genetic dissection of a reproductive isolation gene, *zygotic hybrid rescue*, of *Drosophila melanogaster*. Jpn J Genet 70:223-232

Schäfer U (1979) Viability in *Drosophila hydei* × *D. neohydei* hybrids and its regulation by genes located in the sex heterochromatin. Biol Zentbl 98:153-161

Schartl M (1995) Platifish and swordtails: a genetic system for the analysis of molecular mechanisms in tumor formation. Trends Genet 11:185-189

Schwab M (1987) Oncogenes and tumor suppressor genes in *Xiphophorus*. Trends Genet 3:38-42

Short RV (1997) An introduction to mammalian interspecific hybrids. J Hered 88:355-357

Singh N, Barbour KW, Berger FG (1998) Evolution of transcriptional regulatory elements within the promoter of a mammalian gene. Mol Biol Evol 15:312-325

Slifer EH (1942) A mutant stock of *Drosophila* with extra sex-combs. J Exp Zool 90:31-40

Sturtevant AH (1920) Genetic studies on *Drosophila simulans*. I. Introduction. Hybrids with *Drosophila melanogaster*. Genetics 5:488-500

Sturtevant AH (1946) Intersexes dependent on a maternal effect in hybrids between *Drosophila repleta* and *D. neorepleta*. Proc Natl Acad Sci USA 32:84-87

Sturtevant AH, Dobzhansky Th (1936) Observations on the species related to new forms of *Drosophila affinis* with descriptions of seven. Am Nat 70:574-584

Taketo T, Saeed J, Nishioka Y, Donahoe PK (1991) Delay of testicular differentiation in the B6.Y^{DOM} ovotestis demonstrated by immunocytochemical staining for Müllerian inhibiting substance. Dev Biol 146:386-395

Taketo-Hosotani T, Nishioka Y, Nagamine CM, Villalpando I, Merchant-Larios H (1989) Development and fertility of ovaries in the B6.Y^{DOM} sex-reversed female mouse. Development 107:95-105

Tamarina NA, Ludwig MZ, Richmond RC (1997) Divergent and conserved features in the spatial expression of the *Drosophila pseudoobscura esterase-5B* gene and the *esterase-6* gene of *Drosophila melanogaster*. Proc Natl Acad Sci USA 94:7735-7741

Ting CT, Tsaur SC, Wu ML, Wu CI (1998) A rapidly evolving homeobox at the site of a hybrid sterility gene. Science 282:1501-1504

True JR, Weir BS, Laurie CC (1996) A genome-wide survey of hybrid incompatibility factors by the introgression of marked segments of *Drosophila mauritiana* chromosomes into *Drosophila simulans*. Genetics 142:819-837

Tsaur SC, Wu CI (1997) Positive selection and the molecular evolution of a gene of male reproduction, *Acp26Aa* of *Drosophila*. Mol Biol Evol 14:544-549

Tsaur SC, Ting CT, Wu CI (1998) Positive selection driving the evolution of a gene of male reproduction, *Acp26Aa*, of *Drosophila*: II. Divergence versus polymorphism. Mol Biol Evol 15:1040-1046

Tucker PK, Lundrigan BL (1993) Rapid evolution of the sex determining locus in Old World mice and rats. Nature 364:715-717

Virdee SR, Hewitt GM (1992) Postzygotic isolation and Haldane's rule in a grasshopper. Heredity 69:527-538

Wade MJ, Johnson NA (1994) Reproductive isolation between two species of flour beetles, *Tribolium castaneum* and *T. freemani*: variation within and among geographical populations of *T. castaneum*. Heredity 72:155-162

Wade MJ, Johnson NA, Wardle G (1994) Analysis of autosomal polygenic variation for the expression of Haldane's rule in flour beetles. Genetics 138:791-799

Wallace H, Langridge WHR (1971) Differential amphiplasty and the control of ribosomal RNA synthesis. Heredity 27:1-13

Wang D, Marsh JL, Ayala FJ (1996) Evolutionary changes in the expression pattern of a developmentally essential gene in three *Drosophila* species. Proc Natl Acad Sci USA 93:7103-7107

Watanabe TK (1979) A gene that rescues the lethal hybrids between *Drosophila melanogaster* and *D. simulans*. Jpn J Genet 54:325-331

Weis S, Schartl M (1998) The macromelanophore locus and the melanoma oncogene *Xmrk* are separate genetic entities in the genome of *Xiphophorus*. Genetics 149:1909-1920

Weiss MC, Green H (1967) Human-mouse hybrid cell lines containing partial complements of human chromosomes and functioning human genes. Proc Natl Acad Sci USA 58:1104-1111

Whitfield LS, Lovell-Badge R, Goodfellow PN (1993) Rapid sequence evolution of the mammalian sex-determining gene *SRY*. Nature 364:713-715

Wilson AC, Carlson SS, White TJ (1977) Biochemical evolution. Annu Rev Biochem 46:573-639

Wittbrodt J, Adam D, Malitschek B, Mäueler W, Raulf F, Telling A, Robertson SM, Schartl M (1989) Novel putative receptor tyrosine kinase encoded by the melanoma-inducing *Tu* locus in *Xiphophorus*. Nature 341: 415-421

Wu CI, Davis AW (1993) Evolution of postmating reproductive isolation: the composite nature of Haldane's rule and its genetic bases. Am Nat 142:187-212

Wu CI, Palopoli MF (1994) Genetics of postmating reproductive isolation in animals. Annu Rev Genet 28:283-308

Wu CI, Johnson NA, Palopoli MF (1996) Haldane's rule and its legacy: why are there so many sterile males? Trends Ecol Evol 11:281-284

Zakian SM, Kulbakina NA, Meyer MN, Semenova LA, Bochkarev MN, Radjabli SI, Serov OL (1987) Non-random inactivation of the X-chromosome in interspecific hybrid voles. Genet Res 50:23-27

Zakian SM, Nesterova TB, Cheryaukene OV, Bochkarev MN (1991) Heterochromatin as a factor affecting X-inactivation in interspecific female vole hybrids (Microtidae, Rodentia). Genet Res 58:105-110

Zouros E (1989) Advances in the genetics of reproductive isolation in *Drosophila*. Genome 31:211-220

2
A Molecular Perspective on the Conservation of Diversity

Craig Moritz

Department of Zoology and Entomology and the Cooperative Research Centre for Tropical Rainforest Ecology and Management, The University of Queensland, Qld, 4072, Australia

Abstract

The need to protect genetic variation, along with other hierarchical levels of biological diversity, is identified in numerous international conventions and national policies, but the concepts and information needed to achieve this goal are deficient. In a constantly changing world, conservation planning should seek to maintain evolutionary processes and the ecological function of systems essential to achieve this. For the purposes of assessment of conservation values of areas, I suggest that genetic diversity should be partitioned into adaptive and historical components. Surrogates such as morphological differentiation (e.g. congruent distributions of subspecies) tend to reflect the adaptive rather than the historical component of genetic diversity. Subject to ongoing viability of populations, the former is replaceable, whereas the latter is not. It follows that the identification of historically isolated communities through comparative phylogeography or, perhaps, environmental modelling is an important component of conservation assessments. This paper illustrates the application of comparative phylogeography to conservation assessments through two case studies; rainforest fauna in the wet tropics of Australia, and forest-dwelling herpetofauna in South-east Queensland. In both systems, mtDNA diversity is strongly partitioned among populations in a manner broadly consistent with environmental modelling, although they differ in the extent to which historical vicariance defines congruent distributions of diversity. Areas prioritised for conservation should encompass these historically isolated communities, together with strong environmental gradients within each.

Key words. genetic diversity, phylogeography, conservation priorities, rainforests, mitochondrial DNA (mtDNA), adaptation, historical biogeography, environmental gradients, vertebrates, reptiles, amphibians, Evolutionarily Significant Units, evolutionary processes, historical isolation, Australia

1 Introduction

The need to protect genetic diversity, as well as that of species and ecosytems is widely recognised (Noss 1990) and has been acknowledged in international agreements (e.g. the Convention on Biological Diversity) as well as policies of many national governments. However, appropriate measures and strategies for protecting genetic diversity remain ill-defined and this, together with the lack of baseline information for most systems, is a substantial impediment to practical efforts to set conservation priorities. At one level genetic diversity can be considered as the total information content represented by a suite of species or higher taxa, leading to the use of various phylogenetic measures of distinctiveness in weighting schemes (Vane-Wright et al. 1991; reviewed in Humphries et al. 1995; Crozier 1997). At the other extreme is consideration of the level of genetic diversity within populations and the relationship between diversity and the potential response to selection (Frankel and Soule 1981).

Some clarity can be achieved by considering the underlying conservation goal — what are we trying to protect and what is the most efficient means of doing so? In an eloquent paper, Frankel (1974) stressed the need to maintain evolutionary processes in increasingly modified systems and advocated the development of an evolutionary ethic in conservation and, more generally, in society. We can define the following overarching goal:

> *To maintain evolutionary processes and the ecological viability of populations and landscapes necessary to achieve this.*

This acknowledges that both genetic diversity and ecological processes are necessary to maintain continuing evolution. However, while viability of populations and ecosystems is necessary, it may not be sufficient. It follows that successful conservation strategies will draw on understanding of both ecological and evolutionary processes, combined with socio-economics.

2 Components of Genetic Diversity and Conservation Strategy

In recent years, there has been increasing debate over concepts and criteria for recognising intraspecific units for conservation. Traditionally, conservation managers have sought to protect taxonomically recognised entities — species and subspecies. However, surveys of molecular variation have often revealed a lack of congruence between subspecies defined on morphological criteria and neutral molecular markers (Avise and Ball 1990; O'Brien and Mayr 1991), the former presumably reflecting the joint action of selection and history, and the latter history alone. The concept of an "Evolutionarily Significant Unit" (ESU) was developed in order to provide an objective approach to prioritising intraspecific units for protection (Ryder 1986) and was subsequently refined for applications under the USA Endangered Species Act (Waples 1991, 1995).

There remains, however, disagreement over the criteria for recognising ESUs (Nielsen and Powers 1995). I have argued that molecular signatures of substantial historical isolation, such as reciprocal monophyly of mtDNA alleles (in animals) should be the primary criterion (Moritz 1994; Moritz et al. 1995), whereas others suggest that both molecular and phenotypic measures of divergence should be considered (e.g. Dizon et al. 1992; Vogler and DeSalle 1994; Waples 1995). The term "Evolutionarily Significant Unit" itself is, perhaps, unfortunate as it suggests that we can predict the evolutionary potential of individual populations from their genetic makeup. Other than for very specific selective factors (e.g., insecticide resistance), this is unlikely to be the case.

The view that both genetic and phenotypic criteria should be used for defining ESUs confounds two different evolutionary processes — historical isolation and adaptation (Fig. 1). Historical isolation, combined with genetic drift or divergent selection will generate unique and irreplaceable combinations of genotypes, which may or may not be manifest as differences in phenotype (e.g., see below). This process of isolation and divergence (but only rarely selection) can be studied readily using molecular makers through analyses of phylogeography (Avise et al. 1987). By contrast, the process of adaptation and divergent selection will generate differences in phenotype among populations, but because it can occur in the presence of gene flow (Endler 1977) there need not be significant phylogeographic structure. Importantly, recent experimental studies with natural populations have demonstrated that divergence of phenotypes through selection can occur rapidly and re-

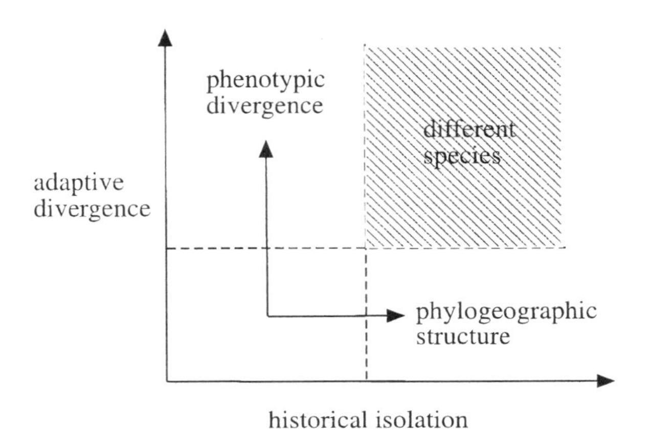

Fig. 1. Illustration of two major components of genetic diversity within species. Adaptive divergence, e.g. that along gradients of selection, is expected to result in phenotypic difference among populations and does not require complete suppression of gene flow. By contrast, historical isolation is expected to result in phylogeographic structure at neutral molecular markers, but does not necessarily result in phenotypic difference. Where both processes operate, the resulting phenotypically and phylogenetically distinct populations can be recognised as distinct taxonomic entities (Avise and Ball 1990)

peatedly, subject to population viability (e.g., Malhotra and Thorpe 1991; Reznick et al. 1997; Losos et al. 1997) and the same conclusion can be drawn from phylogenetic analysis of recent adaptive radiations (e.g., Givnish 1997; Losos et al. 1998). This leads to the conclusion that phenotypic diversity among populations is potentially replaceable, whereas the genetic diversity due to historical isolation is not.

Bearing the above in mind, what conservation strategy is appropriate to meet our goal of protecting evolutionary processes and potential? I would suggest a binary approach, wherein we first identify historically isolated sets of population within a species (ESUs *sensu* Moritz 1994) and then seek to maximise the potential for adaptive evolution within these. The latter would include, for example, ensuring that the areas managed for each ESU encompass heterogeneous landscapes or environmental gradients relevant to phenotypic evolution and functional diversity of the species concerned. The maintenance of population and metapopulation viability and of functional landscapes (see the chapter by Harrison, this volume) is integral to this strategy.

Finally, the weight given to the two dimensions of diversity may vary depending on context. An emphasis on historical isolation (ESUs) is most appropriate to species undergoing gradual evolution in relatively stable systems (e.g. see below). For lineages undergoing rapid adaptive radiation and speciation the historical axis of diversity may be less important and that concerned with adaptation more so. Insofar as selectively driven speciation can occur without total genetic isolation (Rice and Hostert 1993; Grant 1998) and over short time periods (e.g. Meyer et al. 1990), we should not apply the ESU concept blindly, but rather consider the ecological features of the species and their environment that promote this process.

3 Extension from Species to Communities

Quite correctly, there are concerns that by focussing on individual ESUs within individual species we are “fiddling while Rome burns”. Realistically, government policies mandate attention towards threatened species, but the real gains are to made through a proactive approach to conservation at the level of landscapes and ecosystems. This is being acknowledged increasingly in strategies for locating nature reserves (e.g. Sattler 1994) and managing diversity in multiple use landscapes. Priorities for such efforts typically are determined by selecting areas that optimise the representation of vegetation communities and species in the total set, and there has been substantial progress in developing algorithms to achieve this (Pressey et al. 1993, 1994; Lombard et al. 1997). The challenge now is to develop concepts that permit consideration of genetic diversity and, more importantly, evolutionary processes in the management of landscapes.

The concept of ESUs can be extended to communities through comparative analyses of phylogeography for species with similar dispersal capability and ecological requirements (Avise 1992; Moritz and Faith 1998). Through the analysis of a few putative indicator taxa, typically species with low vagility, the aim is to identify geographic areas within which populations of species comprising ecological communities have had evolutionary histories separate from those elsewhere in the range. Where such a history of vicariance occurs, independently evolving communities can be recognised. Further, priorities for management can be set according to the degree of congruence among species, the magnitude of divergence, or some combination of these (Moritz and Faith 1998). This approach to recognising the historical component of genetic diversity can, in principle, be combined with assessments based on diversity of species and communities to identify a set of areas that are representative of all three tiers of diversity. Within these areas, management should then ensure that functional, ecologically and environmentally heterogeneous landscapes are retained in order to protect viability and promote continued adaptive evolution. In the following, I illustrate the basis and application of this approach using data from rainforest faunas of the subtropical and tropical coast of Australia.

4 Evolution and Conservation of Rainforest Faunas

4.1 The Vertebrates of the Wet Tropics of Australia

The tropical rainforests of north-east Australia are small in extent (< 1M ha), yet harbour a high diversity of species, many of which are endemic (Commonwealth of Australia 1986; Nix and Switzer 1991). Besides the high endemicity, other attributes that make this a useful system with which to explore evolutionary processes and strategies for conserving genetic diversity are; (i) a strong background of paleo-ecology, providing evidence for fluctuations in the distribution and composition of rainforests during the climatic fluctuations of the Quaternary (Kershaw 1994; Hopkins et al. 1993), (ii) a reasonably stable alpha taxonomy and good information on species distributions (e.g. Webb and Tracey 1981; Nix 1991; Williams et al. 1996), and (iii) accumulating data from mtDNA on patterns of phylogeography within endemic species of vertebrates (Moritz et al. 1993; Joseph and Moritz 1994; Joseph et al. 1995; Schneider et al. 1998; Schneider and Moritz 1999).

The analyses of phylogeography in several species of bird, frog and lizard have revealed strong and geographically congruent separation of populations to the north and south of a historical disjunction of rainforests predicted by paleoclimatological modelling ("Black Mountain Corridor", BMC, Fig. 2). Levels of sequence divergence at mtDNA protein coding genes vary from <1% to 2% in birds to between 3% and 12% in frogs and lizards. For most of the herpetofauna examined, this implies several million years of separation, whereas more recent (Pleistocene) separations are inferred for the birds. In general, the magnitude of separation is inversely proportional to presumed vagility and positively related to the degree of specialisation to rainforests. In addition to the general vicariance pattern about the BMC, there are additional more species-specific breaks in phylogeography at other

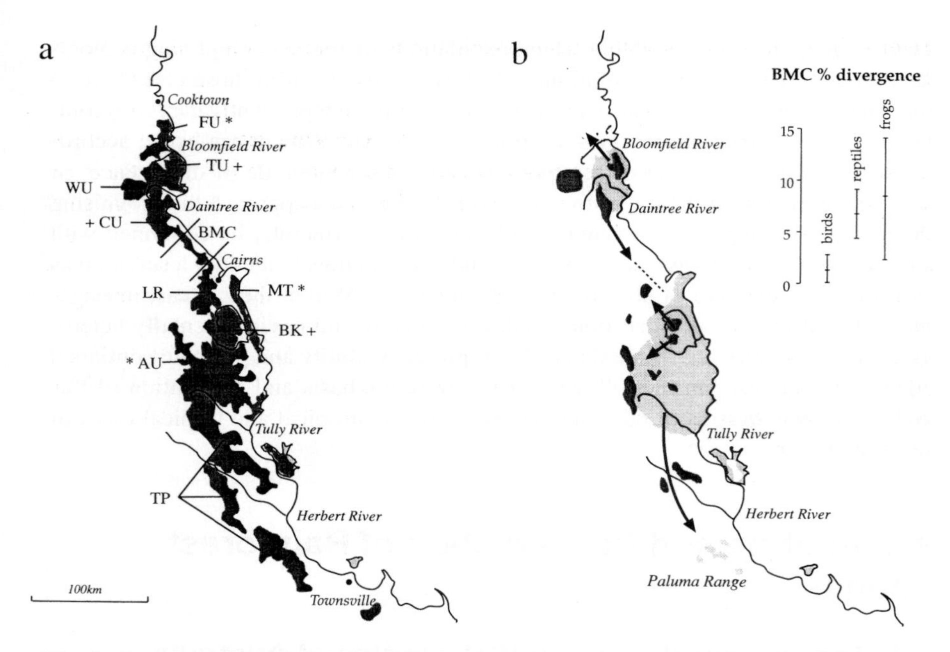

Fig. 2. Current and predicted historical distributions of rainforests. **a**. Current distribution of rainforest in the Wet Tropics of north-east Queensland, Australia, with key locations and conservation priorities indicated. FU, Finnegan Uplands; TU Thornton Uplands; CU Carbine Uplands; BMC, Black Mountain Corridor; BK, Bellenden Ker Range; MT, Malbon Thompson Range; AU Atherton Uplands. * Locations identified as being significant to retain historically isolated communities. + Locations most important for retaining diversity of endemic species (see text). **b.** Predicted historical distribution of rainforests from paleoclimatological modelling (stippled, Nix 1991) and plant biogeography (dark, Webb and Tracey 1981). The arrows indicate hypothesised routes of expansion during the early-mid Holocene. The bars to the right represent the mean and limits of mean sequence divergence observed across major phylogeographic breaks in birds, frogs and lizards (from Joseph et al. 1995; Schneider et al. 1998 and unpublished data)

locations (Schneider et al. 1998). More detailed analysis of genetic structure in three species of lizard provided evidence for contraction of populations to refugia within each of the rainforest areas north and south of the BMC, with subsequent expansions, presumably during periods of the early Holocene when the cooler and wetter conditions were optimal for these elevation species (Schneider and Moritz 1999). In particular, it appears that the vertebrates of the southernmost rainforest areas represent a totally recolonised fauna (Fig. 2; Schneider et al. 1998; see also Williams and Pearson 1997).

For the wet tropics vertebrates, the major component of historical diversity occurs across the BMC, with lesser breaks at three other locations (Fig. 2). Interestingly, this is not evident from the current distribution of rainforest — the rainforest connection across the BMC is at least 8,000 years old (Hopkins et al. 1993), whereas the recolonised area to the south is currently separated by a relatively dry and warm river valley which has probably been unsuitable for cool adapted rainforest vertebrates for much of the past 5,000 years (Nix 1991; Winter 1997).

To consider the adaptive axis of genetic diversity, we can use as a surrogate patterns of morphological variation. Comparisons of several traits, including ecomorphological characters and those typically used by taxonomists, in three species of lizard revealed no significant divergence across the BMC other than for a single scale character in one species (Schneider and Moritz 1999). The same lack of morphological divergence across the BMC is evident for frogs (M. Cunningham, unpublished data) and is inferred for birds from the lack of recognised sub-species distributed to either side of the BMC. This observation casts doubt on the role of isolation and drift in promoting divergence and is inconsistent with the refuge model of speciation as originally proposed for rainforest vertebrates (Haffer 1969; Schneider and Moritz 1999). It also provides a vivid illustration of the need to consider separately the historical and adaptive axes of diversity. This is emphasised further by a recent comparison of molecular and morphological variation in another species of lizard endemic to the wet tropics (C. Schneider, T. Smith, B. Larrison and C. Moritz, unpublished data). In this species, *Carlia rubigularis*, morphological divergence across the BMC was trivial, despite sequence divergence of 12%. In contrast, highly significant shifts in body size and shape was observed between populations in adjacent rainforest and open forest despite high levels of gene flow. This echoes recent observations of increased morphological divergence of birds across ecotones relative to that within rainforests (Smith et al. 1997) and points to the need to protect diverse habitats within areas in order to maintain adaptive diversity among populations.

From the above information, it is clear that to protect the historical axis of genetic diversity in wet tropics vertebrates, we should prioritise areas to the north and south of the BMC. To maintain the process of adaptive divergence, we should ensure that maximum diversity of vegetation communities are protected within each area. A more detailed analysis (Moritz and McDonald, in press) using the quantitative approach described by Moritz and Faith (1998) identified as key areas for protecting historical diversity the Finnegan Uplands to the north, the Atherton Uplands to the south and the geographically intermediate Malbon Thompson range (Fig. 2). By contrast, analyses of patterns of complementarity of endemic species of vertebrate across the same area identified a different set of priority areas, the northern Thornton and Carbine Uplands and the Bellenden Ker Range, in particular (Moritz and McDonald, in press; Fig. 2). This illustrates that genetic and species components of diversity are not equivalent, presumably because of distinct determinants operating on different spatial and temporal scales. Both should be considered in conservation planning. However, once the set of areas that adequately capture both species and (historical) genetic diversity are identified, the notion that habitat heterogeneity should be maximised within areas is consistent with both maintaining adaptive diversity and maximising the diversity of species (Williams 1997; Williams and Pearson 1997) and ecosystems.

4.2 Forest Fauna from South-East Queensland

The wet tropics represents the major area of species diversity within a chain of high elevation, mesic "islands" located along the east coast of Queensland (Nix 1991). Rainforests are distributed within and across this "mesotherm archipelago" (mesotherm = 18-21°C mean annual temperature, Nix 1982), depending on the presence of a combination of relatively fertile (mostly basaltic) soils, high rainfall and appropriate temperature regimes (Adam 1992). Towards south-east Queensland (SEQ), and into northern New South Wales (NSW), subtropical rainforest occurs patchily within an increasingly continuous cool (mesotherm) environment (Fig. 3). These rainforests and adjacent dry forests support a diverse vertebrate fauna, although with lower endemicity and, in mammals, fewer rainforest restricted species (Winter 1988). They also occur within a region of overlap of southern (mostly temperate) and northern (tropical) faunas (Adam 1992).

We have begun to examine patterns of phylogeography in forest-dependent lizards and frogs to see whether the climate-driven vicariance of forest faunas observed for the wet tropics occurs also in SEQ and to contribute to current assessments of conservation values in the region. Analysis of a wet-forest (wet sclerophyll and rainforest) restricted species of tree frog (*Litoria pearsoniana*) revealed a substantial disjunction, with 3% sequence divergence between populations in the north-

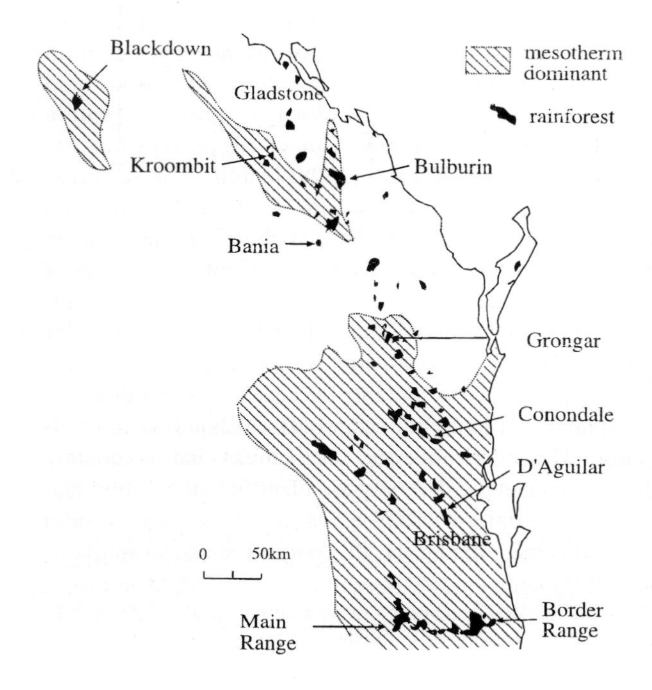

Fig. 3. Map of south-east Queensland, Australia, showing the distribution of rainforest patches (dark) in relation to approximate location of mesotherm climates (shaded) in south-east Queensland. Modified from Adam (1992) and Nix (1993)

ern Conondale (CO) and D'Aguilar Ranges (DA) and those to the south [Main (MR) and Border Ranges (BR) and northern NSW; McGuigan et al. 1998]. The pattern of genetic divergence was consistent with paleo-climatological modelling (as for the wet tropics) and also with area relationships suggested by distributions of rainforest restricted herpetofauna, suggesting the possibility of a general pattern. However, subsequent (and mostly preliminary) analyses have revealed a more complex picture (Figs. 4 and 5). Phylogeographic analysis of two species of rainforest restricted skinks from SEQ (*Saproscincus rosei* and *Eulampius murrayi;* D. O'Connor, R. Sadlier and C. Moritz, unpublished data) again revealed strong population subdivision (being stronger in the high altitude species, *S. rosei*), but the area relationships were idiosyncratic.

Preliminary data for dry forest species sampled from across SEQ reveal substantial population subdivision (>80% of diversity among populations), except for *Carlia pectoralis*, a species of skink located in the drier end of the forest spectrum for which only a northern isolate is divergent (Fig. 4). Again there are various

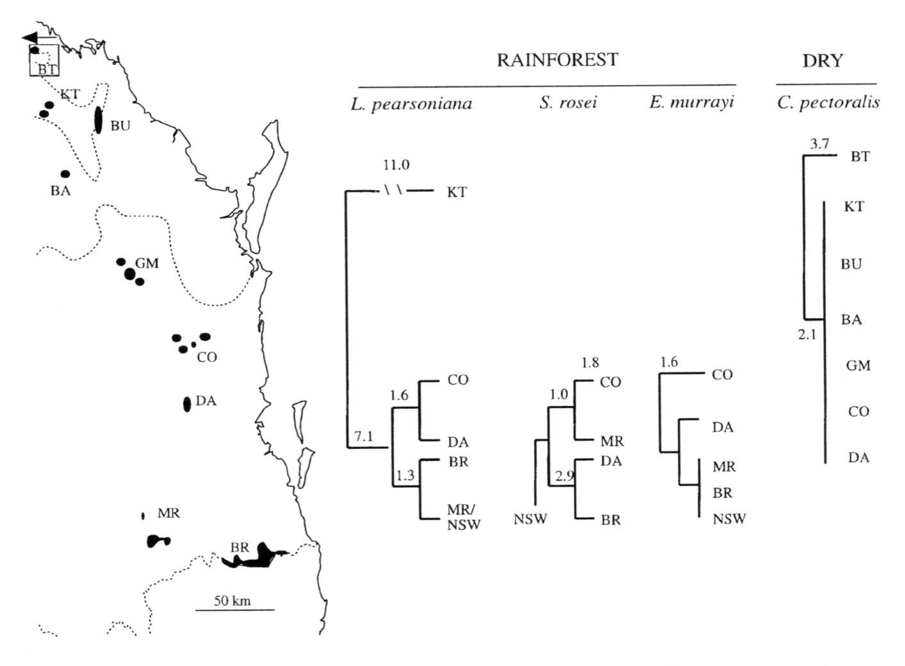

Fig. 4. Diagrammatic representations of the distribution of mtDNA phylogeographies from three species of rainforest herpetofauna and one dry forest species from south-east Queensland, Australia. In each case the trees are simplified from neighbour joining analyses of Kimura 2 parameter distances obtained from sequence analysis of c. 500 bp from 5–10 individuals per locality. Locations containing monophyletic alleles are grouped into a single branch and only distances of >1% are shown. *BA*, Bania; *BR*, Border Range; *BT*, Blackdown Tableland; *BU*, Bulburin; *CO*, Conondale; *DA*, D'Aguilar; *GM*, Grongar-Marodian; *MR*, Main Range; *KT*, Kroombit Tops; *NSW*, New South Wales. Data from McGuigan et al. 1998 and C. Moritz, D. O'Connor, R. Sadlier and C. Hoskins (unpublished)

relationships apparent among the southern areas (CO to BR, Fig. 5), although isolation by distance cannot be ruled out. More interesting is the increasing level of sequence divergence among northern populations (i.e. BU, BA, KT and BT), these representing disjunct pockets of the mesotherm archipelago. Whereas sequence divergence levels within both rainforest and dry forest species are typically <4% among southern sites, the northern populations are frequently separated from others by 5% to >10%. This same general pattern is repeated among populations of several species of cameanid snails that occur in the wet and dry forests of the region (A. Hugall, J. Stanisic and C. Moritz, unpublished data).

The available evidence for the SEQ fauna indicate there is no deep overriding vicariance event as observed across the BMC in the Wet Tropics. However, there are two clear outcomes from the analysis. First, as might be expected from the current disjunct distribution of their habitat, the rainforest species show strong population subdivision among areas, such that each area (CO, DA, MR and BR) contains putative ESUs for one or more taxa. Second, divergence among populations of all species tends to increase towards the northern sites, coinciding with increasing fragmentation of the mesotherm climate. Accordingly, to protect the

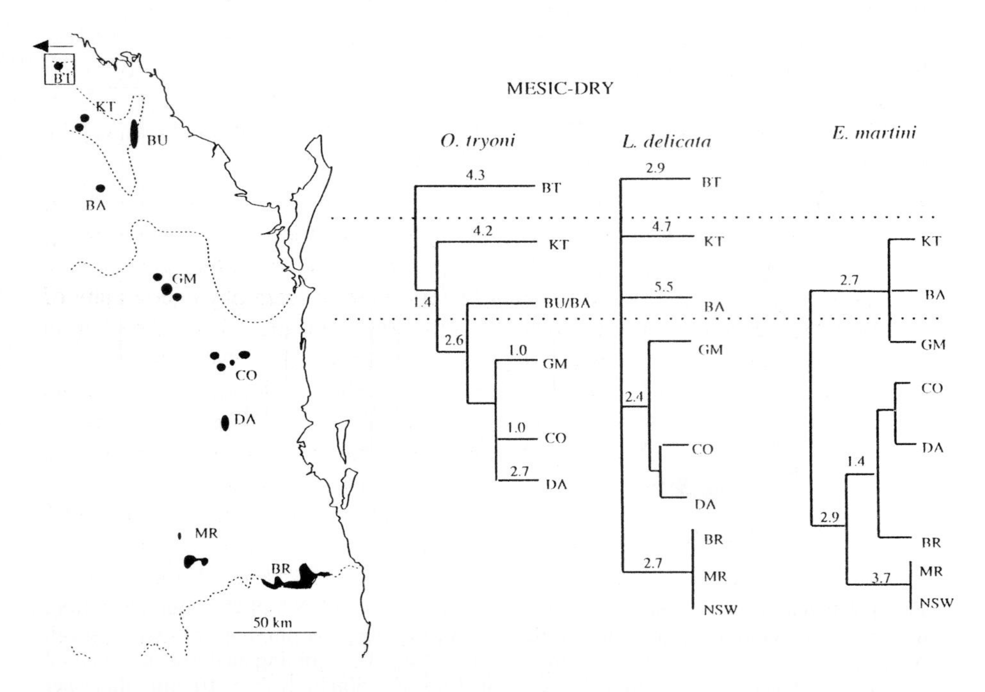

Fig. 5. Diagrammatic representation of mtDNA phylogeography for three species of lizard (*Oedura tryoni*; *Lampropholis delicata*; *Eulampus martini*) from mesic to dry forests from south-east Queensland, Australia. The horizontal dashed lines indicate the major disjunctions of the mesotherm climate (see Fig. 3). Other details as for Fig. 4, except that sample sizes for some species/locality combinations are sometimes less than 5

historical component of diversity, it will be important to protect each of the major rainforest areas, with particular emphasis on those in the north. One notable feature is that the increasing population isolation towards the north applies to both dry and mesic adapted species, so that protection of these mesotherm outliers should extend beyond rainforests to adjacent areas of wet and dry sclerophyll habitats. There has not yet been a formal evaluation of the correspondence between priorities derived from the species and genetic data for SEQ. However, the general recommendations from the latter seem consistent with the presence of locally endemic species of herpetofauna in the rainforest areas (see McGuigan et al. 1998) and the presence of distinctive combinations of bird species in both wet and dry forests in the mesotherm isolates to the north (Nix 1993).

5 Future Challenges

The brief review of evidence from the Wet Tropics and SEQ fauna serve to illustrate the application of key concepts — the separation of genetic diversity due to historical isolation from that due to adaptation and the extension of the ESU approach to species assemblages — and also point the way to combining conservation assessment of genetic diversity with other hierarchical levels; species and ecosystems. Rather than debating which of these is most important, we should seek to consider all three in prioritising and implementing management for conservation.

Many questions remain to be answered before the strategy outlined here can be implemented in any general sense. These include:

- To what extent can modelling of environmental conditions, both current and historical, predict the broad geographic pattern of historical isolation among populations within species? The correspondance between the paleoclimate modelling of rainforest and the area history inferred from phylogeography of Wet Tropics vertebrates and that between the current mesotherm distribution and patterns of genetic diversity in SEQ are encouraging. However, more sophisticated modelling and further molecular analysis are needed before we can use environmental data as a surrogate for this aspect of genetic diversity.
- To what extent do historically structured communities exist in other biomes, particularly those in more topographically uniform or recently colonised areas (e.g., papers on comparative phylogeography in Molecular Ecology 7(3), 1998)?
- How does natural selection operate across heterogeneous environments and what landscape and metapopulation structures are necessary to maintain this process?
- Does genetic diversity within populations limit the response to selection, especially in stressful or ecologically marginal environments?

Answers to these questions, as well as further refinement of a statistical framework and algorithms for explicitly incorporating measures of genetic diversity into area planning will enhance our efforts to conserve biological diversity in a manner that recognises the dynamics of the evolutionary process.

Acknowledgements

The research described here was supported by grants from the Australian Research Council, the Cooperative Research Centre for Tropical Rainforest Ecology and Management and the Queensland Department of Environment and Heritage. I am grateful to my colleagues from the Molecular Zoology lab of the University of Queensland for discussion and permission to cite unpublished data. Thanks to L. Kelemen, L. Pope and A. Hugall for assistance with preparation of figures.

References

Adam P (1992) Australian rainforests. Oxford Univ Press, Oxford, London

Avise J (1992) Molecular population structure and the biogeographic history of a regional fauna: a case history with lessons for conservation. Oikos 63:62-76

Avise JC (1994) Molecular markers, natural history and evolution. Chapman & Hall, New York

Avise JC, Ball RM (1990) Principles of genealogical concordance in species concepts and biological taxonomy. Oxford Surv Evol Biol 7:45-68.

Commonwealth of Australia (1986) Tropical rainforests of north Queensland—their conservation significance. Australian Govt Publishing Service, Canberra

Crozier RH (1997) Preserving the information content of species: genetic diversity, phylogeny, and conservation worth. Annu Rev Ecol Syst 28:243-268

Dizon AE, Lockyer C, Perrin WF, Demaster DP, Sisson J (1992) Rethinking the stock concept: a phylogeographic approach. Conserv Biol 6:24-36

Endler J (1977) Geographic variation, speciation and clines Princeton Univ Press, Princeton

Frankel OH (1974) Genetic conservation: our evolutionary responsibility. Genetics 78:53-65

Frankel OH, Soule ME (1981) Conservation and Evolution. Cambridge Univ Press, Cambridge

Givnish TJ (1997) Adaptive radiation and molecular systematics: issues and approaches. In: Givnish TJ, Systma KJ (Eds) Molecular evolution and adaptive radiation, Cambridge Univ Press, Cambridge, pp 1-54

Grant PR (1998) Speciation and hybridisation of birds on islands. In: Grant PR (Ed) Speciation on islands, Oxford Univ Press, Oxford, pp 142-162

Haffer J (1969) Speciation in amazonian forest birds. Science 165:131-137

Hopkins MS, Ash J, Graham AW, Head J, Hewett RK (1993) Charcoal evidence of the spatial extent of the Eucalyptus woodland expansions and rainforest contractions in North Queensland during the late Pleistocene. J Biogeogr 20:59-74

Humphries CJ, Williams PH, Vane-Wright WI (1995) Measuring biodiversity value for conservation. Annu Rev Ecol Syst 26:93-111

Joseph L, Moritz C (1994) Mitochondrial DNA phylogeography of birds in eastern Australian rainforests: first fragments. Aust J Zool 42:385-403

Joseph L, Moritz C, Hugall A (1995) Molecular support for vicariance as a source of diversity in rainforest. Proc Roy Soc Lond B 260:177-182

Kershaw AP (1994) Pleistocene vegetation of the humid tropics of northeastern Queensland, Australia. Palaeogeog Palaeoclimatol Palaeoecol 109:399-412

Lombard AT, Cowling RM, Pressey RL, Mustart PJ (1997) Reserve selection in a species rich and fragmented landscape on the Agulhas plain, South Africa. Conserv Biol 11:1101-1116

Losos JB, Warheit KI, Scheoner TW (1997) Adaptive differentiation following experimental island colonisation in Anolis lizards. Nature 387:70-72

Losos JB, Jackman TR, Larson A, de Quiroz K, Rodriguez-Schettino L (1998) Contingency and determinism in replicated adaptive radiations of island lizards. Science 279:2115-2118

Malhotra A, Thorpe RS (1991) Experimental detection of rapid evolutionary response in natural lizard populations. Nature 353:347-348

McGuigan K, McDonald K, Parris K, Moritz C (1998) Mitochondrial DNA diversity and historical biogeography of a wet forest restricted frog (*Litoria pearsoniana*) from mid-east Australia. Mol Ecol 7:175-186

Meyer A, Kocher TD, Basasibwaki P, Wilson AC (1990) Monophyletic origin of Lake Victoria cichlid fishes suggested by mitochondrial DNA sequences. Nature 347:550-553

Moritz C (1994) Defining evolutionarily significant units for conservation. Trends Ecol Evol 9:373-375

Moritz C, Faith D (1998) Comparative phylogeography and the identification of genetically divergent areas for conservation. Mol Ecol 7:419-429

Moritz C, McDonald KR (in press) Evolutionary approaches to the conservation of tropical rainforest vertebrates. In: Moritz C, Bermingham E (Eds) Rainforests: past and future. Chicago Univ Press, Chicago

Moritz C, Joseph L, Adams M (1993) Cryptic genetic diversity in an endemic rainforest skink (*Gnypetoscincus queenslandiae*). Biodiv Conserv 2:412-425

Moritz C, Lavery S, Slade R (1995) Using allele frequency and phylogeny to define units for conservation and management. Amer Fish Soc Symp 17:249-262

Nielsen JL, Powers D (1995) Evolution and the Aquatic Ecosystem: Defining unique units in population conservation. American Fisheries Society, Bethesda, WA

Nix HA (1982) Environmental determinants of biogeography and evolution in Terra Australis. In: Barker WR, Greenslade PJM (Eds) Evolution of the flora and fauna of arid Australia. Peacock Publications, Adelaide, pp 47-66

Nix HA (1991) Biogeography: patterns and process. In: Nix HA, Switzer M (Eds) Rainforest animals. Atlas of vertebrates endemic to Australia's wet tropics Australian National Parks and Wildlife Service, Canberra, pp 11-39

Nix HA (1993) Bird distributions in relation to imperatives for habitat conservation in Queensland. In: Catterall CP, Driscoll PV, Hulsman K, Muir D, Taplin A (Eds) Birds and their habitats: status and conservation in Queensland. Queensland Ornithological Society Inc., Brisbane, pp 12-21

Nix HA, Switzer M (1991) Rainforest animals. Atlas of vertebrates endemic to Australia's wet tropics. Australian National Parks and Wildlife Service, Canberra

Noss RF (1990) Indicators for monitoring biodiversity: a hierarchical approach. Conserv Biol 4:355-364

O'Brien SJ, Mayr E (1991) Bureaucratic mischief: recognizing endangered species and subspecies. Science 251:1187-1188

Pressey RL, Humphries CJ, Margules CR, Vane-Wright RI, Williams PH (1993) Beyond opportunism: key principles for systematic reserve selection. Trends Ecol Evol 8:124-128

Pressey RL, Johnson IR, Wilson PD (1994) Shade of irreplaceability: towards a measure of the contribution of sites to a reservation goal. Biodiv Conserv 3:242-262

Reznick DN, Shaw FH, Rodd FH, Shaw RG (1997) Evaluation of the rate of evolution in natural populations of guppies (*Poecilia reticulata*). Science 275:1934-1937
Rice WR, Hostert EE (1993) Perspective: laboratory experiments on speciation: what have we learned in forty years. Evolution 47:1637-1653
Ryder OA (1986) Species conservation and systematics: the dilemma of subspecies. Trends Ecol Evol 1:9-10
Sattler P (1994) Towards a nationwide biodiversity strategy: the Queensland contribution. In: Moritz C, Kikkawa J (Eds) Conservation Biology in Australia and Oceania. Surrey Beatty & sons, Sydney, pp 313-325
Schneider CJ, Cunningham M, Moritz C (1998) Comparative phylogeography and the history of endemic vertebrates in the Wet Tropics rainforests of Australia. Mol Ecol 7:487-498
Schneider CJ, Moritz C (1999) Refugial isolation and evolution in the wet tropics rainforests of Australia. Proc Roy Soc Lond B 266:191-196
Smith T, Wayne R, Girman D, Bruford M (1997) A role for ecotones in generating rainforest biodiversity. Science 276:1855-1857
Vane-Wright RI, Humphries CJ, Williams PH (1991) What to protect — systematics and the agony of choice. Biol Conserv 55:235-254
Vogler AP, DeSalle R (1994) Diagnosing units of conservation management. Conserv Biol 8:354-363
Waples RS (1991) Pacific salmon, *Oncorhynchus* spp., and the definition of "species" under the Endangered Species Act. Marine Fisheries Rev 53:11-22
Waples RS (1995) Evolutionarily significant units and the conservation of biological diversity under the Endangered Species Act. In: Nielsen J, Powers D (Eds) Evolution and the aquatic ecosystem. American Fisheries Society, Bethesda, Maryland, pp 8-27
Webb L, Tracey J (1981) Australian rainforests: pattern and change. In: Keast JA (Ed) Ecological biogeography of Australian. W. Junk, The Hague, pp 605-694
Williams SE (1997) Patterns of mammalian species richness in the Australian tropical rainforests: are extinctions during historical contractions of the rainforest primary determinants of current regional patterns in biodiversity? Wildlife Res 24:513-530
Williams S, Pearson R (1997) Historical rainforest contractions, localized extinctions and patterns of vertebrate endemism in the rainforests of Australia's wet tropics. Proc Roy Soc Lond B 264:709-716
Williams SE, Pearson RG, Walsh PJ (1996) Distributions and biodiversity of the terrestrial vertebrates of Australia's wet tropics: a review of current knowledge. Pacific Conserv Biol 2:327-362
Winter JW (1988) Ecological specialization of mammals in Australian tropical and subtropical rainforest: refugial or ecological determinism. In: Kitching R (Ed) The ecology of Australia's wet tropics. Surrey Beatty and Sons, Sydney, pp 127-138
Winter JW (1997) Responses of non-volant mammals to late Quaternary climatic changes in the wet tropics region of north-eastern Australia. Wildlife Res 24:493-511

3 Mitochondrial Molecular Clocks and the Origin of Euteleostean Biodiversity: Familial Radiation of Perciforms May Have Predated the Cretaceous/Tertiary Boundary

YOSHINORI KUMAZAWA [1], MOTOOMI YAMAGUCHI [2,3], AND MUTSUMI NISHIDA [2,3]

[1] Department of Earth and Planetary Sciences, Nagoya University, Furo-cho, Chikusa-ku, Nagoya 464-8602, Japan
[2] Department of Marine Bioscience, Fukui Prefectural University, 1-1 Gakuen-machi, Obama 917-0003, Japan
[3] Current Address: Ocean Research Institute, University of Tokyo, 1-15-1 Minamidai, Nakano-ku, Tokyo 164-8639, Japan; e-mail: mnishida@ori.u-tokyo.ac.jp

Abstract

Euteleostean fishes, especially those in the order Perciformes, can represent vertebrate biodiversity. Fossil evidence suggests an apparently explosive radiation of perciform families in the early Cenozoic, raising the question of how freshwater-adapted perciforms like cichlids could distribute over the well-separated landmasses of Gondwanaland origin. In order to gain insights into the time and mode of the perciform radiation by molecular methods, complete mitochondrial genes for NADH dehydrogenase subunit 2 and cytochrome *b* were sequenced for 13 perciforms, a characiform and a basal neopterygian *Amia*. Phylogenetic trees constructed from concatenated amino acid sequences of these genes together with published sequences for other taxa were mostly in agreement with current views on ichthyological classification. The relative rate tests indicated approximate homogeneity of molecular evolutionary rates among many of the diverse fish groups. Gamma-corrected distances based on the mitochondrial protein sequences were shown to provide a good estimate of pairwise distances even for divergences of a few hundred million years ago in time or 0.5–1.0 substitutions per site in pairwise distance. The molecular evolutionary rate for the fishes could be calibrated consistently using the diver-

---

Key words. Teleostei, Perciformes, Cichlidae, freshwater fishes, mitochondrial DNA, cytochrome *b*, phylogenetic tree, fossil records, molecular clock, vertebrate evolution, biodiversity, adaptive radiation, biogeography, divergence time, Cretaceous/Tertiary boundary

gence of African and neotropical cichlids at the time of continental breakup, actinopterygian vs. sarcopterygian divergence, and bony fish vs. chondrichthyan divergence. An alternative but unlikely assumption of the cichlid divergence in the Cenozoic caused clear inconsistency with the latter calibration points. All pairwise distances among the examined perciform families exceeded those between African and neotropical cichlids and diversification of the perciforms was estimated to be substantially older than that deducible from the first occurrence evidence of the corresponding fossil records. Thus, the explosive radiation of perciform families in the early Cenozoic was not supported by the mitochondrial sequence data. Molecular methods are expected to provide new insights into macroevolutionary history of vertebrates which has been established on the basis of fossil evidence.

1 Introduction

Teleosts are a modern group of fishes with well over 22 000 living species which are classified into 40 orders (Nelson 1994). They are characterized by various morphological characteristics with regard to the feeding and locomotor apparatus (Gosline 1971; Carroll 1988; Benton 1997). Osteoglossomorphs (arowanas and their relatives), Elopomorphs (eels, tarpons, and notacanths), Clupeomorphs (herrings and anchovies), and Euteleostei (the remaining teleosts) were recognized to be four basal teleost groups (Nelson 1994). Paleontological records suggest that the first teleosts (pholidophorids and ichthyokentemids) occurred in the early Late Triassic approximately 230 million years ago (MYA), and that various orders of Teleostei were diversified during Jurassic and Cretaceous times (Goody 1969; Benton 1993, 1997). Fossil records for euteleost orders start to occur only from the beginning of the Cretaceous (Benton 1993).

The predominant euteleostean order Perciformes, which contains 18 suborders, 148 families, and the largest number (>9000) of species of any vertebrate orders, shows the greatest amount of morphological diversity (Nelson 1994). Perciforms currently dominate in vertebrate ocean life and are the dominant fish group in (sub)tropical freshwaters. Paleontological records suggest that perciforms originated in the early Late Cretaceous (approximately 90 MYA) and that many of perciform families rapidly radiated in the Paleocene or early Eocene after the Cretaceous/Tertiary boundary (K/T boundary) at 65 MYA (Benton 1993; Carroll 1997). These diversified perciforms contained a number of genera which look very much like extant forms now dominating the marine vertebrate fauna (Carroll 1988). This radiation appeared so explosive that it was occasionally compared to the apparently explosive radiation of mammals at about the same time with a speculation that the perciform radiation may have been related to global changes in climate or ecosystems that led to the extinction of most archaic actinopterygian groups that flourished throughout the Mesozoic (Carroll 1988).

In this study, molecular phylogenetic analyses were conducted in order to evaluate the fossil-based history of perciform radiation in the early Cenozoic. We did not assume *a priori* rates of the molecular clock for bony fishes. Rate calibration was conducted using three independent data points, two of which are external calibration points based on reliable paleontological and molecular evidence. The other is an internal calibration point reasonably assuming that cichlids could not disperse between well-separated African and South American landmasses. Cichlids are secondary freshwater fishes that have undergone remarkable adaptive radiation on landmasses of Gondwanaland origin (see, e.g., Nishida 1997; Stiassny and Meyer 1999). Recent phylogenetic studies of cichlids based on molecular (Sültmann et al. 1995; Zardoya et al. 1996; Streelman and Karl 1997; Streelman et al. 1998; Takahashi et al. 1998; Kanie and Nishida, manuscript in preparation) and morphological (Stiassny 1991) evidence consistently indicate that African and neotropical cichlids are monophyletic relative to each other, and that Indian and Malagasy species make outgroups of the African + neotropical clade. This phylogenetic framework is in good agreement with a currently well-accepted separation process of Gondwanaland (Smith et al. 1994; The Plates Project 1998), supporting the close association between the continental-clade formation of cichlids and the Gondwanaland breakup. These three calibration points are congruent with each other and serve to set a rate calibration for fishes, which may be useful for rough estimation of divergence times among diverse fish groups.

2 Materials and Methods

2.1 Samples and Sequence Determination

Fifteen fish species sequenced in this study are listed in Table 1. Three species were selected from each of African and neotropical cichlids so as to represent deep lineages among each group, in light of previous work (Kocher et al. 1995; Sültmann et al. 1995; Zardoya et al. 1996; Nishida 1997; Streelman et al. 1998; Takahashi et al. 1998). Representatives of four out of six families in the suborder Labroidei (Nelson 1994) were employed (i.e. Cichlidae, Embiotocidae, Pomacentridae, and Labridae). Fish samples for African cichlids and non-cichlid perciforms were captured in native localities of individual species, while the others were obtained from local shops. Total DNA was extracted from a small amount of muscle tissues or fins according to Kocher et al. (1989) and used for the amplification by the polymerase chain reaction (PCR) of genes for NADH dehydrogenase subunit 2 (ND2) and cytochrome *b* (cytb). They are two of the best represented genes for fishes among mitochondrial protein genes that can provide clearly orthologous sets of sequences.

The sequences and positions of the PCR primers used for the amplification are shown in Table 2 and Fig. 1, respectively. Amplification of ND2 genes was carried out either with L4296 and H5635 or with L4157m and H5934m. Amplification of cytb genes was done with L14724m and H15990 except for *T. polylepis* cytb gene that was amplified with L14724m and fcytb-2, and with fcytb-1 and H15915. PCR reactions consisted of 30 cycles of denaturation at 92°C for 40 s, annealing at 45–

Table 1. Scientific names and current classification of taxa sequenced in this study

Species	Classification[a]
African cichlids	
Tylochromis polylepis	Cichlidae, Labroidei, Perciformes
Oreochromis niloticus	Cichlidae, Labroidei, Perciformes
Tropheus moorii	Cichlidae, Labroidei, Perciformes
South American cichlids	
Cichlasoma citrinellum	Cichlidae, Labroidei, Perciformes
Geophagus jurupari	Cichlidae, Labroidei, Perciformes
Astronotus ocellatus	Cichlidae, Labroidei, Perciformes
others	
Ditrema temmincki	Embiotocidae, Labroidei, Perciformes
Chrysiptera cyanea	Pomacentridae, Labroidei, Perciformes
Pseudolabrus japonicus	Labridae, Labroidei, Perciformes
Trachurus japonicus	Carangidae, Percoidei, Perciformes
Apogon semilineatus	Apogonidae, Percoidei, Perciformes
Scomber japonicus	Scombridae, Scombroidei, Perciformes
Rhinogobius giurinus	Gobiidae, Gobioidei, Perciformes
Phenacogrammus interruptus	Characidae, ———, Characiformes
Amia calva	Amiidae, ———, Amiiformes

Although complete ND2 gene sequences were reported for *T. polylepis*, *O. niloticus*, *T. moorii*, and *C. citrinellum* (Kocher et al. 1995), we determined both ND2 and cytb gene sequences for these species using our specimens.

[a] Classification is based on Nelson (1994). Names for the family, suborder, and order are shown from left to right. No suborder name is given to *P. interruptus* and *A. calva*.

55°C for 1 min, and extension at 72°C for 1–3 min. Amplified products were purified with QIAquick PCR purification kit (QIAGEN) and used directly as template for sequence determination with the Applied Biosystems 373A DNA Sequencer. Complete nucleotide sequences for the ND2 and cytb genes were determined unambiguously by sequencing both strands from internal sequencing primers listed in Table 2 (or some taxon-specific internal primers, data not shown) as well as those used for the initial amplification. Database accession numbers for the sequences determined in this study are AB018970–AB018999.

Table 2. Primers used for PCR amplification and sequencing

Name	Sequence (5' to 3')
PCR and sequencing primers	
L4157m	CGTCGGGGATCCTACCCACGATTCCGNTAYGAYCA
L4296	ACGTAGGGATCACTTTGATAG
H5635	AGGTCTTAGCTTAATTAAAG
H5934m	CCCGACGCTGCAGGGTGCCAATGTCTTTRTGRTT
L14724m	TGACTTGAAAAACCAYCGTTG
H15915	ACCTCCGATCTYCGGATTACAAGAC
H15990	AGTTTAATTTAGAATCYTGGCTTTGG
Internal sequencing primers	
L4437	AAGCTATCGGGCCCATACC
H5540	CCGCTGAGGGCTTTGAAGGC
fND2-1	GCCCACCTAGGVTGAATAAT
fND2-2	ATTATTCASCCYAGGTGGGC
fND2-3	TCMACCTGACARAAACT
fND2-4	ATTATTCAKCCWAGGTG
fcytb-1	CGATTCTTYGCMTTCCA
fcytb-2	GAKCCKGTTTCGTGNAGGAA
fcytb-3	TMGTMCAATGAATCTGAGG
fcytb-4	TTKGAKCCRGTTTCGTG

Primers L4157m and H5934m were modified, respectively, from L4160m and H5937m (Kumazawa and Nishida 1993), and L14724m and H15915 were modified, respectively, from L14724 and H15915 (Irwin et al. 1991).

2.2 Phylogenetic Analyses

The determined nucleotide sequences for ND2 and cytb genes were converted to amino acid sequences and aligned by eye with the corresponding sequences for trout (Zardoya et al. 1995), cod (Johansen and Bakke 1996), carp (Chang et al. 1994), loach (Tzeng et al. 1992), bichir (Noack et al. 1996), coelacanth (Zardoya and Meyer 1997), and sharks (Martin et al. 1992; Naylor et al. 1997). The alignment is obtainable from Y.K. upon request. Phylogenetic analyses were conducted using concatenated amino acid sequences of ND2 and cytb after unalignable parts in the C-terminus of each gene, as well as gap sites, were removed (719 sites in total). The neighbor-joining (NJ) tree (Saitou and Nei 1987) was obtained with the njboot program included in Lintre package (Takezaki et al. 1995) with the option of amino Poisson-gamma distance and a gamma-parameter estimated from the data

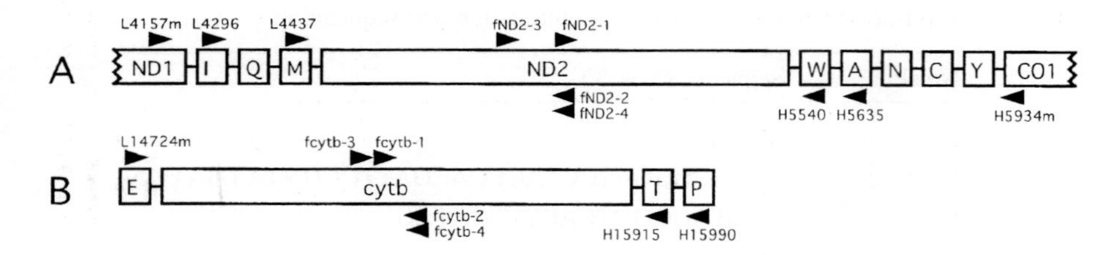

Fig. 1. Position of primers used for amplification and/or sequencing of (**A**) ND2 and (**B**) cytb genes. Refer to Table 2 for their sequence. See text for actual combinations of primers used for amplification

set using Takezaki's gmaes program. The relative rate test was conducted with the tpcv program included in Lintre (the two-cluster test). The maximum-likelihood (ML) distances among taxa were estimated with PUZZLE version 4.0 (Strimmer and von Haeseler 1996). The mtREV24 model with the amino acid frequency estimated from the data set (Adachi and Hasegawa 1996) was used as a model of amino acid substitutions. Gamma-distributed rates of amino acid substitutions among sites (Ota and Nei 1994) were used with the gamma-parameter estimated from the data set with PUZZLE.

3 Results

3.1 Phylogenetic Tree and Rate Constancy Test

Figure 2 shows an NJ tree obtained from the concatenated amino acid sequences of ND2 and cytb genes. The tree topologies among teleost (super)orders are largely in agreement with previous morphological (or paleontological) studies (see, e.g., Rosen 1982; Lauder and Liem 1983; Nelson 1994) and molecular studies (see, e.g., Lydeard and Roe 1997; Streelman et al. 1998) with respect to early divergence of ostariophysan taxa (carp, loach, and characin in this case) and the monophyly of perciform species examined. Although the relationships among a protacanthopterygian (trout), a paracanthopterygian (cod), and acanthopterygians (perciforms), as well as among perciform families were not resolved with high bootstrap probabilities, several plausible nodal relationships were supported with significant or high bootstrap values; i.e., monophyly of teleosts (80%), perciforms (97%), cichlids (76%), neotropical cichlids (96%), and African cichlids (93%). These nodal relationships supported with significant or high bootstrap values were unchanged when the same sequence data were analyzed with a different method (e.g., ML method, and NJ method based on the ML distances) or model (e.g., no assumption of gamma-distribution of rates among sites).

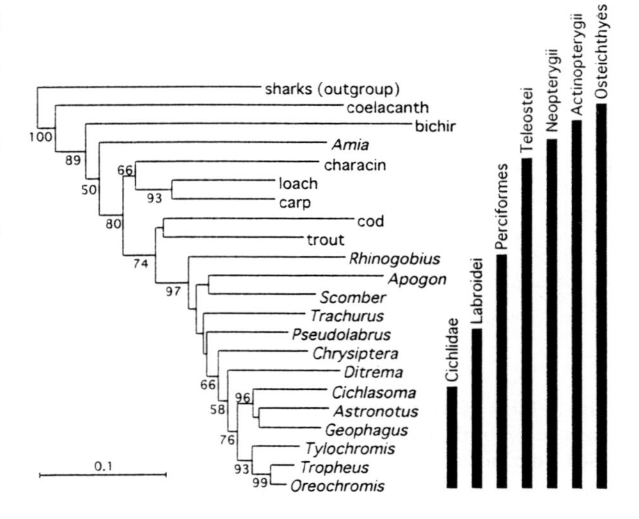

Fig. 2. Neighbor-joining tree constructed from ND2 and cytb amino acid sequences as described in Materials and Methods. Bootstrap values from 1000 replications (only those that are above 50%) are shown on the corresponding branches. Sharks used as outgroups are *I. paucus*, *I. oxyrinchus*, *C. carcharias*, *S. tiburo*, *S. lewini*, *C. plumbeus*, *N. brevirostris*, and *G. cuvier* (Martin et al. 1992; Naylor et al. 1997)

In order to use molecular sequence data as the molecular clock, the homogeneity of modes and rates of sequence evolution among lineages should be examined. Because the ND2/cytb sequence data of bichir were found to have a significantly (5% chi-square test with PUZZLE) different amino acid frequency from the average frequency among all the taxa, bichir was removed for subsequent analyses to estimate divergence times. Because the two-cluster test (Takezaki et al. 1995) using sharks as an outgroup suggested a molecular evolutionary rate on a lineage leading to *Apogon* to be significantly ($p<0.01$) faster than that on the other lineages, *Apogon* was also removed for subsequent analyses. No significant heterogeneity of the evolutionary rate was detected with this criterion among the remaining 19 bony fish species. When the ND2/cytb sequence data were rooted with sea lamprey (Lee and Kocher 1995), the two-cluster test showed that evolutionary rates are not significantly ($0.05<p$) different between actinopterygian lineages (teleosts and *Amia*) and chondrichthyan lineages (sharks) (data not shown), implying rough rate-homogeneity among diverse groups of fishes such as sharks and teleosts, as discussed in more detail in the following sections.

3.2 Performance of ND2 and cytb Sequences as Molecular Clocks

Rapidly evolving mtDNA sequences are generally considered to be useful for analyzing closely related taxa. However, we found that amino acid sequences of mitochondrial protein genes may be useful as a molecular clock for taxa with diver-

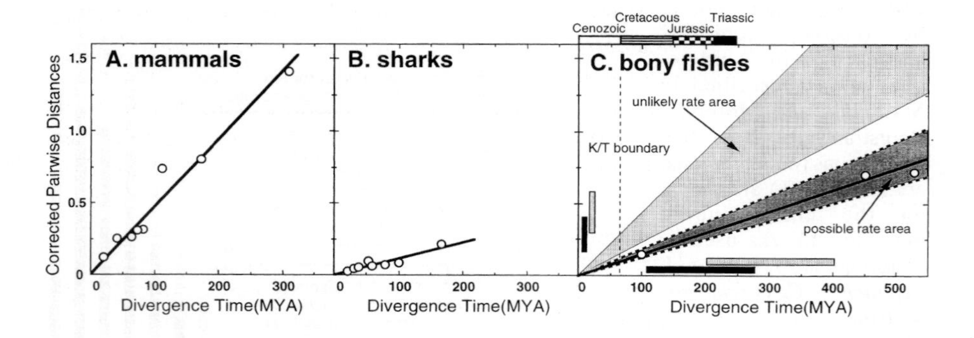

Fig. 3. Plots of pairwise distances against estimated divergence times among (**A**) 15 mammals and 3 birds, (**B**) sharks, and (**C**) bony fishes (plus sharks in part). Pairwise distances used were the gamma-corrected ML distances from the ND2 and cytb amino acid sequences. In *A*, the data consist of 716 sites (the gamma parameter α=0.38). Estimated divergence times used are those from Kumar and Hedges (1998), which seem to be currently the most reliable molecular time estimate based on multiple nuclear protein sequences and are largely in agreement with the Cenozoic and Paleozoic fossil evidence (Kumar and Hedges 1998). Divergences plotted are; 19.6 MYA for cow vs. goat, 40.7 MYA for mouse vs. rat, 64.7 MYA for cow/goat vs. whales, 74.0 MYA for cat/seals vs. rhinos/horse/donkey, 83.0 MYA for cow/goat/whales vs. cat/seals/rhinos/horse/donkey, 112 MYA for the ferungulates vs. the rodents, 173 MYA for the placental mammals vs. opossum/wallaroo, and 310 MYA for the mammals vs. chicken/ostrich/rhea. A regression line through the origin (R=0.99) using StatView (Abacus Concepts, Inc) gave a rate estimate for mammals (2.4×10^{-3} substitutions/site/million year (SSMY)). See Kumazawa and Nishida (1999) for more details. In *B* and *C*, the data consist of the ND2/cytb sequences for 11 sharks (8 species listed in the legend of Fig. 2 plus *A. superciliosus*, *C. taurus*, and *L. nasus*) and 19 bony fishes (all the species shown in Fig. 2 except bichir and *Apogon*) (683 sites, α=0.31). Estimated divergence times used in *B* are from fossil records (Martin et al. 1992; Benton 1993). A regression line through the origin (R=0.98) gave a rate estimate for sharks (6.0×10^{-4} SSMY). Estimated divergence times used in *C* are from the probable divergence time for African and neotropical cichlids (100 MYA), and from an independent molecular study (Kumar and Hedges 1998) for sarcopterygian vs. actinopterygian divergence (450 MYA) and for bony fishes vs. sharks divergence (528 MYA). The *thick line* is a rate calibration for bony fishes assuming the cichlid divergence at 100 MYA (7.3×10^{-4} SSMY). Possible ranges for the rate are shown by the *dark-shaded area bordered by dotted lines*, which are calibrations assuming the intercontinental cichlid divergence at 80-120 MYA ($6.1–9.1 \times 10^{-4}$ SSMY) (see legend of Fig. 4). *Light-shaded areas* represent unlikely ranges of the rate assuming cichlid divergence in the early Cenozoic (35–65 MYA). *Black and grey bars* show possible ranges for the cross-family and cross-order divergences, respectively, by assuming the rate 7.3×10^{-4} SSMY. These ranges cover divergences of all 51 cross-family species-pairs among seven perciform families and all 69 cross-order species-pairs among five teleost orders

gence times of hundreds of million years ago, when distances are appropriately estimated using the gamma-correction of the rate heterogeneity over sites (Fig. 3; Kumazawa and Nishida 1999). In Figs. 3A and 3B, we examined the performance of the ND2/cytb sequences as the molecular clock for long periods of time using mammals (plus some birds) and sharks considered to have relatively reliable fossil records. For mammals, reliable divergence-time estimates are also available from an independent molecular data set using multiple nuclearly-encoded protein sequences, and these time estimates are largely in agreement with Cenozoic and Paleozoic fossil records (Kumar and Hedges 1998).

As shown in Fig. 3A, gamma-corrected ML distances among the mammalian (plus avian in part) ND2/cytb sequences accumulate almost linearly (R=0.99) by time. The linear relationship was similarly observed when gamma-corrected Poisson distances were plotted, but not in the case of simple ML or Poisson distances (Kumazawa and Nishida 1999). The evolutionary rate of ND2/cytb amino acid sequences for sharks (Fig. 3B) appears to be much slower (approximately a fourth) than that of mammals (Fig. 3A). The reduction of molecular evolutionary rates in sharks is in agreement with previous work (Martin et al. 1992) using DNA sequences. Taken together, these results support the assumption that ND2/cytb amino acid sequences can perform as a molecular clock even for divergences of a few hundreds of million years ago in time or 0.5–1.0 substitutions per site in pairwise distance. These genes were therefore expected to be useful for estimating divergence times among teleosts.

3.3 Rate Calibration in Bony Fishes

Fragmentation processes of Gondwanaland have been well-elucidated from recent geological evidence (Smith et al. 1994; The Plates Project 1998). Figure 4 shows three possible models for the evolutionary history of cichlids based on both the molecular and geological evidence. For the reasoning described in the legend of Fig. 4, the vicariance or dispersal model leads to the reasonable assumption that extant African and neotropical cichlids separated from each other at approximately 100 MYA (80–120 MYA). When molecular evolutionary rates for bony fishes were calibrated under this assumption (Fig. 3C), the line calibrated at 100 MYA nearly runs through another data point for the divergence between coelacanth and actinopterygians (teleosts plus *Amia* in this case) plotted against an independent time estimate using multiple nuclear protein sequences (450 MYA, Kumar and Hedges 1998). Although the other data point for the divergence between bony fishes and sharks appears to deviate slightly from this calibration line, it is placed well within the possible range of calibration (see the dark-shaded area in Fig. 3C). If cichlids had dispersed across the paleo-Atlantic or originated from labroid ancestor(s) at a time which is less incompatible with the fossil record (i.e., 35–65 MYA from the Eocene to the Paleocene), the molecular evolutionary rate for bony fishes would have to be increased by at least 50% from the estimate calibrated at 100 MYA, causing clear incompatibility with the other calibration points (see the light-shaded area in Fig. 3C).

The molecular evolutionary rate for bony fishes was thus calibrated using both internal and external data points. The rate calibrated at 100 MYA (Fig. 3C) is nearly the same as or only slightly faster than that for sharks (Fig. 3B), which is in agreement with the results of the two-cluster test mentioned above. It should be noted that, due to slower rates of molecular evolution in fishes than in mammals (Thomas and Beckenbach 1989; Martin et al. 1992; Adachi et al. 1993; Cantatore et al. 1994; Fig. 3 of this study), pairwise distances among bony fishes used in this study are within a range (<0.71) where linearity in calibration was observed for mammals (Fig. 3A).

Table 3. Preliminary divergence-time estimates from the ND2/cytb sequences

Divergences at each node of Fig. 2	Pairs[a]	Time (MYA) Γ-ML	Time (MYA) Γ-Poisson[b]
African vs. neotropical cichlids	9	100	100
Cross-family divergences among perciforms			
Cichlidae vs. Embiotocidae	6	130	137±16
Cichlidae/Embiotocidae vs. Pomacentridae	7	125	127±14
Cichlidae/Embiotocidae/Pomacentridae vs. Labridae	8	138	130±15
Labroidei vs. Carangidae	9	157	148±17
Labroidei/Carangidae vs. Scombridae	10	171	171±19
Labroidei/Carangidae/Scombridae vs. Gobiidae	11	229	210±23
Cross-order divergences among teleosts			
Salmoniformes vs. Gadiformes	1	249	258±32
Cypriniformes vs. Characiformes	2	266	250±29
Salmoniformes/Gadiformes vs. Perciformes	24	265	258±25
Salmoniformes/Gadiformes/Perciformes vs. Cypriniformes/Characiformes	42	296	284±28
deeper divergences			
Teleostei vs. Amiiformes	17	404	367±40
Actinopterygii vs. Sarcopterygii	18	455	490±54

Divergence times based on gamma-corrected ML and Poisson distances were obtained using PUZZLE (Strimmer and von Haeseler 1996) and tpcv of the Lintre package (Takezaki et al. 1995), respectively, with the rate calibrated using the divergence of African and neotropical cichlids at 100 MYA (see Fig. 3C and text) and a gamma parameter (α=0.28) estimated from the data set. These gamma-corrected distances were shown to give a good estimate of distances for deep divergences (Fig. 3; Kumazawa and Nishida 1999).

[a]Number of species pairs used for the divergence-date estimation.

[b]Means of divergence times among the corresponding species-pairs are shown with one standard error.

3.4 Divergence Date Estimates Among Teleosts

The calibrated rate was used to estimate divergence times among teleost orders and among perciform families (Table 3). Estimated divergence times among five teleost orders dated back to the Paleozoic (mostly Permian). A somewhat broad range for the possible rates (the dark-shaded area in Fig. 3C) may reduce accuracy of the estimation, and we are aware that the sequence data should be further expanded in the numbers of both sites and taxa in order to obtain very accurate phylogeny and divergence times. We thus consider the time estimates in Table 3 to be tentative approximation. However, it seems obvious that the range of estimated times is considerably older than the estimates deducible from fossil records. Fossils for these euteleost orders are not known before the Cretaceous (Benton 1993), whereas the molecular time estimates for the interordinal divergences even predate the first occurrence record of Teleostei in the early Late Triassic.

Estimated divergence times among seven perciform families were broadly distributed in the Mesozoic (mostly from the Jurassic to the early Cretaceous; Table 3). They are also considerably older than the fossil-based estimates which supported the notion of rapid radiation of various perciform families after the K/T boundary (Benton 1993). Although many of the 148 perciform families were not represented in the present study, four out of six existing families in the suborder Labroidei were sampled and all the divergence dates among these families predated that between African and South American cichlids (see Table 3). The apparent discrepancy between molecular and fossil-based estimates becomes most conspicuous by the fact that no fossils clearly assignable to acanthomorphs (an advanced euteleost group including the Perciforms) are known before the Cenomanian stage (90–97 MYA) (Patterson 1993). Taken together, there is a considerable gap between divergence-date estimates for teleosts from molecular evidence and those based on the first occurrence evidence of fossil records.

4 Discussion

4.1 Intercontinental Distribution of Cichlids

Some previous researchers (see, e.g., Bănărescu 1990; Stiassny 1991) considered that cichlids inhabited Gondwanaland and were separated into African and neotropical lineages upon its breakup (the vicariance model in Fig. 4A). However, fossil records of the Cichlidae are known only from the Cenozoic (Oligocene, Benton 1993) while the breakup of Gondwanaland occurred much earlier in the early or middle Cretaceous. This gap led other researchers (see, e.g., Lundberg 1993; Briggs 1995) to consider an alternative explanation that euryhaline cichlids could disperse across the paleo-Atlantic ocean over a modest saltwater gap during the late Cretaceous or possibly the early Cenozoic because some (but a very limited number of) extant cichlid species are adapted to brackish water (Nelson 1994). Maisey (1993) referred to possible tectonic settings for the long-distance dispersal.

A. the vicariance model

B. the dispersal model

C. the seawater-origin model

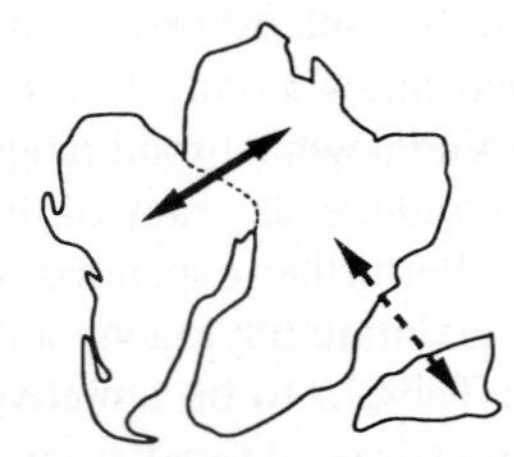

Before separation of African and South American landmasses (Solid and broken arrows mean migration possible and impossible, respectively)

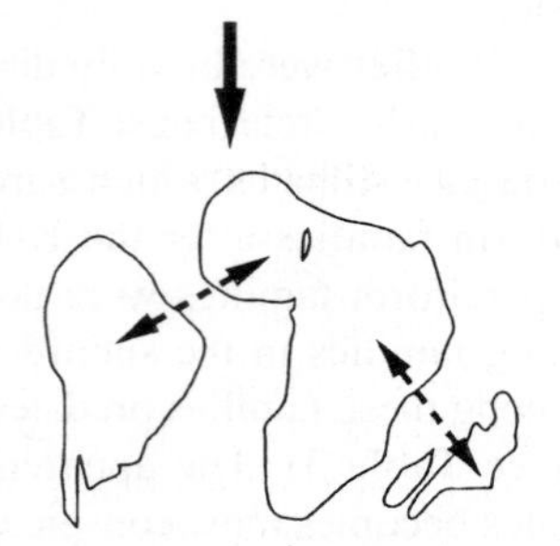

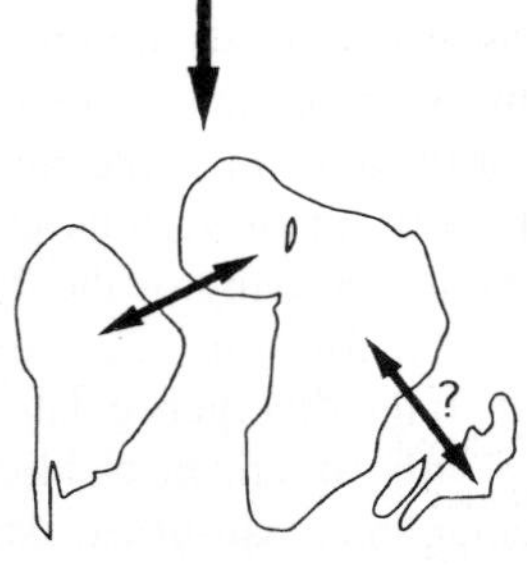

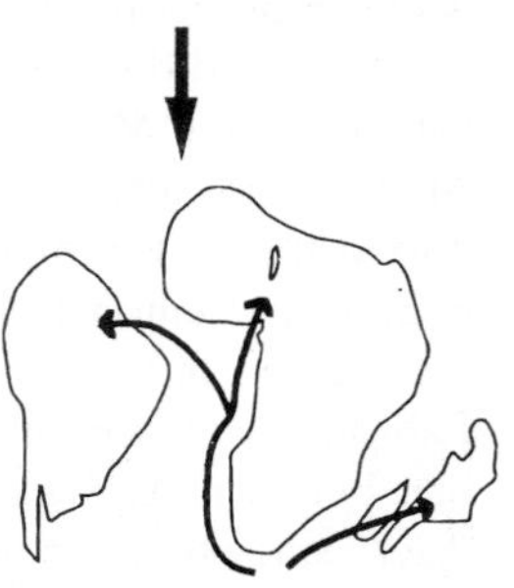

After separation of African and South American landmasses

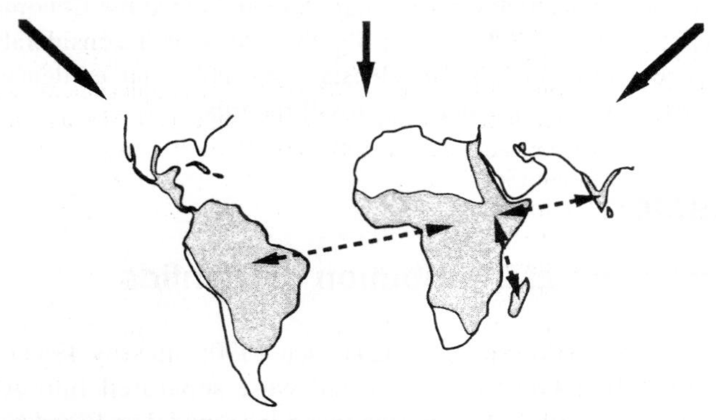

Extant cichlids inhabit shaded areas of the world

However, there is no direct evidence for the actual occurrence of the special tectonic settings, without which it would have been virtually impossible for even the euryhaline cichlids to travel between well-separated African and South American landmasses. Since cichlids are primarily fishes of lowland tropics, it is also unlikely that they dispersed through freshwater systems in high-latitude areas. Because the Cretaceous earth had a relatively warm climate and high sea-levels throughout the period (Briggs 1995; refs. therein), it seems unlikely that African and South American landmasses became temporarily connected through a landbridge after their complete disconnection at 100 MYA. We therefore deduced that dispersal, if any, could have happened only when the separated landmasses were relatively close to each other (the dispersal model in Fig. 4B).

Among the three possible models for the intercontinental distribution of cichlids (Fig. 4), the vicariance model seems to explain most naturally the phylogenetic relationship among continental cichlid groups, but this model as well as the dispersal model have to assume that old lineages were replaced by new ones by remarkable adaptive radiations (see the legend of Fig. 4). Cichlids in the African rift lakes are known for their explosive adaptive radiation and this may have occurred frequently in the Cretaceous, too. The seawater-origin model remains possible but

Fig. 4. Models for the evolutionary history of cichlids that may reconcile the molecular and geological data. (**A**) the vicariance model, (**B**) the dispersal model, and (**C**) the seawater-origin model. In **A**, multiple freshwater cichlid lineages were already widespread on Gondwanaland before its breakup. The Indian-Malagasy landmass was first separated from Gondwanaland and old lineages now inhabiting India and Madagascar were geographically isolated. African and South American landmasses were then separated from each other and cichlids could not disperse between them. Remarkable adaptive radiation of new lineages and their replacement of old ones shaped extant African and neotropical cichlids to be monophyletic to each other. The vicariance model may predict that extant African and neotropical cichlid lineages separated from each other at 100–120 MYA. 100 MYA is the age of complete separation of African and South American landmasses (Smith et al. 1994; The Plates Project 1998). However, the lineage separation may have somewhat preceded the complete separation of the two landmasses. Fossil evidence on taxa such as ostracods, foraminiferans, and nannoplanktons strongly favored a marine north-south connection between the two landmasses as early as the Lower Albian (approximately 110 MYA) (Briggs 1995; refs. therein). Geological evidence shows that the two landmasses were connected only through narrow equatorial regions at 120 MYA (Smith et al. 1994; The Plates Project 1998). In **B**, after the main landmasses became disconnected, some freshwater cichlids readapted to brackish water could disperse in short distances between them. They replaced preexisting lineages (if any) on each landmass and gave rise to extant lineages. The dispersal model may predict that extant African and neotropical lineages separated from each other at 80-100 MYA. Even euryhaline cichlids readapted to brackish water may not have been able to disperse between well-separated African and South American landmasses after 80 MYA. In *C*, after Gondwanaland became well-fragmented, freshwater-adapted cichlids leading to extant lineages arose from labroid ancestor(s) that could travel across paleo-oceans. The seawater-origin model does not, by itself, constrain the lineage separation time between landmasses. However, Fig. **3C** supports the premise that the lineage separation time between African and neotropical cichlids based on this model cannot have been much younger than 80 MYA seems to lack positively supportive evidence. At present we have little evidence with which to deduce what the labroid ancestor(s) of cichlids were. Phylogenetic relationships among labroid families should be investigated further by using more taxa from each family.

Although the molecular data collected in this study do not specify one of the three models, they can constrain conditions under which the models hold. Figure 3C suggested that the dispersal or origination of cichlids could not have happened in the early Cenozoic. The gap in the fossil record may not become strong evidence of a Cenozoic dispersal, because the incompleteness of fossil records for long periods is now likely for perciforms in general (Table 3).

4.2 Apparent Discrepancies Between Molecular and Fossil Chronologies

Macroevolutionary history of organisms through time has been traditionally reconstructed on the basis of fossil records. In principle, the divergence time for two lineages can be deduced from the date of the earliest fossil records clearly attributable to either of the two lineages (Marshall 1990; Benton 1993). However, in general, fossil evidence is incomplete in various aspects . It may be lacking or fragmentary for some taxa with a small population size or with less preservable components, biased by various environmental factors for fossilization, or obscured by the arbitrariness of researchers' morphological interpretations (Marshall 1990; Carroll 1997; Benton 1998). It thus seems significant to evaluate the fossil-based history using independent approaches such as molecular evolutionary methods.

The present study shows the possibility of a long-term lack of fossil records for not only cichlids but also many other families of the Perciformes. If this is really the case, how can it have happened? We provide two explanations that may stand simultaneously. First, fossils of fishes are not necessarily considered to be preserved very well. A number of perciform families are completely lacking in their fossil record and many of the first occurrences for the other families having recognizable fossil records are based only on otoliths (Benton 1993; Patterson 1993). Moreover, due to the less preservable nature of fish fossils in general, interpretation of occurrence and disappearance of fish lineages tends to be influenced by a limited number of fossil localities of exceptional preservation (the Lagerstätten effect; see, e.g., Goody 1969; Patterson 1993; Benton 1998). These problems may cause strong bias in inferring evolutionary history of fishes, leading to considerable underestimation of origination or divergence dates, and apparently rapid cladogenesis or mass extinction (Benton 1998).

Second, fossil records in general tend to be influenced by population size of the corresponding species and various taphonomical conditions for fossilization (Carroll 1997; Benton 1998). It may therefore be possible to deduce that perciforms had a much smaller population size in the Mesozoic than in the Cenozoic and/or they occupied much more confined ecological niches than today. It is noteworthy that perciforms are not recognized even in the Santana Formation that produces ample marine fish fossils of exceptional preservation for the Aptian–Albian Age (~110 MYA) (Maisey 1993), when perciforms may have already radiated according to the molecular evidence presented in this study (Table 3).

4.3 Implications

The present study suggested that euteleosts have had a much longer history than previously thought. This notion seems so different from the previous view on this issue that it should be further evaluated in the future by collecting longer sequence data for more perciform taxa and/or by using more calibration points for the molecular clock. Possible other candidates for the rate calibration may include characiforms (Ortí 1997), cyprinodontiforms (Murphy and Collier 1997), and geminate species pairs that most likely speciated by the rise of the Panamanian Isthmus (Bermingham et al. 1997). At present, the molecular evolutionary rates reported in the latter two papers may not be directly comparable to ours due to, e.g., the differences in sequenced genes and/or order of time scale for the calibration.

It has been pondered why perciforms appear to have evolved various morphotypes rapidly in the early Cenozoic, while they have been largely conservative since then (Carroll 1988). Our results may partly explain this question by relaxing the time period during which the original morphological changes could occur. According to molecular evidence presented herein, radiation of perciforms may not have been related to global changes in climate and ecosystems that led to the extinction of most archaic actinopterygian groups that flourished throughout the Mesozoic. However, as discussed above, it is still possible that their population size was dramatically increased and/or that they occupied various niches upon the global changes after the K/T boundary The long-term lack of fossil records, as well as the apparent uncoupling between lineage radiation of new groups and extinction of old ones have also been implicated for terrestrial vertebrates such as mammals and birds by some molecular studies (Janke et al. 1994; Hedges et al. 1996; Cooper and Penny 1997; Kumar and Hedges 1998).

Occurrence of a similar phenomenon for organisms living in different environments (on land and in the sea) may imply its ubiquitousness for more diverse groups of organisms that lived in different geological ages. Vertebrates are considered to have relatively well-documented fossil records and there have been no substantial changes in the accepted broad-scale pattern of vertebrate evolution since the 19th century (Benton 1998). However, this pattern may not be unchangeable or, at least, it may be worth questioning and reexamining with newer data sets from modern methods such as molecular evolutionary ones.

Acknowledgments

We thank N. Takezaki for sending us an earlier version of her programs now contained in the Lintre package, and instructions on the use of these programs. We also thank M. Miya and an anonymous reviewer for many valuable comments on an earlier version of the manuscript and K. Tamaki, Y. Terai and N. Okada for sending us a manuscript before publication. This work was supported by a grant from the Ministry of Education, Science and Culture of Japan to Y.K. (No. 09214102) and M.N. (No. 10660189).

References

Adachi J, Hasegawa M (1996) Model of amino acid substitution in proteins encoded by mitochondrial DNA. J Mol Evol 42:459-468

Adachi J, Cao Y, Hasegawa M (1993) Tempo and mode of mitochondrial DNA evolution in vertebrates at the amino acid sequence level: rapid evolution in warm-blooded vertebrates. J Mol Evol 36:270-281

Bănărescu P (1990) Zoogeography of fresh waters, Vol 1. AULA-Verlag, Wiesbaden

Benton MJ (1993) The Fossil record 2. Chapman and Hall, London

Benton MJ (1997) Vertebrate palaeontology. Chapman and Hall, London

Benton MJ (1998) The quality of the fossil record of the vertebrates. In: Donovan SK, Paul CRC (Eds) The adequacy of the fossil record. John Wiley and Sons, New York, pp 269-303

Bermingham E, McCafferty SS, Martin AP (1997) Fish biogeography and molecular clocks: perspectives from the Panamanian Isthmus. In: Kocher TD, Stepien CA (Eds) Molecular systematics of fishes. Academic Press, London, pp 113-128

Briggs JC (1995) Global biogeography. Elsevier, Amsterdam

Cantatore P, Roberti M, Pesole G, Ludovico A, Milella F, Gadaleta MN, Saccone C (1994) Evolutionary analysis of cytochrome b sequences in some perciformes: evidence for a slower rate of evolution than in mammals. J Mol Evol 39:589-597

Carroll RL (1988) Vertebrate paleontology and evolution. WH Freeman, New York

Carroll RL (1997) Patterns and processes of vertebrate evolution. Cambridge Univ Press, New York

Chang Y-S, Huang F-L, Lo T-B (1994) The complete nucleotide sequence and gene organization of carp (*Cyprinus carpio*) mitochondrial genome. J Mol Evol 38:138-155

Cooper A, Penny D (1997) Mass survival of birds across the Cretaceous-Tertiary boundary: molecular evidence. Science 275:1109-1113

Goody PC (1969) The relationships of certain upper Cretaceous teleosts with special reference to the myctophoids. Bull Br Mus Nat Hist (Geol), London 7:3-255

Gosline WA (1971) Functional morphology and classification of teleostean fishes. Univ Hawaii Press, Honolulu

Hedges SB, Parker PH, Sibley CG, Kumar S (1996) Continental breakup and the ordinal diversification of birds and mammals. Nature 381:226-229

Irwin DM, Kocher TD, Wilson AC (1991) Evolution of the cytochrome *b* gene of mammals. J Mol Evol 32:128-144

Janke A, Feldmaier-Fuchs G, Thomas WK, von Haeseler A, Pääbo S (1994) The marsupial mitochondrial genome and the evolution of placental mammals. Genetics 137:243-256

Johansen S, Bakke I (1996) The complete mitochondrial DNA sequence of Atlantic cod (*Gadus morhua*): relevance to taxonomic studies among codfishes. Mol Marine Biol Biotechnol 5:203-214

Kocher TD, Thomas WK, Meyer A, Edwards SV, Pääbo S, Villablanca FX, Wilson AC (1989) Dynamics of mitochondrial DNA evolution in animals: amplification and sequencing with conserved primers. Proc Natl Acad Sci USA 86:6196-6200

Kocher TD, Conroy JA, McKaye KR, Stauffer JR, Lockwood SF (1995) Evolution of NADH dehydrogenase subunit 2 in East African cichlid fish. Mol Phylogenet Evol 4:420-432

Kumar S, Hedges SB (1998) A molecular timescale for vertebrate evolution. Nature 392:917-920

Kumazawa Y, Nishida M (1993) Sequence evolution of mitochondrial tRNA genes and deep-branch animal phylogenetics. J Mol Evol 37:380-398

Kumazawa Y, Nishida M (1999) Molecular phylogenetic analysis of vertebrate radiations. In: Iwatsuki K (Ed) IIAS Reports. No. 1999, International Institute for Advanced Studies, Kyoto, Japan, in press [online text will be available: http://www.iias.or.jp/home.html and http://www2.ori.u-tokyo.ac.jp/~mnishida/ publication/IIAS_Report.html]

Lauder GV, Liem KF (1983) The evolution and interrelationships of the actinopterygian fishes. Bull Mus Comp Zool 150:95-197

Lee W-J, Kocher TD (1995) Complete sequence of a sea lamprey (*Petromyzon marinus*) mitochondrial genome: early establishment of the vertebrate genome organization. Genetics 139:873-887

Lundberg JG (1993) African-South American freshwater fish clades and continental drift: problems with a paradigm. In: Goldblatt P (Ed) Biological relationships between Africa and South America. Yale Univ Press, New Haven, pp 156-198

Lydeard C, Roe KJ (1997) The phylogenetic utility of the mitochondrial cytochrome *b* gene for inferring relationships among actinopterygian fishes. In: Kocher TD, Stepien CA (Eds) Molecular systematics of fishes. Academic Press, London, pp 285-303

Maisey JG (1993) Tectonics, the Santana Lagerstätten and the implications for late Gondwanan biogeography. In: Goldblatt P (Ed) Biological relationships between Africa and South America. Yale Univ Press, New Haven, pp 435-454

Marshall CR (1990) The fossil record and estimating divergence times between lineages: maximum divergence times and the importance of reliable phylogenies. J Mol Evol 30:400-408

Martin AP, Naylor GJP, Palumbi SR (1992) Rates of mitochondrial DNA evolution in sharks are slow compared with mammals. Nature 357:153-155

Murphy WJ, Collier GE (1997) A molecular phylogeny for aplocheiloid fishes (Atherinomorpha, Cyprinodontiformes): the role of vicariance and the origins of annualism. Mol Biol Evol 14:790-799

Naylor GJP, Martin AP, Mattison EG, Brown WM (1997) Interrelationships of Lamniform sharks: testing phylogenetic hypotheses with sequence data. In: Kocher TD, Stepien CA (Eds) Molecular systematics of fishes. Academic Press, London, pp 199-218

Nelson JS (1994) Fishes of the world. John Wiley and Sons, New York

Nishida M (1997) Phylogenetic relationships and evolution of Tanganyikan cichlids: a molecular perspective. In: Kawanabe H, Hori M, Nagoshi M (Eds) Fish communities in Lake Tanganyika. Kyoto Univ Press, Kyoto, pp 1-24

Noack K, Zardoya R, Meyer A (1996) The complete mitochondrial DNA sequence of the bichir (*Polypterus ornatipinnis*), a basal ray-finned fish: ancient establishment of the consensus vertebrate gene order. Genetics 144:1165-1180

Ortí G (1997) Radiation of characiform fishes: evidence from mitochondrial and nuclear DNA sequences. In: Kocher TD, Stepien CA (Eds) Molecular systematics of fishes. Academic Press, London, pp 219-243

Ota T, Nei M (1994) Estimation of the number of amino acid substitutions per site when the substitution rate varies among sites. J Mol Evol 38:642-643

Patterson C (1993) An overview of the early fossil record of acanthomorphs. Bull Mar Sci 52:29-59

Rosen DE (1982) Teleostean interrelationships, morphological function and evolutionary inference. Am Zool 22:261-273

Saitou N, Nei M (1987) The neighbor-joining method: a new method for reconstructing phylogenetic trees. Mol Biol Evol 4:406-425

Smith AG, Smith DG, Funnell BM (1994) Atlas of Mesozoic and Cenozoic coastlines. Cambridge Univ Press, New York

Stiassny MLJ (1991) Phylogenetic intrarelationships of the family Cichlidae: an overview. In: Keenleyside MHA (Ed) Cichlid fishes: behaviour, ecology and evolution. Chapman and Hall, London, pp 1-35

Stiassny MLJ, Meyer A (1999) Cichlids of the rift lakes. Sci Am 280:64-69

Streelman JT, Karl SA (1997) Reconstructing labroid evolution using single-copy nuclear DNA. Proc Roy Soc Lond Biol Sci 264:1011-1020

Streelman JT, Zardoya R, Meyer A, Karl SA (1998) Multilocus phylogeny of cichlid fishes (Pisces: Perciformes): evolutionary comparison of microsatellite and single-copy nuclear loci. Mol Biol Evol 15:798-808

Strimmer K, von Haeseler A (1996) Quartet puzzling: a quartet maximum-likelihood method for reconstructing tree topologies. Mol Biol Evol 13:964-969

Sültmann H, Mayer WE, Figueroa F, Tichy H, Klein J (1995) Phylogenetic analysis of cichlid fishes using nuclear DNA markers. Mol Biol Evol 12:1033-1047

Takahashi K, Terai Y, Nishida M, Okada N (1998) A novel family of short interspersed repetitive elements (SINEs) from cichlids: the patterns of insertion of SINEs at orthologous loci support the proposed monophyly of four major groups of cichlid fishes in Lake Tanganyika. Mol Biol Evol 15:391-407

Takezaki N, Rzhetsky A, Nei M (1995) Phylogenetic test of the molecular clock and linearized trees. Mol Biol Evol 12:823-833

The Plates Project (1998) Atlas of paleogeographic reconstructions. Plates Progress Report No 215, Univ Texas Institute for Geophysics Technical Report No 181, 88 pp

Thomas WK, Beckenbach AT (1989) Variation in salmonid mitochondrial DNA: evolutionary constraints and mechanisms of substitution. J Mol Evol 29:233-245

Tzeng C-S, Hui C-F, Shen S-C, Huang PC (1992) The complete nucleotide sequence of the *Crossostoma lacustre* mitochondrial genome: conservation and variations among vertebrates. Nucleic Acids Res 20:4853-4858

Zardoya R, Meyer A (1997) The complete DNA sequence of the mitochondrial genome of a "living fossil", the coelacanth (*Latimeria chalumnae*). Genetics 146:995-1010

Zardoya R, Garrido-Pertierra A, Bautista JM (1995) The complete nucleotide sequence of the mitochondrial DNA genome of the rainbow trout, *Oncorhynchus mykiss*. J Mol Evol 41:942-951

Zardoya R, Vollmer DM, Craddock C, Streelman JT, Karl S, Meyer A (1996) Evolutionary conservation of microsatellite flanking regions and their use in resolving the phylogeny of cichlid fishes (Pisces: Perciformes). Proc Roy Soc Lond Biol Sci 263:1589-1598

4
Molecular α-Taxonomy of a Morphologically Simple Fern *Asplenium nidus* Complex from Mt. Halimun National Park, Indonesia

Noriaki Murakami[1], Yoko Yatabe[1], Hisako Iwasaki[1], Dedy Darnaedi[2], and Kunio Iwatsuki[3]

[1] Department of Botany, Graduate School of Science, Kyoto University, Kitashirakawa-Oiwake-cho, Sakyo-ku, Kyoto 606-8502, Japan
[2] Botanical Gardens, Bogor, Jl. Ir. H. Juanda 13, Bogor 16122, Indonesia
[3] Faculty of Science, Rikkyo University, 3-34-1 Nishi-Ikebukuro, Toshima-ku, Tokyo 171-0021, Japan

Abstract

Asplenium nidus is one of the most common epiphytic fern species in the Old World tropics. It has simple leaves without any particular appendages, and the lack of good species recognition might be due to the lack of good taxonomic characters. Recently, it has become easier to collect DNA nucleotide sequence data from wild plants using PCR and direct sequencing techniques. This kind of molecular information has mostly been used for phylogenetic analyses, but can also be useful for α-taxonomy (recognition of naturally existing species). In this work, we made a detailed analysis of the genetic variation of *A. nidus* in one locality, Mt. Halimun National Park, West Java, Indonesia. We collected leaf materials from 25 individuals, all identified as *A. nidus* according to the most recent monograph of the group by Holttum (1974). We found three types of *rbcL* sequences and there was variation at 30 sites out of 1,227 bp. Moreover, we found good correlation among *rbcL* types, some morphological characters (leaf shape and size, sorus length) and ecological traits (habitats and altitudes). DNA nucleotide sequence information might be useful as a primary key even for α-taxonomy, especially for morphologically simple organisms like ferns.

Key words. α-taxonomy, *Asplenium* sect. *Thamnopteris*, *Asplenium nidus*, fern, habitat, Halimun National Park, molecular taxonomy, *rbcL* - species complex

1 Introduction

The bird-nest fern *Asplenium nidus* L. (Fig. 1) is one of the most common and well-known epiphytic fern species with large simple leaves in the Old World tropics. *Asplenium* sect. *Thamnopteris* was first established by Presl (1836) as consisting of the sole species *A. nidus*. Smith (1841a,b) raised the section to a new genus *Neottopteris* including other related species. This group was defined by a synapomorphic character peculiar to Aspleniaceae, submarginal veins connecting lateral veins. Thus, it is easily recognizable and its monophyly is almost clear. In spite of its clear monophyly as a whole group, species delimitation within the section is very confusing. Holttum (1974) monographed 15 species of sect. *Thamnopteris* using gross morphological characters such as frond, scale and rhizome shapes, but neither their naturalness *sensu* Copeland (1929) or cohesion *sensu* Templeton (1989) as species is clear, because their morphology is too simple to find good qualitative taxonomic characters for species recognition.

In our previous work (Murakami et al. 1999), we examined variations in *rbcL* nucleotide sequences of Japanese species of *Asplenium* sect. *Thamnopteris,* as well as materials from Thailand, Vietnam, China, Australia and New Caledonia. We found a very large amount of variation in this section, which is comparable to those in some angiosperm families. Three species of sect. *Thamnopteris, A. antiquum* Makino, *A. australasicum* (J. Smith) Hooker and *A. nidus* are distributed in Japan. We inferred that *A. antiquum* is a species of ancient origin with 40–50 nucleotide

Fig. 1. Typical epiphytic growth habit of *Asplenium nidus* in Mt. Halimun National Park

differences in 1,194 bp (base pairs) sequences of the gene, though morphologically it is not very different from the other species of the sect. *Thamnopteris*. It was also discovered that the so called "*A. australasicum*" in Japan has a very different *rbcL* sequence from *A. australasicum sensu* Holttum distributed in Australia and South Pacific Islands. Based on these molecular data, we described the Japanese "*A. australasicum*" as a new species, *Asplenium setoi* N. Murak. et Seriz. However, it is sometimes difficult to recognize species as realistic biological units by examining allopatric populations. Comparisons are required among sympatric populations of the plants with various *rbcL* types which may or may not be able to be reproductively isolated depending on their genetic differences.

In the present work, we examined *rbcL* variations of sympatric populations in Mt. Halimun National Park where a large amount of biodiversity can be expected. We used as materials, the plants that were all identified as one species, *A. nidus sensu* Holttum (1974). Mt. Halimun National Park is located in west Java, Indonesia, and has primarily well preserved tropical lowland and mountain forests (1000–1800 m altitude). Relatively large morphological variations of *A. nidus* especially in leaf shape and size, soral morphology and position, and the existence or absence of a keel on the undersurface of the rachis and petiole, were observed. We found a very large amount of *rbcL* sequence variation in the sympatric populations. Furthermore, we compared the *rbcL* sequences of the plants from Mt. Halimun with those from other paleotropical localities, including its neighboring areas on Java Island such as Mt. Gede and Bogor, in order to clarify their phylogenetic position. We also compared differences in morphological characters and ecological habitats between different *rbcL* types in order to recognize natural biological units in sect. *Thamnopteris* distributed in Mt. Halimun.

2 Materials and Methods

2.1 Plant Materials

Fresh green leaves of 20 plants of *A. nidus* representing a wide spectrum of morphological variation were collected in a natural primary mountain forest (1000–1800 m alt.) near Chikaniki station of Mt. Halimun National Park. For comparison, we also collected plant materials from Mt. Gede (3 individuals) and Bogor Botanical Garden (naturally growing plants, 1 individual), which are 50-60 km apart from Mt. Halimun. Usually two leaves were collected from each individual, and one was used for DNA extraction and the other as voucher specimen. When only one leaf was available, a small portion was used for molecular analysis, and the remainder was deposited as a voucher specimen. Voucher specimens were kept in the herbaria of the Graduate School of Science, Kyoto University (KYO); the Graduate School of Science, The University of Tokyo (TI); and Aichi Kyoiku University (AICH). Morphological comparison was made using these voucher specimens. Voucher information is shown in Table 1.

Table 1. Taxa and vouchers for the plant materials sequenced

Plant samples	Voucher	Locality	DDBJ No.
A. griffithianum (outgroup)			
	NM J93-001 (TI)	Yaku Is., Kagoshima Pref., JAPAN	AB013252
A. antiquum			
JP-Yakushima	JY 5151 (TI)	Suzunoko River, Yaku Is., Kagoshima Pref., JAPAN	AB013237
A. aff. *antiquum*			
VN-Sapa	KI et al. 94-	Sapa, Hoang Lien Son Prov.,	AB013244
V259 (TI)	VIETNAM		
A. australasicum			
AU-Brisbane	NS s.n. Aug. 7, 1997 (KYO)	Brisbane, AUSTRALIA	AB013249
NEW CALEDONIA	NM 97-N014 (KYO)	Mt. Mou, Is. Grande Terre, NEW CALEDONIA	AB013250
A. setoi			
JP-Okinawa	SS 71596 (AICH)	Urazoe, Okinawa Is., Okinawa Pref., JAPAN	AB013243
JP-Daitoh	NM 96-J101 (TI)	Kita-Daitoh Is., Daitoh Islands, Okinawa Pref., JAPAN	AB013241
A. nidus			
Halimun A*	YY 98-I14 (KYO)	Mt. Halimun, West Java, INDONESIA	AB023500
Halimun B*	KI 97-I03 (KYO)	Mt. Halimun, West Java, INDONESIA	AB023501
Halimun C*	KI 97-I23 (KYO)	Mt. Halimun, West Java, INDONESIA	AB013245
Gede (1,400 m)*	NM 97-I31 (KYO)	Mt. Gede (1,400 m), West Java, INDONESIA	AB013245
Gede (2,400 m)*	YY 98-I151 (KYO)	Mt. Gede (2,400 m), West Java, INDONESIA	AB023502
Bogor*	NM 97-I35 (KYO)	Bogor Botanical Garden, West Java, INDONESIA	AB023508
JP-Amami	NM 94-J022 (TI)	Mt. Yuwan, Amami Is., Kagoshima Pref., JAPAN (cultivated in Univ. of Tokyo)	AB013239
CH-Hekou*	NM & XC 95-2851 (TI)	Hekou, Yunnan Prov., CHINA	AB023503
TH-Suthep	NF et al. 94-T382 (TI)	Mt. Suthep, Chiang Mai Prov., THAILAND	AB013247
VN-CatBa	KI et al. 94-V374 (TI)	Cat Ba Island, Hai Phong Prov., VIETNAM	AB013246
VN-Concuong1*	NF et al. 95-V2416 (TI)	Concuong, Vinh Prov., VIETNAM	AB023504
VN-Concuong2*	NF et al. 95-V2440 (TI)	Concuong, Vinh Prov., VIETNAM	AB023505

Table 1. *Continued*

Plant samples	Voucher	Locality	DDBJ No.
VN-Concuong3*	NF et al. 95-V2443 (TI)	Concuong, Vinh Prov., VIETNAM	AB023506
VN-Dalat*	KI et al. 98-V518 (KYO)	Dalat, Dalat Prov., VIETNAM	AB023507
VN-TamDao	KI et al. 94-V339 (TI)	Tam Dao, Vinh Phu Prov., VIETNAM	AB013248
VN-Bavi	KI et al. 94-V320 (TI)	Mt. Bavi, Ha Noi Prov., VIETNAM	AB013251

JY, Jun Yokoyama; KI, Kunio Iwatsuki; NF, Nobuyuki Fukuoka; NM, Noriaki Murakami; NS, Norio Sahashi; SS, Shunsuke Serizawa; UH, Ujang Hapid; XC, Xiao Cheng; YY, Yoko Yatabe.
DDBJ, DNA Data Bank of Japan.
* Nucleotide sequences of *rbcL* were analyzed for the first time in this study.

2.2 DNA Extraction and *rbcL* Sequencing

Total DNA was extracted using 2X CTAB (Hexadecyl trimethyl ammonium bromide) solution according to Doyle and Doyle (1987). If necessary, the DNA was purified using Qiagen Column Tip-20 (Qiagen GmbH, Hilden, Germany) according to the manufacturer's instructions. PCR (Polymerase Chain Reaction) amplification of *rbcL* fragments followed Hasebe et al. (1994) except that we used original primers OT-NP1(5'-TATCCATTGGACCTTTTTGAAGAAGGTTC-3') and OT-2PR (5'-TCTCTTTCTCCTTCTAGTTTACCTACTAC-3') instead of their NP1 and 2PR primers. The PCR products were purified using a GENE CLEAN II kit (BIO 101, Vista, California, USA) after electrophoresis in 1.0% agarose gel, and then used as templates for direct sequencing. Sequencing reactions were prepared using a Big Dye terminator cycle sequencing kit (Perkin Elmer Applied Biosystems, Foster, CA, USA) . The reaction mixtures were analyzed on an Applied Biosystems Model 377 automated sequencer (Perkin Elmer Applied Biosystems). Sequences were aligned using the Sequence Navigator program (Perkin Elmer Applied Biosystems).

2.3 Phylogenetic Analyses

The nucleotide sequence data obtained in this study as well as our previous work (Murakami et al. 1999) were used for molecular phylogenetic analyses (Table 1). The *rbcL* sequences were analyzed using PAUP version 3.1 (Swofford 1993). A branch and bound search was conducted to find the most parsimonious trees. We used *A. griffithianum* as an outgroup to root the tree. Its relevance as an outgroup was shown by the more comprehensive molecular work as a whole for Aspleniaceae (Murakami 1995). A bootstrap analysis with 10 000 replications was performed in order to estimate the reliability for various clades.

2.4 Ecological Observations

One leaf each was collected from 30 individuals of *A. nidus* along the loop trails near Chikaniki Station, Mt. Halimun National Park, using a 6.5 m long fishing pole with a small sickle on top. For each individual collected, their altitude, position on the tree (height from the ground), and degree of sun shade were recorded. After returning to our laboratory at Kyoto University, a small portion of each leaf was used for DNA extraction to determine their *rbcL* types. About 500 bp sequences of *rbcL* were examined to identify the three types. Based on the recorded ecological information and the obtained molecular data, the ecological preference of each *rbcL* type was inferred.

3 Results

3.1 Variations in *rbcL* Sequence

We determined 1194 bp nucleotide sequences of *rbcL*. Three types of *rbcL* sequences were found from the 25 plants of *A. nidus* from Halimun. We called them type A , B and C (Fig. 2). Among these three types, 30 sites of the *rbcL* sequences varied.

We also determined the *rbcL* nucleotide sequences of the plants from Mt. Gede and Bogor Botanical Garden. The type C was identified from the middle elevations (1400–1900 m alt.) of Mt. Gede, but the plants from its higher elevations (1900–2400 m alt.) and Bogor Botanical Garden (300 m alt.) had very different *rbcL* from any of the three types from Mt. Halimun.

3.2 Molecular Phylogenetic Analyses

The two equally most parsimonious trees (length = 146 steps; Consistency Index = 0.801; Retention Index = 0.851) were obtained by the Wagner parsimony method, and their strict consensus tree is shown in Fig. 3. These three types were found to be not closely related, when we included the plants of sect. *Thamnopteris* from the other localities of the Old World. The type A is closely related to *A. nidus* from Yunnan, China, and central Vietnam. The type C belongs to the same clade as type A, but it was the most basally diverged from the other members. The type B made a clade with *A. australasicum sensu* Holttum (1974) from Australia and New Caledonia. We also analyzed the plants identified as *A. nidus* from Mt. Gede and

```
                                                              50                                                          100
HalimunA  CCCGAATAC  AAGACCTTAG ATACCGATAT CTTGGCAGCT TTCCGGATGA CCCCACAACC CGGAGTACCA GCTGAAGAAG CCGGAGCTGC AGTAGCTGCAG
HalimunB  .........  ......AA.. .......... ...A...... .......... .......... T......... .......... .......... .........G.
HalimunC  .........  ......AA.. ..C....... .......... .......... .......... .......... .......... .......... ...........
                                                             150                                                          200
HalimunA  AGTCCTCTA  CAGGTACGTG GACCACTGTG TGGACAGACG GTTTGACCAG TCTTGACCGT TACAAGGGCC GATGCTACGA TATTGAGCCC GTTGCTGGAGA
HalimunB  .........  .......... .......... .......... .......... .......... .......... .......... .......... ...........
HalimunC  .........  .......... .......... .......... .......... .......... .......... .......... .......... ...........
                                                             250                                                          300
HalimunA  AGAAAATCA  GTATATTGCG TACGTAGCTT ACCCATTGGA CCTCTTTGAA GAAGGTTCTG TAACCAACAT GTTAACTTCA ATTGTAGGTA ACGTTTTTGGA
HalimunB  G........  ......C... .......... .T........ ...T...... .......A.. .C........ .......... .......... ...........
HalimunC  G........  .......... .......... .T........ ...T...... .......... .......... .......... .......... ...........
                                                             350                                                          400
HalimunA  TTTAAAGCT  CTACGTGCCC TACGGTTGGA AGACCTTCGA ATCCCCCCCG CTTACTCTAA AACTTTCAAC GGACCGCCAC ATGGTATTCA GGTTGAAAGGG
HalimunB  .........  .......... .......... .......... .......... .......... .......... .......... .......... ...........
HalimunC  .........  .......... .......... .......... .......... .......... .......... .......... .......... ...........
                                                             450                                                          500
HalimunA  ATAAGTTAA  ACAAATACGG ACGCCCCTTA TTGGGATGTA CAATCAAGCC AAAATTAGGT CTGTCTGCCA AAAATTATGG TAGAGCCGTT TATGAATGCCT
HalimunB  .........  .......... .......... .......... .......... .......... ..A....... ......C... .......... ...........
HalimunC  .........  .......... .......... .......... .......... .......... ..A....... .......... .......... ...........
                                                             550                                                          600
HalimunA  TCGTGGTGG  ACTTGATTTT ACAAAAGATG ACGAAAACGT AAATTCCCAA CCATTCATGC GTTGGAGAGA TCGTTTCTTA TTTGTTGCAG AAGCTCTTTTC
HalimunB  .........  .......... .......... .......... .......... .......... .C........ .......... .......... ...........
HalimunC  .........  .......... .......... .......... .......... .......... .C........ .......... .......... ...........
                                                             650                                                          700
HalimunA  AAATCCCTG  TCTGAAACAG GCGAGATTAA GGGGCATTAC TTAAACGCTA CTGCAGGTAC GTGTGAAGAA ATGATGAAAA GAGCTGTTTT TGCGAGAGAAT
HalimunB  ......A..  .......... ....A..... .......... .......... .......... A......... .......... .......... ...........
HalimunC  ....T.A..  .......... .......... .......... .......... .......... A......... .......... .......... ...........
                                                             750                                                          800
HalimunA  TGGGCGCAC  CAATTGTCAT GCATGACTAC CTAACTGGAG GGTTTACTGC AAATACTAGT TTAGCTTACT ATTGCAGAGA TAATGGGTTG CTTCTTCATAT
HalimunB  .........  .......... .......... ..G....... .......... .......... ......C... .......... .......... ...........
HalimunC  .........  .T........ .......... ..G....... .......... .......... ......C... .......... .......... ...........
                                                             850                                                          900
HalimunA  TCATCGCGC  GATGCATGCT GTAATTGATA GACAACGAAA TCACGGTATG CATTTTCGTG TATTGGCCAA GGCTTTGCGC ATGTCCGGCG GGGATCACATA
HalimunB  .........  .......... .......... .......... .......... .......... .......... ...A...... .......... ...........
HalimunC  .........  .......... .......... .......... .......... .......... .......... .......... .......... ...........
                                                             950                                                         1000
HalimunA  CACGCTGGA  ACTGTAGTAG GTAAACTAGA AGGAGAAAGG GAAGTCACTT TGGGTTTTGT CGATTTACTT CGCGACGATT ATATCGAGAA AGATCGTTCAC
HalimunB  .........  .......... .......... .......... .......... .......... .......... .......... .......... .......G...
HalimunC  .........  .......... .......... .......... .....T.... .......... .......... .......... .......... .......G.G.
                                                            1050                                                         1100
HalimunA  GCGGTGTTT  ATTTCACTCA AGATTGGGTA TCTATGCCGG GTGTAATCCC AGTAGCTTCG GGGGGTATCC ACGTATGGCA CATGCCAGCT CTAACCGAAAT
HalimunB  .T...A...  .......... .......... .......... .......... .......... .......... .......... .......... ...........
HalimunC  .........  .......... .......... .......... .......... .......... .......... .......... .......... ...........
                                                            1150
HalimunA  CTTTGGGGA  CGATGCTGTC TTACAGTTCG GTGGAGGAAC CTTGGGGCAT CCTTGGGGAA ATGCACCCGG TGCGGTAGCC AACCGAGTCG CATTA
HalimunB  .........  .......... .......... .......... ...A...... ..C....... .......... .......... .......... .....
HalimunC  .........  .......... .......... .......... .......... .......... .......... .......... .......... .....
```

Fig. 2. Alignment of the three types of *rbcL* sequences from *Asplenium nidus* in Mt. Halimun National Park

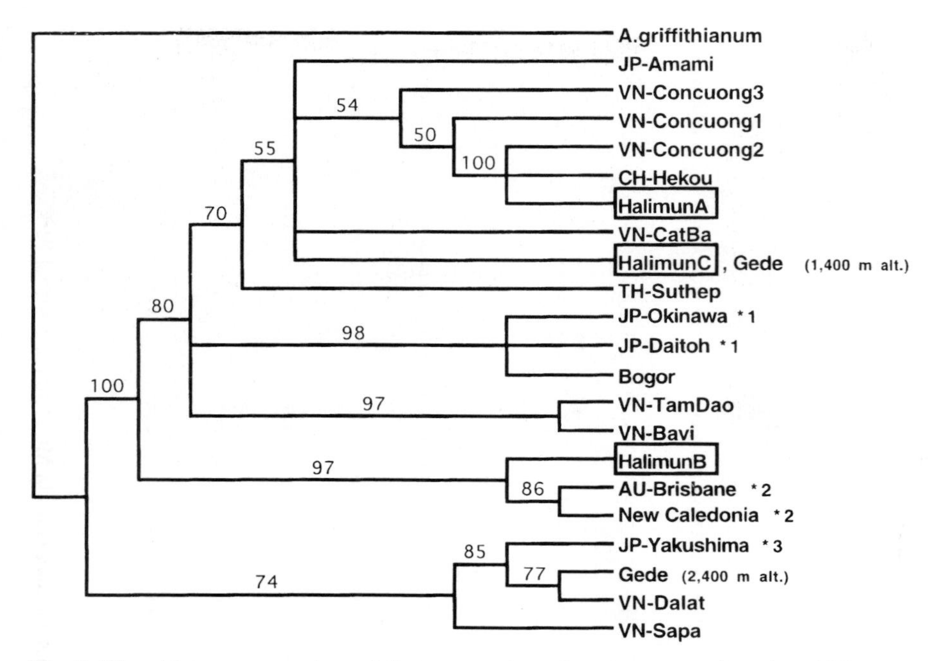

Fig. 3. The strict consensus tree of the two most parsimonious trees (length = 126 steps; consistency index = 0.801; retention index = 0.851) from the analysis of *rbcL* sequence data of *Asplenium nidus* and its related species from various localities (see Table 1) in the Old World tropics. Numbers above the branches are bootstrap percentages. AU, Australia; CH, China; JP, Japan; TH, Thailand; VN, Vietnam. *1, identified as *A. setoi;* *2, *A. australasicum;* *3, *A. antiquum* (see Murakami et al. 1999)

Bogor Botanical Garden, which are only 50–60 km apart from Mt. Halimun, but they were not closely related to any of the three *rbcL* types from Halimun. The plants from Mt. Gede (2400 m altitude) made a clade with *A. antiquum* from Japan and its relatives from Vietnam, and those from Bogor with *A. setoi* from Japan. Thus, at least five *rbcL* types are found to be distributed within a small area of West Java.

3.3 Morphological Comparison

Generally, the three *rbcL* types were morphologically easily distinguishable. Typical plants of the type A (Fig. 4a) have relatively narrow leaves of medium size, long and relatively scattered sori (sori do not occur along all leaf lateral veins) and a markedly keeled petiole and rachis. The type B (Fig. 4b) has broader and larger leaves with short and dense sori, and a slightly keeled petiole. The type C has narrow leaves of medium size similar to those of the type A, but different in that it has shorter sori, and a non-keeled petiole and rachis (Fig. 4c).

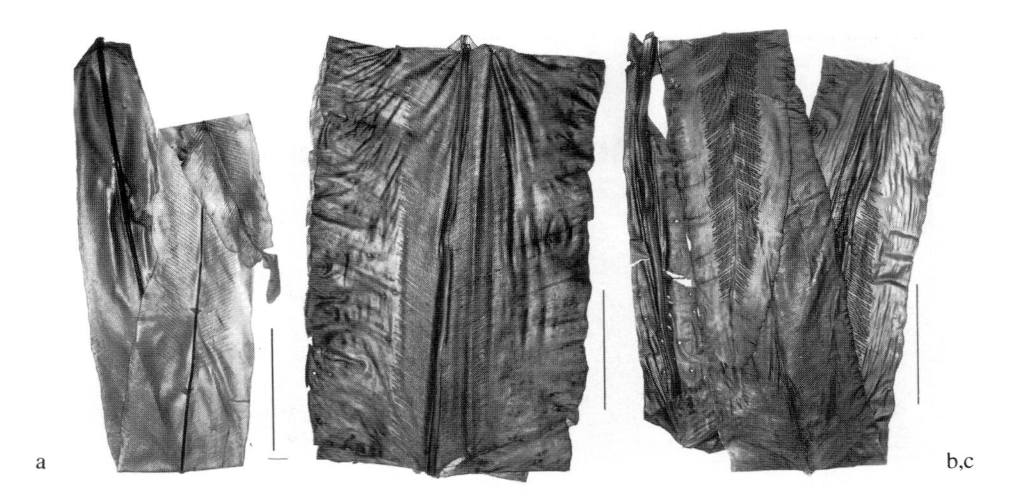

Fig. 4 a-c. Typical leaves of the three *rbcL* types of *Asplenium nidus* in dried condition. **a** Type A; **b** Type B; **c** Type C

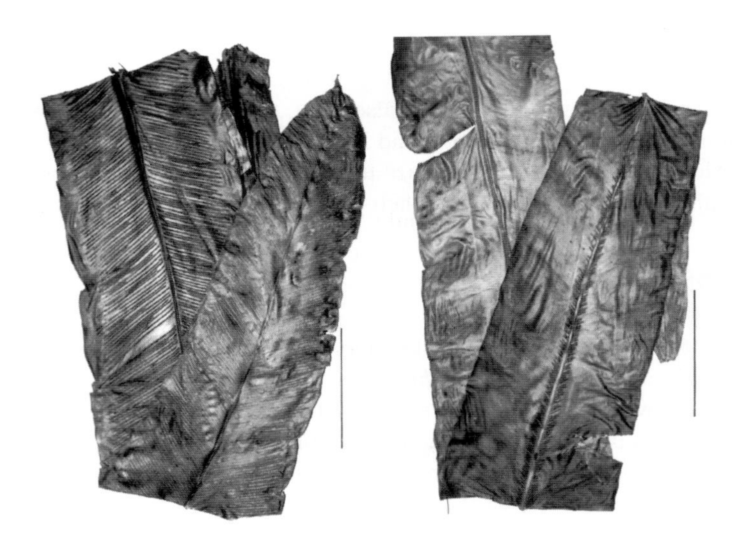

Fig. 5 a,b. Intermediate leaves between the three *rbcL* types of *Asplenium nidus*. **a** Intermediate form between types A and B. Its *rbcL* sequence is of type A; **b** Intermediate form between types B and C. Its *rbcL* sequence is of type B

a b

Fig. 6 a,b. Typical habitats of types A and B of *Asplenium nidus*. **a** Type A usually growing on lower sections of tree trunks in deep shade; **b** Type B usually growing on higher sections of tree trunks in open to half-shaded places

However, some intermediate forms also exist, such as the plants shown in Fig. 5a (an intermediate form between A and B) and Fig. 5b (intermediate between B and C). Thus, it is difficult to recognize these three *rbcL* types only by leaf morphology, although we must examine their anatomical features carefully.

3.4 Ecological Habitats

There are differences in the ecological preferences of types A and B in Mt. Halimun National Park, although these two *rbcL* types grow together at 1000–1250 m alt. The type A grows on relatively low sections (1–6 m high from the ground) of tree trunks where it is deeply shaded (Fig. 6a). In contrast, the type B grows on higher sections of tree trunks where it is more open, or only half shaded (Fig. 6b). This habitat differentiation between the types A and B was more obvious when they grew on the same tree trunks. For example on one tree trunk, six individuals of the type A were identified up to about 6 m high, and over 6 m high, only the type B plants were epiphytic (Fig. 7).

The type C was not observed mixed with the type B, and very seldom with the type A. The type C usually grows separately at higher elevations over 1140–1700 m alt. It prefers tree trunks in even more shaded forests on drier ridges than the type

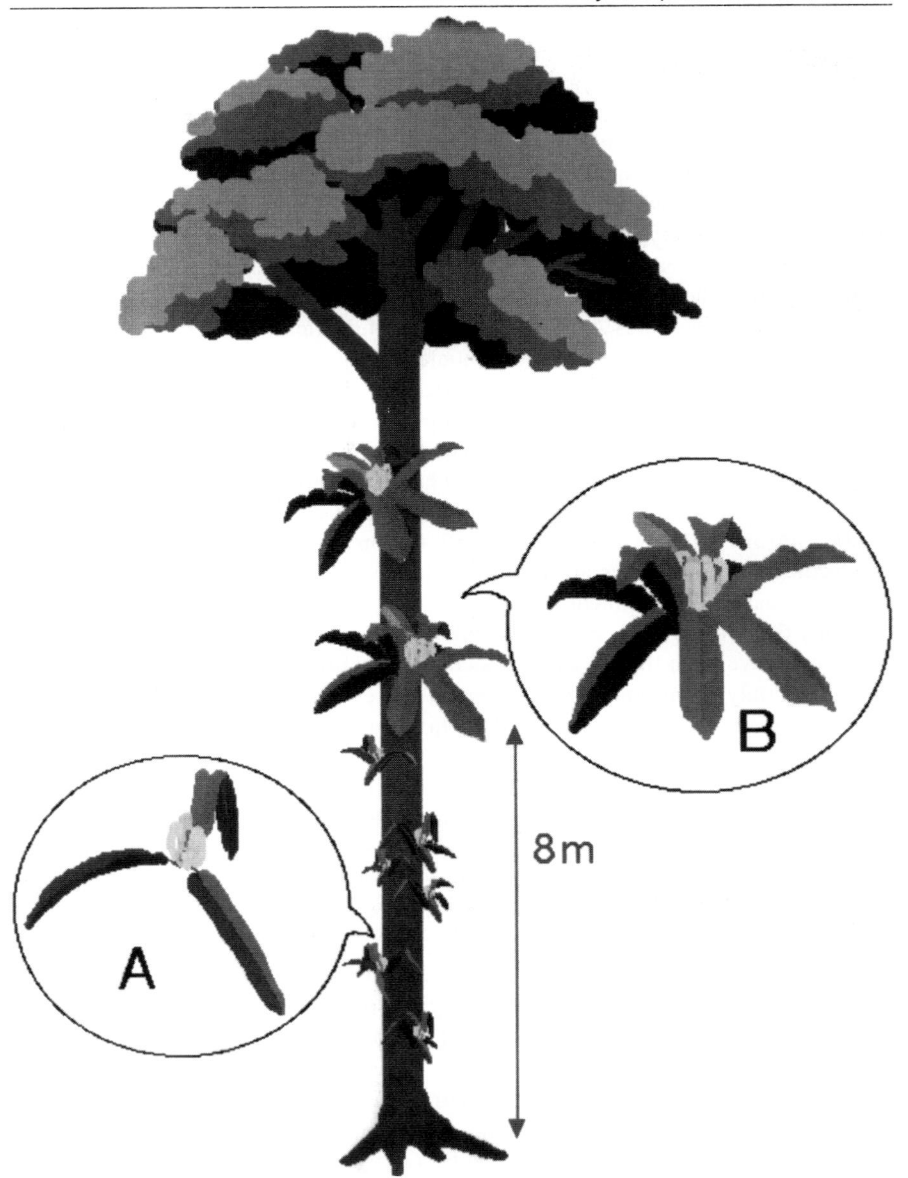

Fig. 7. Growing positions of types A and B of *Asplenium nidus* on a tree trunk in Haımun. Plants of typical morphology are drawn, although they are not always recognizable morphologically

A, whereas the types A and B grow on trees of riparian wet forests. The type C grew on similar sections of tree trunks as the type A.

4 Discussion

We found three *rbcL* sequence types among plants from a small area of Mt. Halimun National Park which were all identified as *A. nidus* according to Holttum's (1974) classification. Based on the molecular phylogenetic tree obtained, it was determined that these three *rbcL* types are distantly related to each other. Thus, their origin is certainly old, and it is not likely that they recently speciated in Mt. Halimun.

In this work, the three *rbcL* types were found to have different ecological preferences as well. The types A and B mixed on Mt. Halimun, but even in sympatric populations, their habitats were differentiated as to their relative position on tree trunks. The type C occurs mostly parapatrically, with distribution at higher elevations. Distances among the populations were within a range of a few km, and therefore gene exchange through spores between the populations of C and the other two *rbcL* types should happen frequently. There was no better evidence of their dispersability than the fact that the type C was also distributed at mid-elevations (1400–1900 m) of Mt. Gede, which is 50–60 km apart from Mt. Halimun. Still, their ecological preferences are recognizable, and we can conclude that the three *rbcL* types have different niches.

Thus, these three *rbcL* types are both genetically and ecologically differentiated. They have been treated as conspecific so far because it is difficult to distinguish them only by morphology. However, it is evident that biologically they belong to different entities, and should be taxonomically treated as three separate species.

In the present study, we showed sequence variations of the *rbcL* gene even in a fern species complex which is morphologically hard to distinguish. Based on this information, we concluded that the three *rbcL* types found in Mt. Halimun belong to different biological units. It is becoming easier to determine partial sequences of particular genes. Such molecular techniques are useful even for monographic works which try to delimit species, especially morphologically similar taxa like ferns.

Ecological (habitat) comparison is also a powerful method to recognize different biological units with similar morphology, especially when they are sympatrically distributed. Murakami and Moran (1993) recognized four species of the *Hymenasplenium obtusifolium* complex (*H. obtusifolium*, *H. riparium*, *H. repandulum*, *H. volubile*) based on their differentiated habitats (subaquatic, lithophytic, epiphytic on large tree trunks, and epiphytic on thin trees, respectively) in addition to morphological and molecular (restriction fragment length polymorphism analysis of whole chloroplast DNA) data of their mixed populations in Panama. Murakami et al. (1998) distinguished a new species *H. inthanonense* from the most closely related species, *H. cheilosorum,* using habitat differentiation in addition to morphological and molecular (*rbcL* nucleotide sequences) information. If populations defined by molecular markers belong to different biological units, they should also have differentiated niches. Otherwise, they would not be able to coexist for a long time in nature. Searching for coincidence between molecular and ecological information is sometimes a better strategy for α-taxonomy than trying to find morphological discontinuity, at least for morphologically simple ferns such as the *A. nidus* complex.

We are also examining intraspecific variations of *rbcL* sequences in other distantly related fern groups, such as *Stegnogramma pozoi* (Thelypteridaceae), *Osmunda cinnamomea*, *Os. claytoniana* (Osmundaceae), *Ophioglossum pendulum*, *Op. petiolatum* (Ophioglossaceae) and *Cheiropleuria bicuspis* (Cheiropleuriaceae) (unpublished data except for *S. pozoi:* Yatabe et al. 1998). In each species, we found a large amount of *rbcL* sequence variation. Thus, it is highly probable that the same strategy as we used for *A. nidus* complex can be applicable for various fern groups. Our method might be useful for fern α-taxonomy in general.

Acknowledgments

This work was supported by a grant from The New Technology Development Foundation and by a Grant-in-Aid for Scientific Research (B) from the Japan Society for the Promotion of Science to NM. This work was also supported by a grant from the Japan Bio-industry Association for KI and NM, especially for the collection of plant materials from Indonesia. We thank Mr. U. Hapid, Mr. Bayu Adjie, Ms. Julisasi Tri Hadiah, Mr. Esti Endah Ariyanti, Bogor Botanical Gardens, Dr. M. Ito, Chiba University and the staff members of Indonesian Institute of Science and Mt. Halimun National Park for their kind assistance in our field work and collection of plant materials. We also thank Ms. K. Matsuda for her assistance in the sequencing experiments. We are grateful to Dr. N. Fukuoka, Shoei Junior College, for his instruction in using the fishing pole to collect epiphytic materials, and Mr. K. Hirai (University of Tokyo) and Ms. Y. Shibao (Kyoto University) for their cultivation of our plant materials in their green houses.

References

Copeland EB (1929) The Oriental genera of Polypodiaceae. Univ Calif Publ Bot 16:45-128

Doyle JJ, Doyle JL (1987) A rapid DNA isolation procedure for small quantities of fresh leaf material. Phytochem Bull 19:11-15

Hasebe M, Oumori T, Nakazawa M, Sano T, Kato M, Iwatsuki K (1994) *rbcL* gene sequences provide new evidence for the evolutionary lineages of leptosporangiate ferns. Proc Natl Acad Sci USA 91:5730-5734

Holttum RE (1974) *Asplenium* Linn., sect. *Thamnopteris* Presl. Gardens' Bull Singapore 27:143-154

Murakami N (1995) Systematics and evolutionary biology of the fern genus *Hymenasplenium* (Aspleniaceae). J Plant Res 108:257-268

Murakami N, Moran RC (1993) Monograph of the Neotropical species of *Asplenium* sect. *Hymenasplenium* (Aspleniaceae). Ann Missouri Bot Gard 80:1-38

Murakami N, Yokoyama J, Iwatsuki K (1998) *Hymenasplenium inthanonense* (Aspleniaceae), a new fern species from Doi Inthanon, and its phylogenetic status. Thai Forest Bull (Bot) 26:40-52

Murakami N, Watanabe M, Yokoyama J, Yatabe Y, Iwasaki H, Serizawa S (1999) Molecular taxonomic study and revision of the three Japanese species of *Asplenium* sect. *Thamnopteris*. J Plant Res 112:15-25

Presl CB (1836) Tentamen pteridographiae, seu genera filicacearum. Filiorum Theophili Haase, Prague

Smith J (1841a) Enumeratio filicum Philippinarum: a systematic arrangement of the ferns collected by Mr. Cuming, Esq., F.L.S., in the Philippine Islands and the Peninsula of Malacca, between the years 1836 and 1840. J Bot 3:392-422

Smith J (1841b) An arrangement and definition of the genera of ferns, with observations on the affinities of each genus. J Bot 4:38-70, 147-198

Swofford D (1993) Phylogenetic analysis using parsimony, Version 3.1

Templeton AR (1989) The meaning of species and speciation: a genetic perspective. In: Otto D, Endler JA (Eds) Speciation and its consequences. Sinauer Associates, Sunderland, pp 3-27

Yatabe Y, Takamiya M, Murakami N (1998) Variation in the *rbcL* sequence of *Stegnogramma pozoi* subsp. *mollissima* (Thelypteridaceae) in Japan. J Plant Res 111:557-564

5
Adaptive Radiation, Dispersal, and Diversification of the Hawaiian Lobeliads

THOMAS J. GIVNISH

Department of Botany, University of Wisconsin, Madison, WI 53706, USA

Abstract

Molecular data provide several key insights into the origin and diversification of the Hawaiian lobeliads, which comprise one-ninth of the flora of the most isolated archipelago on earth. Soon after colonizing the Hawaiian chain from elsewhere in the Southern hemisphere, this group radiated into four major clades, adapted to bogs and other open, wet high-elevation habitats, to inland outcrops and rock walls, to sea cliffs and dry forest, and to rain- and cloud-forest edges and interiors. Woodiness and bird pollination arose long before the lobeliads arrived in Hawaii; fleshy fruits evolved autochthonously in moist to wet forests, and were lost on sea cliffs. Limited seed dispersal (associated with fleshy fruits in forest-interior plants) appears to have triggered substantial speciation in the genus *Cyanea,* and helped generate convergent radiations in flower length, elevational distribution, and plant height on each of the four major islands. Some of the processes that accelerated speciation in *Cyanea* – such as limited seed dispersal, narrow ranges, and highly specialized flowers – also appear to have increased the likelihood of extinction. The roles of adaptive radiation vis-à-vis limited dispersal and sexual selection in promoting speciation in other groups (African rift-lake cichlids, coral reef fish, lilies and their relatives) are also discussed, with special reference to the phenomena of convergent adaptive radiations and of visual selection. Natural enemies may exert stronger density-dependent mortality on their hosts under rainier, more humid, warmer, and less seasonal conditions, and help create strong ecological gradients in plant species diversity within the tropics. Adaptive radiation may thus have interacted synergistically with several processes to generate the taxonomic richness and ecological diversity of the largest clade in the Hawaiian flora.

Key words. convergent adaptive radiation, molecular systematics, geographic speciation, Hawaii, insular evolution, Lobeliaceae, fleshy fruits, species richness, ecological determinants of speciation and extinction, rift-lake cichlids, coral-reef fish, tropical forest diversity, natural enemies, density-dependent mortality, whole-plant compensation point

1 Introduction

Adaptive radiation – the rise of a diversity of ecological roles and attendant adaptations within a lineage, as exemplified by Darwin's finches (Grant 1986), the African rift lake cichlids (Stiassny and Meyer 1999), and the Hawaiian silversword alliance (Baldwin 1997) – is one of the most important processes bridging ecology and evolution, with profound implications for the origin of adaptations and for the genesis and maintenance of biological diversity (Givnish 1997). Ecological divergence among the members of a radiation is often thought – and in three recent cases (Schluter 1994; Turner et al. 1996; Rainey and Travisano 1998) has been directly shown – to reflect selection to avoid competition with close relatives. Adaptive radiation is often apparent in organisms that have invaded oceanic islands or similarly isolated lakes or mountaintops (e.g., Losos et al. 1998; Albertson et al. 1999). Few groups are able to colonize such remote areas, eliminating competition from many lineages, while increasing the importance of competition with close relatives. Although less easy to recognize, adaptive radiation can also occur on continents and in oceans – often after a group invades a previously unoccupied adaptive zone, through the origin of a key innovation or extinction of competing groups (Simpson 1953; Clark and Johnston 1996) – as part of the grand diversification of life on earth at all taxonomic levels (Foote 1996; Goldblatt et al. 1995; Jernvall et al. 1996; Givnish et al. 1997, 1999; Hapeman and Inoue 1997; Kirsch and Lapointe 1997; Smith and Littlewood 1997; Springer et al. 1997; Bond and Opell 1998; Johnson et al. 1998; Price et al. 1998; Chiba 1999; Sato et al. 1999).

In this paper, I examine the importance of adaptive radiation and other processes in promoting the remarkable diversification of the Hawaiian lobeliads. Molecular systematics provides an essential tool for evolutionary studies of this and similar groups: plants that have undergone extensive adaptive radiation often have diverged so dramatically from putative ancestral groups that it can be difficult to ascertain relationships based on morphology alone (Givnish et al. 1994, 1995, 1997; Sang et al. 1994; Baldwin and Robichaux 1995; Böhle et al. 1996; Francisco-Ortega et al. 1997; Givnish 1998). In the lobeliads, limited dispersal and adaptive radiation appear to have interacted synergistically to generate more species than might be expected from either process alone, and has helped generate a number of convergent radiations on different islands. Paradoxically, some of the processes that may accelerate speciation in these tropical plants also may increase their likelihood of extinction. The roles of limited dispersal and sexual selection in promoting speciation are also discussed in the context of other groups, with a focus on how these processes might intersect with adaptive radiation. Finally, I consider how interactions with natural enemies and mutualists – long considered to be crucial factors favoring high plant diversity in tropical forests (Janzen 1970; Connell 1971) – may also help generate strong ecological gradients in plant species diversity within the tropics.

2 Adaptive Radiation in Hawaiian Lobeliads

The Hawaiian lobeliads (Lobeliaceae) have long been viewed as one of the most spectacular cases of adaptive radiation in plants on oceanic islands (Rock 1919; Carlquist 1965, 1974; Lammers 1990; Givnish et al. 1994, 1995; Givnish 1998). They include alpine rosette shrubs, succulent cliff plants, rain-forest and bog shrubs, trees, and treelets, and even a few epiphytes and vine-like species. They are the angiosperm family with the largest number of taxa endemic to the Hawaiian Islands, with over 110 species in 6 genera, comprising one-ninth of the native flora (Wagner et al. 1990; Lammers et al. 1994). Most species were pollinated by honeycreepers, and many were dispersed by these or other groups of native birds (Rock 1919; Lammers and Freeman 1986; Givnish et al. 1995).

The three endemic genera with fleshy fruits (*Cyanea, Clermontia, Delissea*) encompass some 95 species and – based on gross morphology – are often considered the largest group of Hawaiian plants derived from a single immigrant, although the source of this immigrant is debatable (Rock 1919; Mabberley 1974, 1975; Lammers 1990). *Cyanea* (65 spp) is the largest angiosperm genus endemic to Hawaii, and is comprised mainly of unbranched trees and treelets (1–18 m tall) of mesic- and wet-forest interiors. Its tubular corollas are whitish to purplish and 15–80 mm in length, with a curvature strongly resembling the bills of the nectar-feeding honeycreepers that pollinated them. Its fruits are orange or purple in color, and generally are less than 15 mm in diameter. Leaves vary from a few mm to 0.25 m in width; nearly one-third of the species have toothed, lobed, or highly dissected margins, especially in the juvenile stage, accompanied by prickles along the leaf veins and shoot axes.

Clermontia (22 spp) differs from *Cyanea* in having a branched, often shrubby habit, and occurs mainly in forest edges and gaps rather than shaded understories. Its orange fruits are often more than 25 mm in diameter; its leaves are undivided and unarmed; a few species (e.g., *C. peleana*) are mainly epiphytic. *Delissea* (9 spp) bears a narrow crown atop a tall stem, and occurs mainly in partially open dry and mesic forests. Its flowers and fruits are similar to those of *Cyanea;* its petals have one or more knob-like projections.

Among the capsular taxa, *Lobelia* sect. *Galeatella* includes four species of rosette shrubs of alpine bogs and wet openings; all have massive woody stems and spectacular terminal inflorescences of large, tubular flowers, white or crimson in color. *Lobelia* sect. *Revolutella* includes nine species of smaller shrubs of rock walls, cliffs, ridges, and open forests; they bear smaller flowers with recurved petals that are tubular, and magenta to blue in color. *Trematolobelia* (4 spp) bears a mop of linear leaves atop a short stem, and frequents wet ridges and openings in wet or boggy forests. It has branched horizontal inflorescences and capsules with a bizarre mechanism of dehiscence, opening by a network of ragged pores in the capsule wall. Finally, *Brighamia* (2 spp) includes some of the most bizarre plants in the world. Two rare species grow on sea cliffs on Kaua`i and Moloka`i, and bear a cabbage-head of fleshy leaves atop an unbranched, succulent woody stem. Unlike the other Hawaiian lobeliads, *Brighamia* has white to yellow flowers with straight, narrow tubular corollas, which are thought to have been pollinated by hawkmoths based on their nectar chemistry.

Based on morphology, the fleshy-fruited genera are usually seen as the product of a single colonization event, while the capsular taxa have been thought to be the products of two to four additional colonizations (Rock 1919; Wimmer 1953; Mabberley 1974, 1975; Lammers 1989, 1990; but see Givnish et al. 1995). Here I will present a molecular phylogeny to show that the Hawaiian lobeliads are instead the product of a single immigration event from elsewhere in the Pacific basin or the Southern Hemisphere, and that this ancestor appears to have been woody, wind-dispersed, bird-pollinated, and adapted to cool, rainy, open, high-elevation habitats.

2.1 Molecular Systematics

Knox et al. (1993) used restriction-site variation and re-arrangements of the chloroplast genome to infer relationships among *Sclerotheca* and a sampling of species in the large, cosmopolitan genus *Lobelia.* They demonstrated that herbaceous diploids like *L. cardinalis* are basal within *Lobelia,* and that woody, montane, mostly tetraploid species from various parts of the tropics form a monophyletic clade. This suggested that *L. cardinalis* would be a suitable outgroup for a cladistic analysis of relationships among the Hawaiian lobeliads and their putative relatives.

Givnish et al. (1994, 1995, and unpubl. data) analyzed relationships within and among the Hawaiian lobeliad genera based on cpDNA restriction site variation, using global parsimony and five outgroups. Virtually all of the extant Hawaiian taxa were included in this survey. Our strict consensus tree (Fig. 1) shows that (1) each endemic genus and each of the endemic sections of *Lobelia* are monophyletic; (2) *Cyanea* is sister to *Clermontia;* (3) the bizarre cliff succulent *Brighamia* with hawkmoth-pollinated flowers and capsular fruits is sister to *Delissea,* with bird-pollinated flowers and fleshy fruits; (4) *Cyanea-Clermontia* is sister to *Brighamia-Delissea;* (5) *Lobelia* sect. *Revolutella* is sister to the latter group; and (6) *Lobelia* sect. *Galeatella* is sister to *Trematolobelia,* forming a clade at the base of the Hawaiian lobeliads. This analysis is consistent with the Hawaiian lobeliads being monophyletic, but lacks many groups of putative relatives.

My colleagues and I therefore sequenced several rapidly evolving spacer regions of the chloroplast genome (*atpB-rbcL, trnL-trnF, psbA-trnH*), as well as the internal transcribed spacer (ITS) and 5.8S ribosomal subunit of nuclear ribosomal DNA for representatives of each of the Hawaiian genera/sections, as well as for representatives of several groups which have been proposed as potential relatives of the Hawaiian lobeliads (Table 1). Our cladistic analysis of these data employed global parsimony and a 1.4:1.0 weighting of transitions to transversions in PAUP* (kindly provided by D. Swofford), using *L. cardinalis* as a super-outgroup. The transition:transversion ratio was estimated as 0.696 (= 1/1.437) using a maximum likelihood approach; a total of 481 informative characters were detected. We ob-

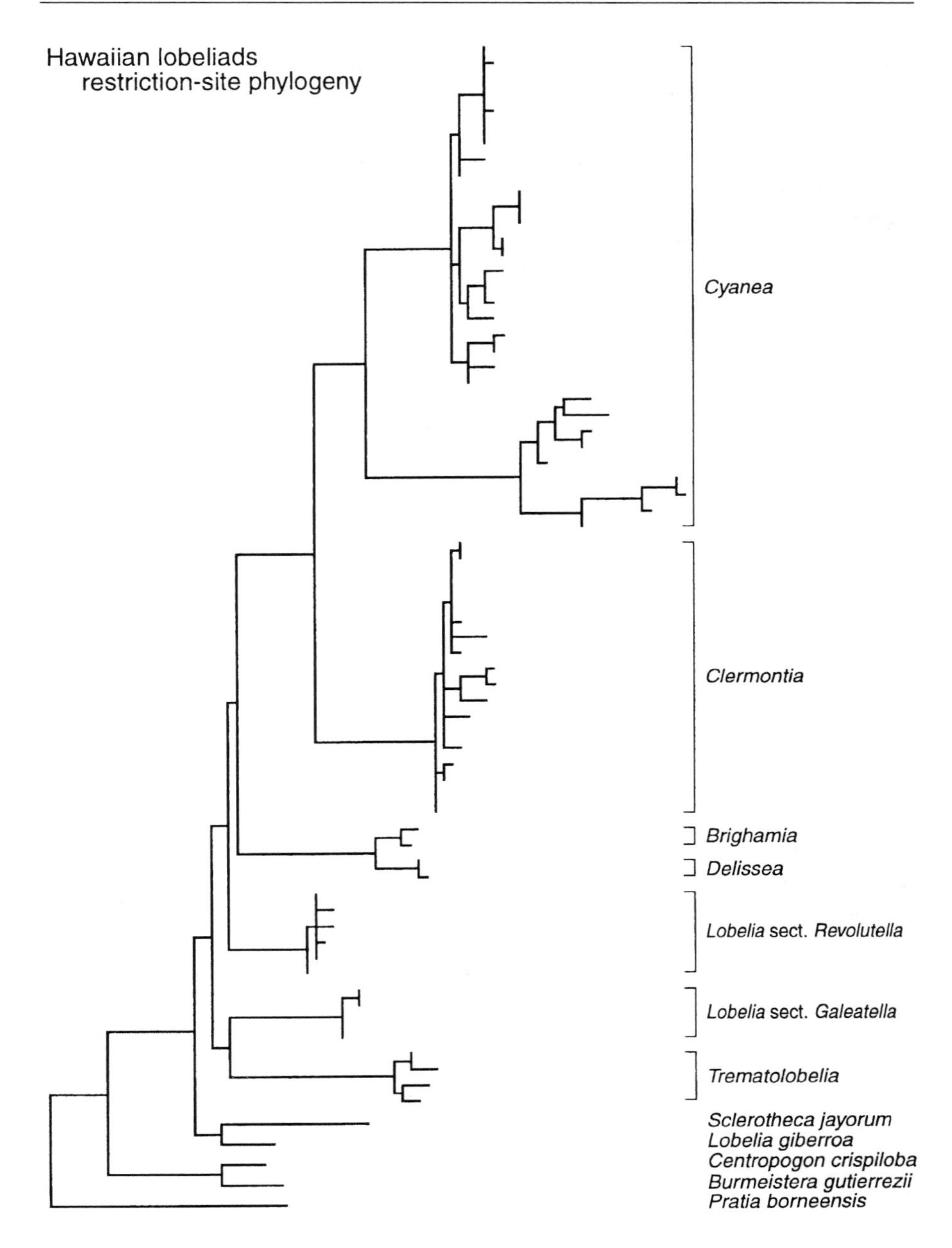

Fig. 1. Phylogram for the Hawaiian lobeliads based on cpDNA restriction-site variation; the length of each horizontal branch is proportional to the number of site mutations inferred between nodes (Givnish et al. 1995, Givnish 1998, and unpubl. data)

Table 1. Proposed relatives of the genera and sections of the endemic Hawaiian lobeliads

Hawaiian taxon	Habitat/habit	Proposed relatives	Source area	Citation
Brighamia (2 spp.)	Sea-cliff succulents	*Sclerotheca/Apetahia*	Rarotonga, Rapa, Tahiti, and Marquesas	Rock 1919; Lammers 1989
		Isotoma	Australia	Rock 1919; Lammers 1989
		Delissea	Hawaii	Givnish et al 1995
Lobelia sect. *Galeatella* (4)	Alpine bogs/grassland rosette shrubs	*Lobelia organensis*	Brazil	Mabberley 1974, 1975; Lammers 1990
		L. boninensis	Bonin Islands	Lammers 1990
		L. columnaris, L. petiolata, and allies	East and West African Highlands	Mabberley 1974, 1975
		L. excelsa and allies	South America	Wimmer 1953; Mabberley 1974, 1975
Lobelia sect. *Revolutella* (9)	Cliff and rock-wall rosette shrubs	*Lobelia nicotianifolia*	Southern Asia	Mabberley 1974, 1975; Lammers 1990
		L. columnaris, L. petiolata, and allies	East and West African Highlands	Mabberley 1974, 1975
		L. excelsa and allies	South America	Wimmer 1953; Mabberley 1974, 1975
Trematolobelia (2)	Subalpine wet-forest treelets	*Sclerotheca/Apetahia*	Rarotonga, Rapa, Tahiti, and Marquesas	Wimmer 1953
		Lobelia sect. *Galeatella*	Hawaii	Mabberley 1974, 1975; Lammers 1990
Clermontia (22)	Mesic and wet forest-edge shrubs and trees			
Cyanea (65)	Mesic and wet forest-interior trees, treelets, and vines	*Burmeistera/Centropogon*	Central /South America	Rock 1919; Carlquist 1970
		Pratia sect. *Colensoa*	Borneo	Lammers 1990
Delissea (9)	Dry forest treelets and trees			

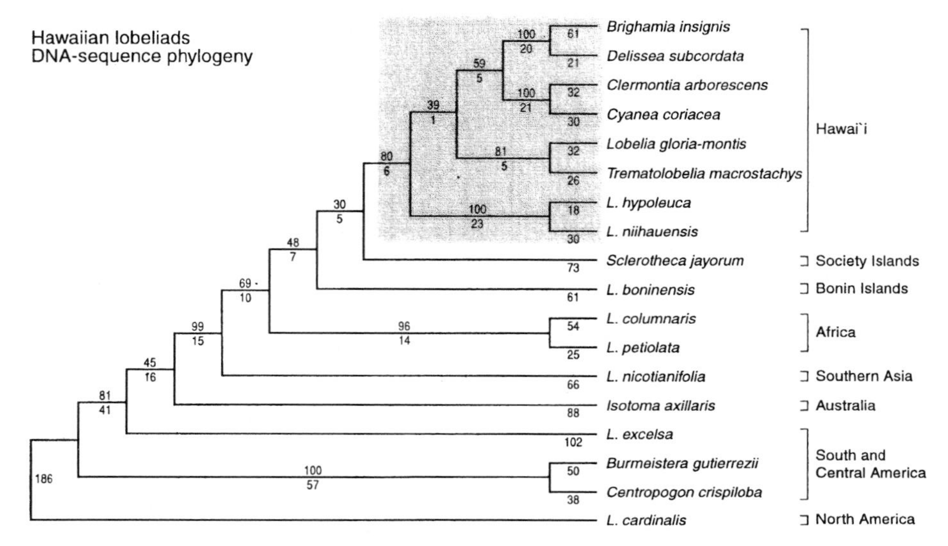

Fig. 2. Phylogram of relationships among representatives of the Hawaiian lobeliads and putative relatives, based on a weighted parsimony analysis of cpDNA and nrDNA sequence variation (see text). Numbers above each node are parsimony-jackknife values, based on 100 iterations under full heuristic search. Numbers below nodes are branch lengths. The consistency index CI was 0.86 (CI' = 0.66, excluding autapomorphies); the standardized consistency index CI^*_9 [normalized to an equivalent value for nine taxa (Givnish and Sytsma 1997)] was 0.94. Based on the equations presented by Givnish and Sytsma (1997), the maximum probability of correct phylogenetic inference was thus 1.00

tained one shortest tree of length 1,239 steps under Fitch parsimony, with a consistency index CI = 0.86 (CI' = 0.66, excluding autapomorphies) (Fig. 2).

2.2 Large-Scale Patterns of Adaptive Radiation

Five principal conclusions can be drawn from this tree. First, contrary to almost all previous views (Rock 1919; Wimmer 1953; Carlquist 1965, 1974; Mabberley 1974, 1975; Lammers 1989, 1990; but see Givnish et al. 1995, Givnish 1998), the Hawaiian lobeliads appear to be monophyletic. This finding – as well as the pattern of relationships among the Hawaiian taxa – is consistent with our restriction-site analysis, except that the latter identifies *Trematolobelia* and *Lobelia* section *Galeatella* as the earliest divergent clade within the Hawaiian group. Our results imply that the lobeliad radiation is one of the largest among plants on any true oceanic island or archipelago.

Second, the closest relatives of the Hawaiian lobeliads appear to be *Sclerotheca* from the Society Islands, and *Lobelia boninensis* from the Bonin Islands. More broadly, the Pacific Basin lobeliads appear to be related to those of Africa or Australasia (Fig. 3A). The single origin of the Hawaiian lobeliads may ultimately

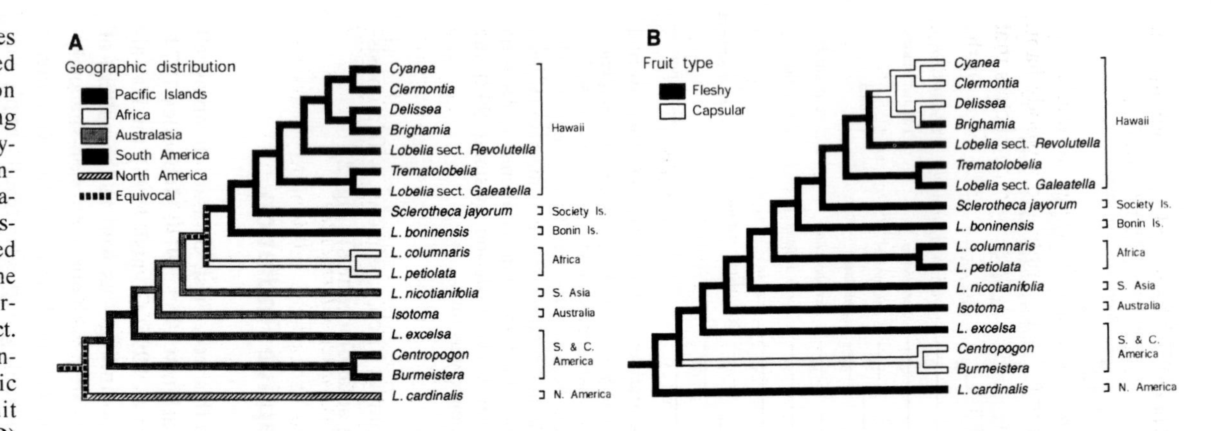

Fig. 3. Character-states of lobeliads and inferred ancestors overlain on the cladogram resulting from a combined analysis of DNA restriction-site and sequence variation. Accelerated transformation was assumed in order to minimize the appearance of convergent gains as an artifact. Characters analyzed include (**A**) geographic distribution, (**B**) fruit type, (**C**) habit, and (**D**) pollination syndrome

be traceable to the small target area of the Hawaiian chain, and the much greater proximity of the Hawaiian Islands to each other than to other moist, montane areas that could support lobeliads in the tropics.

Third, the exclusion of *Centropogon, Burmeistera,* and *Pratia* as close relatives of the three Hawaiian genera that share with them fleshy fruits and a woody habit (Fig. 3B,C) illustrates the danger of assessing phylogeny based on characters that are subject to strong selection pressures. Fleshy fruits arose at least twice in the woody lobeliads (in *Cyanea-Clermontia-Delissea* and *Burmeistera-Centropogon*), each time in moist tropical forests. Fleshy fruits were lost once, in *Brighamia* endemic to dry sea cliffs. Comparative data on the occurrence of different fruit types along ecological gradients, as well as functional arguments, suggest that fleshy fruits are adaptive under moist forest conditions (Givnish 1998). The woody habit evolved long before lobeliads colonized the Pacific, and appears to have first arisen in moist or wet tropical forests (Fig. 3C). Similarly, bird-pollinated flowers appear to have evolved long before lobeliads arrived in Hawaii (Fig. 3D), involving hummingbirds in North and South America, sunbirds in Africa and Asia, and white-eyes, honeyeaters, and honeycreepers in the Pacific. The habitats of bird-pollinated lobeliads are mostly wet and/or cold, in which pollinator thermoregulation would be favored. Both shifts to hawkmoth pollination (in *Brighamia* and *Isotoma*) occurred in dry forest and scrub.

Finally, the habitat occupied by the ancestor of the Hawaiian lobeliads appears to have been wet subalpine openings or bogs – like those now occupied by *Lobelia* section *Galeatella* and *Trematolobelia,* and by *Lobelia nicotianifolia* in southern Asia. There appears to have been a rapid initial radiation into four main lineages in the Hawaiian Islands, involving life in bogs and other moist or wet, relatively open sites (*Trematolobelia-Lobelia* sect. *Galeatella*), moist, inland rock walls and crests (*Lobelia* sect. *Revolutella*), dry forests, scrub, and sea cliffs (*Brighamia-Delissea*) and moist and wet forest interiors and edges (*Cyanea-Clermontia*).

Based on the estimated origin of *Cyanea* at least 8.3 million (M) years ago (Givnish et al. 1995) and the extent of genetic divergence of *Cyanea* from the other Hawaiian lobeliads in this current study, this initial radiation appears to have taken, at most, a few million years. Adaptive radiation in habitat and general growth-form thus appears to have occurred soon after the lobeliads arrived in the Hawaiian chain, involving divergence into four broad adaptive zones. Yet the lobeliads diversified to become the largest family represented in the endemic flora, with roughly 110 species known historically. What factors drove such extensive speciation long after the initial radiation in habit and habitat?

2.3 Seed Dispersal and Geographic Speciation

Givnish et al. (1995) argued that limited seed dispersal, associated with the invasion of moist forest interiors, may have been pivotal. More than half the Hawaiian taxa occur in *Cyanea,* which differs from its sister genus *Clermontia* in having three times as many species (65 vs. 22), a much greater fraction of which are restricted to single islands (91% vs. 59%). Ecologically, *Cyanea* differs from *Clermontia* in being native to forest interiors rather than forest edges, having smaller fruits, and being generally unbranched. Givnish et al. (1995) and Givnish (1998, 1999) argued that fleshy fruits dispersed by forest-interior birds – which are notoriously loathe to cross water- and habitat-barriers elsewhere in the tropics – should be associated with relatively poor dispersal ability. Low dispersal rates should tend to increase genetic divergence between populations, accelerate speciation, and result in most taxa having narrow ecological and geographic ranges. Fleshy fruits dispersed by forest-edge or scrub birds – whose dependence on successional or disturbed vegetation would require frequent movement – should have greater dispersal ability, resulting in lower rates of genetic differentiation and speciation, and broader ranges. Finally, groups with dust-like, wind-dispersed seeds that inhabit open, windswept, high-elevation habitats are likely to disperse over large distances, at least within the Hawaiian archipelago. High dispersal rates would tend to minimize rates of genetic divergence and speciation.

Indeed, high-elevation lobeliad lineages with dust-like seeds have 4.8 ± 3.0 species each; *Clermontia* and *Delissea,* with fleshy fruits in partly closed habitats, have 15.5 ± 9.2 species each; and *Cyanea,* with fleshy fruits in mainly closed forests, has 65 species. Within *Cyanea* itself, the expected pattern in species numbers also seems to hold. The orange-fruited clade (found almost exclusively in closed forest interiors) has 54 species, while the purple-fruited clade [found mainly in forest edges and in more open mesic forests (Givnish et al. 1995)] has only 12 species, despite both groups showing comparable ranges in both flower length and plant height. Thus, the invasion of moist forest interiors appears to have triggered substantially higher rates of speciation, arguably by favoring the evolution of fleshy fruits, decreased dispersability due to reliance on sedentary forest-interior birds, and increased rates of genetic differentiation within species, and leading to increased rates of geographic speciation and narrow species distributions.

2.4 Limited Dispersal and Repeated Small-Scale Radiations

Limited dispersal can trigger extensive speciation and narrow endemism without adaptive differentiation (Givnish 1997), as has apparently happened in many groups of relatively sedentary land snails, fossorial rodents, and marine invertebrates with poorly dispersing larvae (e.g., Gittenberger 1991; Cameron et al. 1996; Poulin and Feral 1996; Cook and Lessa 1998). Moreover, limited dispersal can also act synergistically with selection for ecological divergence to produce multiple, parallel adaptive radiations.

Cyanea provides several apparent examples of this phenomenon. On each of the four major Hawaiian Islands, species have evolved roughly the same range of corolla lengths (Givnish et al. 1995): 17 to 65 mm on Kaua`i, 24 to 75 mm on O`ahu, 18 to 75 mm on Maui, and 22 to 75 mm on Hawai`i (calculations based on assigning each species a corolla length midway between the upper and lower limits in Lammers 1990). Average corolla length is slightly (but significantly) shorter on Kaua`i (37 ± 14 mm) than on the progressively younger islands of O`ahu (54 ± 17 mm), Maui (50 ± 17 mm), and Hawai`i (51 ± 20 mm). Flower length is probably related to the bill length of the honeycreeper(s) and other birds that pollinated each species; interspecific variation in flower length should thus be a measure of the breadth of the pollinator guild partitioned by *Cyanea* taxa on each island. This pattern – combined with a detailed infrageneric analysis that inferred rapid evolutionary increases and decreases in floral length among closely related species – suggests that *Cyanea* underwent convergent radiations in flower length on each major island after a relatively small number of inter-island dispersal events (Givnish et al. 1995). Most of the inferred inter-island dispersal events were from one island to the nearest, younger island in the chain (Fig. 4), as would be expected given constraints on dispersal and the greater likelihood of successful establishment and subsequent radiation by colonists on nearby, newly formed, ecologically unsaturated islands. Such patterns of inter-island dispersal were first documented in *Drosophila* (Carson 1983) based on chromosomal banding patterns, and have now been reported for many other Hawaiian groups, including the silversword alliance (Carr et al. 1989; Baldwin 1992), *Alsininidendron-Schiedea* (Wagner et al. 1995; Sakai

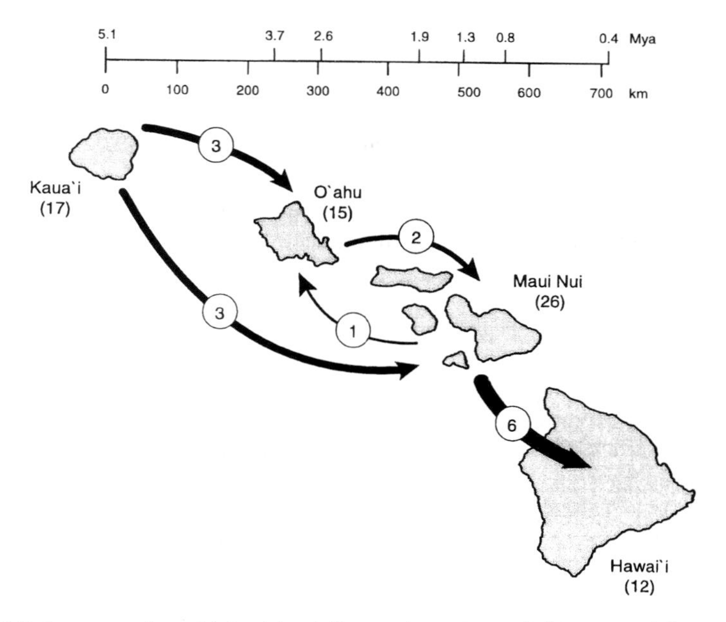

Fig. 4. Minimum number of inter-island dispersal events needed to account for present-day species in *Cyanea,* based on molecular and morphological data (after Givnish et al. 1995); note tendency for species to disperse from one island to the next younger one. Width of each arrow is proportional to the number of dispersal events between the corresponding pair of islands; the number of species found on each island or island group is indicated in parentheses. Mya, million years ago

et al. 1997), *Laupala* crickets (Shaw 1995), and *Tetragnatha* spiders (Gillespie and Croom 1995) based on molecular data, and the plant genera *Hesperomania, Remya, Hibiscadelphus,* and *Kokia* (Funk and Wagner 1995) based on morphology.

Cyanea species appear to partition each island via differences in elevational distribution within moist/wet forests. For example, within the Hardyi clade of purple-fruited *Cyanea,* there appears to have been an evolutionary progression from low to high elevations on Kaua`i, involving *C. coriacea* (150–220 m), *C. hardyi* (425–700 m), and *C. spathulata* (700–1200 m) (Givnish et al. 1995). More broadly, species of *Cyanea* appear to have invaded the full elevational range of appropriate habitats on each major island. Average species elevation (calculated as the mean of the upper and lower limits in Lammers 1990, 1992, 1996, as well as unpubl. pers. obs.) ranges from 185 to 1215 m on Kaua`i, from 320 to 950 m on O`ahu, from 455 to 2130 m on Maui, and from 465 to 1850 m on Hawai`i (Table 2). The much greater elevations achieved on the latter two, young islands partly reflects the greater maximum elevations of Hawai`i (4205 m) and Maui (3055 m) compared with the two older islands of O`ahu (1225 m) and Kaua`i (1598 m), as well as the absence of moist or wet forests above roughly 2200 m and the dominance of bogs and wet shrublands near the crests of O`ahu and Kaua`i (Gagné and Cuddihy 1990).

When species on each island are plotted by their mean corolla length and elevation, there is a roughly even distribution of species along these combined gradients, suggesting an even and strikingly similar partitioning of pollinators and habitats on each island (Fig. 5). Exceptions to this rule are few: (1) *Cyanea humboldtiana* and *C. superba* on O`ahu; (2) *C. copelandii* and *C. mceldowneyi* in wet forests near Waikamoi on East Maui; and (3) *C. degeneriana* and *C. pilosa* on Hawai`i, small-flowered species with pilose foliage which were originally described as subspecies of the same taxon by Rock (1919).

The close similarity of the first pair of species largely disappears if geographic variation in elevational distribution is taken into account. *Cyanea* (formerly *Rollandia* [Lammers et al. 1994]) *humboldtiana* is endemic to the Ko`olau Moun-

Table 2. Range, mean, and standard deviation of average corolla tube length, plant height, and elevation for *Cyanea* taxa occurring on each of the four rain-forested islands of the Hawaiian archipelago (see text). Data compiled from Lammers (1990, 1992, 1996), Rock (1919), and personal observations. Some or all data are missing for *C. glabra, C. linearifolia, C. longissima, C. pinnatifida, C. remyi,* and *C. pycnocarpa*

Island	No. of species	Corolla tube length		Plant height		Elevation	
		Range	Mean ± s.d.	Range	Mean ± s.d.	Range	Mean ± s.d.
Kaua`i	17	17–65	36.7 ± 14.0	1.0–14	4.1 ± 3.6	185–1215	631 ± 328
O`ahu	15	24–75	53.5 ± 17.3	0.5–6.0	2.7 ± 1.6	320–950	562 ± 147
Maui	21	18–75	49.3 ± 17.0	1.5–8.0	4.4 ± 2.3	455–2130	1127 ± 465
Hawai`i	12	22–75	51.1 ± 20.4	1.2–10	4.0 ± 2.8	465–1850	1064 ± 451

tains of eastern O`ahu, while *C. superba* is now limited to the Wai`anae Mountains of western O`ahu. *C. superba* formerly occurred in the Ko`olau range as well, but at a lower elevation [530 m (Rock 1919)] than that cited for the Wai`anae populations [595 m (Lammers 1990)], spacing it nearly 80 m below the mean elevation (608 m) of *C. humboldtiana*. *C. copelandii* is a vine-like species (unique in the genus), while *C. mceldowneyi* is erect and self-supporting. The close ecological similarity

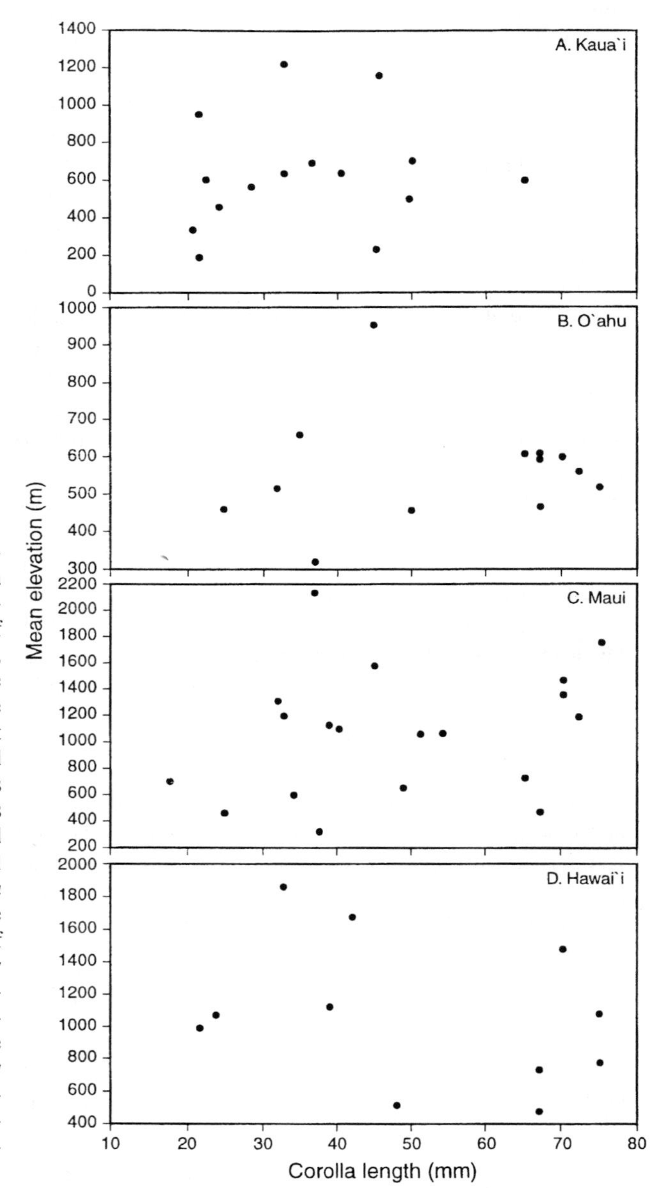

Fig. 5. Convergent adaptive radiations in corolla length and elevational distribution among species of *Cyanea* on the four major, rain-forested islands of the Hawaiian archipelago. The x- and y-values represent the means of upper and lower values for both traits as tabulated by Lammers (1990, 1992, 1996) and supplemented by personal observations for *Cyanea asarifolia* at 600 m in the "Blue Hole" at the foot of Mt. Wai`ale`ale, and by substitution of the mean elevation of the Wai`oli Valley below the falls for the extinct *Cyanea* (formerly *Rollandia*) *parvifolia*, collected there in 1909 without further label information (Rock 1919)

of the last species pair might simply reflect their recent divergence; in this regard, it should be noted that 3 of 12 *Cyanea* taxa on the youngest island of Hawai`i (ca. 0.4 M years old) are considered conspecific with taxa endemic to Maui (Lammers 1990). Three ecologically similar, palm-like species (*C. aculeatiflora, C. hamatiflora, C. macrostegia*) cooccur in wet forests near Waikamoi on East Maui, but diverge somewhat in mean flower length and/or elevation (Fig. 5). *C. aculeatiflora* and *C. macrostegia* are closely related, based on morphology, crossability (Lammers 1990) and molecular data (Givnish et al. 1995), but *C. hamatiflora* is quite distant from both, a member of the purple-fruited clade (Givnish et al. 1995).

These exceptions aside, species of *Cyanea* appear to have undergone largely parallel adaptive radiations in flower length and elevational distribution on Kaua`i, O`ahu, Maui, and Hawai`i – an evolutionary version of the movie "Groundhog Day", in which the genus evolved in much the same way (albeit with a few quirks) on each new island to which it dispersed. This similarity among insular radiations appears to extend to the interspecific range in maximum plant height and the number of species found on each major island as well, with the notable exception that Hawai`i, the youngest island, has fewer species than the older islands, in spite of being much larger in area (Table 2). This shortfall (12 vs. 15–21 spp) probably reflects the short time (0.4 M years) plants and other organisms have had to evolve on Hawai`i, compared with that available on Maui (1.9 M years), O`ahu (3.7 M years), and Kaua`i (5.1 M years). In accord with this hypothesis, a higher fraction of species on Hawai`i (27%) are conspecific with populations on other islands than are species on Maui (24%), O`ahu (14%), or Kaua`i (0%) (tabulated from data in Lammers 1990).

2.5 Extinction Rates in Relation to Adaptive Radiation and Geographic Speciation

Limited dispersal may not only have accelerated speciation and fostered parallel radiations on individual islands in *Cyanea,* but – paradoxically – have increased the likelihood of extinction for individual species as well (Givnish et al. 1995). Reduced dispersal and high rates of speciation appear to have resulted in narrow geographic and elevational ranges: 91% of *Cyanea* species are restricted to single islands, vs. only 58% for species in the sister genus *Clermontia,* while the average elevational range of individual species of *Cyanea* is 438 ± 294 m, vs. 840 ± 367 m for species of *Clermontia* ($P < 0.001$, 2-tailed t-test, 59 d.f.). The narrow geographic and elevational ranges of individual species of *Cyanea,* their limited dispersal, and dependence (at least in some instances) on highly specialized pollinators may have made them more susceptible to extinction (Givnish et al. 1995). Habitat destruction, decimation of pollinators and seed dispersers, and introduction of mammalian herbivores has led to 20% of *Cyanea* species now being thought extinct, with an additional 29% considered endangered (updated from Givnish et al. 1995). By comparison, only 5% of the historically known species of *Clermontia* are extinct, and 13% considered endangered. The principal factors associated with the likelihood of historically known species becoming extinct appear to be (i) initial rarity (see Terborgh and Winter 1980); (ii) occurrence in areas heavily disturbed by humans; and (iii) possession of highly specialized flowers > 45 mm in length (Givnish et al. 1995). Ten of the 13 extinct species of *Cyanea* were known from only one site, compared with only 4 of 52 extant species ($X^2 = 27.1$, $P < 0.0001$ for 1 d.f.). To the extent that data are available, the average elevational breadth of the extinct species was significantly less than that for extant species (102 m vs. 424 m, $P < 0.005$, 2-tailed t-test with 51 d.f.). Finally, the extinct species *C. arborea, C. comata, C. pohaku,* and *C. quercifolia* all occurred on leeward West Maui, in habitats that were heavily logged and/or cleared for pastures or *Eucalyptus* plantations; *C. giffardii* occurred on a single kipuka near Glenwood on Hawai`i that was largely cleared for cattle grazing (Givnish et al. 1995).

3 Limited Dispersal and Convergent Adaptive Radiations as a General Pattern

The general principle that limited dispersal can interact with selection for ecological divergence to produce multiple, parallel adaptive radiations finds support in several other groups. Examples include the parallel radiations of marsupial and placental mammals into a wide range of ecological roles on Australia vs. other continents (Springer et al. 1997); of *Anolis* lizards into a set of ecomorphs that partition the same set of structural habitats on each island of the Greater Antilles (Jackman et al. 1997; Losos et al. 1998); of arboreal, semi-arboreal, and terrestrial forms in different lineages of *Mandarina* land snails, on various islands of the Bonin chain since the Pleistocene (Chiba 1999); and of three floral syndromes in three habitat classes by different clades of the bulbous, heavy-seeded plant genus *Calochortus* (Patterson 1998). Each of these findings belies the claim by Gould (1989) that a "replay of the tape [of life] would lead evolution down a pathway radically different from the road actually taken". They do not support Gould's assertion that historical contingency – the particular lineage undergoing radiation, its genetic and phylogenetic constraints, and the suite of competitors, predators, and mutualists it faces – plays a predominant role in determining the direction of evolution. It must be recognized, however, that evolution within each of these groups has involved diversification based on very similar patterns of development and genetic variation (see Givnish 1997). Historical contingency may be important in determining which lineages undergo adaptive radiation in particular contexts, but ecological factors may be crucial in determining the pattern of adaptive radiation and species diversification that results.

4 Visual Selection, Sexual Selection, Predators, and Species Diversity

Limited dispersal may also have played a key role in the diversification of the African rift-lake cichlids. This group is considered one of the most dramatic cases of adaptive radiation, with species in different lakes evolving a wide range of different feeding strategies and associated dentition, body shape, and coloration (Fryer and Iles 1972; Greenwood 1984; Nishida 1991; Kocher et al. 1993; Meyer et al. 1994; Stiassny and Meyer 1999). Yet, while more than 1,000 species of cichlids have arisen in the seven major lakes, surely there are no more than 20 or 30 adaptive types per lake. The large remaining component of diversity must surely reflect the effects of limited dispersal, at least in part (Givnish 1997). Dispersal between the three largest lakes (Victoria, Tanganyika, Malawi) is minimal and molecular data indicate that parallel, largely independent radiations have occurred in each (Meyer et al. 1990, 1991, 1996; Kocher et al. 1993). In cichlids, extensive intralacustrine speciation is favored by (1) mouth brooding and limited dispersal of young; (2) use of rocks for shelter from predatory species, resulting in philopatry of adults in several groups; (3) the insular nature of suitable rock outcrops around the periphery of each lake; and (4) periodic drought and the resulting dissection of lakes into separate basins (see Mayr 1970; Fryer and Iles 1972; Greenwood 1974, 1978; Meyer et al. 1990, 1996; Johnson et al. 1996; Verheyen et al. 1996; Reinthal and Meyer 1997). Such limited dispersal fosters convergent radiations in individual lakes, and extensive speciation within lakes, associated with low vagility associated with mouth breeding and philopatry associated with isolated areas of rock outcrops.

Recent research by Albertson et al. (1999) supports this perspective. Their molecular phylogeny, based on DNA fingerprinting, resolves relationships among closely related cichlids from Lake Malawi. Adaptive divergence in feeding morphology appears to have occurred early in the history of the Malawi lineage; subsequent diversification has arisen with little change in trophic morphology, suggesting that processes other than adaptive radiation (e.g., limited dispersal) may have played a greater role in recent speciation.

Sexual selection on visual cues used for mate recognition also appears to play a key role in cichlid diversification, with many species displaying some of the most vivid coloration patterns of any freshwater fishes. Seehausen et al. (1997) have recently shown a strong correlation between water clarity and cichlid species diversity within and between the African rift lakes, and have experimentally demonstrated that water turbidity increases mismating by female cichlids. Seehausen et al. (1997, 1999) argue that, under better viewing conditions, sexual selection based on visual cues can generate a greater diversity of pre-mating barriers between species, leading to greater species diversity; greater turbidity led to more mismating and an erosion of cichlid diversity. Indeed, male nuptial coloration (blue vs. red dorsum and/or ventrum) varies rapidly through evolutionary time, especially in species with promiscuous mating systems (Seehausen et al. 1999); mating barriers between species often disappear under dim light or incomplete spectra (Seehausen and van Alphen 1998; Seehausen et al. 1997). I believe that a similar process of sexual selection in brightly lit, clear water may also help account for the astounding diversity and visual splendor of coral reef fish. In both rift lakes and coral reefs, the factors favoring exceptional water clarity – and thus, in part, high species diversity – are likely to be the same: *adjacent deep water basins* (allowing nutrients to purged by gravity from the water column, in the form of sinking plankton, feces, and dead plants and animals); *lack of seasonality* (reducing seasonal turnover of the water column and consequent re-injection of nutrients and particulates into the photic zone); and *high temperatures* (favoring rapid rates of growth – and, ultimately, of nutrient scavenging – by phytoplankton and zooplankton). It should be recognized, however, that high water clarity may not only favor sexual selection based on visual cues, but increase the selective pressures caused by visually orienting predators as well. Those predation pressures are almost surely important causes for the evolution of mouth brooding, strong territoriality, and use of refuge holes in rocky reefs in cichlids – all of which should increase speciation rates by limiting dispersal – and for the evolution of strong territoriality, striking coloration, and the use of refuge holes in coral reefs in coral reef fish.

Interactions with natural enemies might explain some of the remarkable species diversity of many groups of tropical woody plants. Tropical forests not only include the most diverse plant communities on earth – with up to 470 tree and liana species per hectare – they also display well-marked patterns of plant species richness along several ecological gradients (see review by Givnish 1999). Of these, perhaps the most striking is the rise in woody plant diversity with increasing rainfall and decreasing seasonality in the Neotropics (Gentry 1982, 1988; Wright 1992; Clinebell et al. 1995) and Hawaii (Aplet et al. 1998). Interactions with natural enemies may be an important factor helping create this gradient and foster high speciation rates in woody plants native to wet tropical forests (Givnish 1999). Most plant enemies are small, soft-bodied, desiccation-intolerant insects, nematodes, and pathogenic fungi, and are frequently host-specific within the tropics. As argued by Janzen (1970) and Connell (1971) – and decisively supported by the data of Wills et al. (1997) – attacks by host-specific natural enemies can create patterns of density-dependent mortality that can generate and maintain high levels of tropical tree diversity. Givnish (1999) made the Janzen-Connell hypothesis context-specific: wet, humid, aseasonal conditions should put the fewest physical limits on the populations of desiccation-intolerant plant enemies and thereby generate and maintain high levels of tree diversity, and foster high rates of speciation and differentiation by pathogens in tropical woody groups. In addition to the factors discussed by Givnish (1999), high temperatures may elevate the feeding rates of herbivores and thereby increase the potential for density-dependent mortality in their hosts. Indeed, Wilf and Labandeira (1999) recently used the fossil record to show that insect damage to leaves was greater in Wyoming during the Paleocene than during the cooler Eocene. A higher tempo of predation at elevated temperatures may thus be an additional factor promoting high diversity in tropical forests and coral reefs.

High tree diversity on rainy, humid, aseasonal sites in the tropics may also be promoted by the high density of understory stems on such sites (Givnish 1999).

Most woody species in moist tropical forests are understory shrubs or small trees (Whitmore 1975; Gentry 1982; Condit et al. 1996). Greater availability of moisture on rainier, less seasonal sites (and of nutrients on more fertile sites) should reduce whole-plant compensation points, increase shade tolerance (Givnish 1988, 1995; see also Wright 1992; Burslem et al. 1996) and permit more individuals – and, thus, species – to persist in the understory. In fact, there is roughly a doubling of woody stem number along the rainfall gradient in Neotropical forests, almost entirely involving stems < 2.5 cm d.b.h. (Givnish 1999). Given the regular relationship within a given region between stem number and tree species richness (Condit et al. 1996), the observed increase in stem number across the rainfall gradient accounts for only 17% of the observed 8.3-fold increase in tree species richness per 0.1 hectare (Givnish 1999). Both rainfall (a proxy for density-dependent tree mortality) and tree density have significant effects when each is included in a model for tree species richness:

$$\ln \text{species richness} = 0.361 \ln \text{rainfall} + 1.11 \ln \text{individual density} - 4.39$$

($r^2 = 0.735$, $P < 0.0001$ for 48 d.f.). The high density of small-diameter stems in tropical moist and wet forests should account for their high diversity of small-statured trees and treelets per unit area (see Whitmore 1975; Gentry 1982; Condit et al. 1996). The lognormal relationship between species number and stature in a lineage of tropical wet-forest plants such as *Cyanea* (Fig. 6) should be fairly general, and reflect the balance of three processes. First, competition for light should favor the evolution of taller and taller forms, which would tend to extend the species distribution to the right in Fig. 6. Second, shorter species have greater population densities (favoring low extinction rates) and may have lower dispersal capacity (favoring high speciation rates), favoring an accumulation of small-statured species at the left of the species-abundance curve. Strong herbivory pressures caused by high humidity, low seasonality, and high stem abundance among short species of rain-forest understories may select for divergent anti-herbivore defenses, and

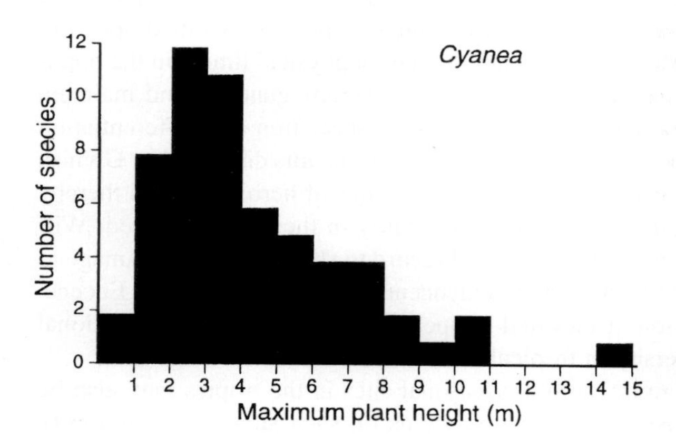

Fig. 6. Histogram of maximum plant height in *Cyanea* in the Hawaiian Islands (after Givnish 1997)

further increase the pool of short-statured species over evolutionary time.

Finally, the proportion of species with fleshy fruits dispersed by animals increases with rainfall in tropical forests (Gentry 1982, 1988). Many of these taxa are understory plants, so their reliance on sedentary forest-interior birds for dispersal should further accelerate speciation and increase the size of the pool of short-statured tree species, even if it has no immediate effect on the number of such species that coexist locally. It is remarkable that many of the most species-rich genera of angiosperms – such as *Chamaedorea, Dipsis,* and *Geonoma* (Arecaceae), *Piper* (Piperaceae), and *Psychotria* (Rubiaceae) – are fleshy-fruited shrubs, treelets, and trees in tropical rain-forest understories (Givnish 1998, 1999). Further research is needed to test whether the exceptional diversity of such groups is mainly a result of limited dispersal ability, density-dependent mortality, and/or adaptive radiation.

Acknowledgments

This research was supported by NSF grant DEB-9509550. Restriction-site and sequence analyses were conducted in collaboration with my colleagues Ken Sytsma, Tom Patterson, Jeff Hapeman, Jim Smith, Bill Hahn, Heather Corliss, Chris Pires, and Aaron Rodriguez; Eric Knox kindly provided access to DNAs of *Isotoma, L. columnaris, L. excelsa, L. nicotianifolia,* and *L. petiolata.* I would like to thank Professor Masahiro Kato and the Ministry of Education, Science, Sports and Culture of Japan for inviting me to participate in this symposium, associated with the awarding of the International Prize in Biology to Otto T. Solbrig. Solbrig's seminal contributions to our understanding of biosystematics, plant adaptation, and the ecological factors constraining plant diversity have long been a personal inspiration.

References

Albertson RC, Markert JA, Danley PD, Kocher TD (1999) Phylogeny of a rapidly evolving clade: the cichlid fishes of Lake Malawi, East Africa. Proc Natl Acad Sci USA 96:5101-5106

Aplet GH, Hughes RF, Vitousek PM (1998) Ecosystem development on Hawaiian lava flows: biomass and species composition. J Veg Sci 9:17-26

Baldwin BG (1992) Phylogenetic utility of the internal transcribed spacers of nuclear ribosomal DNA in plants: an example from the Compositae. Mol Phyl Evol 1:3-16

Baldwin BG (1997) Adaptive radiation of the Hawaiian silversword alliance: congruence and conflict of phylogenetic evidence from molecular and non-molecular investigations. In: Givnish TJ, Sytsma KJ (Eds) Molecular evolution and adaptive radiation. Cambridge Univ Press, New York, pp 103-128

Baldwin BG, Robichaux RH (1995) Historical biogeography and ecology of the Hawaiian silversword alliance (Asteraceae): new molecular phylogenetic perspectives. In: Wagner WL, Funk VA (Eds) Hawaiian biogeography: evolution on a hot spot archipelago. Smithsonian Institution Press, Washington, D. C., pp 259-287

Böhle U-R, Hilger HH, Martin WF (1996) Island colonization and evolution of the insular woody habit in *Echium* L. (Boraginaceae). Proc Natl Acad Sci USA 93:11740-11745

Bond JE, Opell, BD (1998) Testing adaptive radiation and key innovation hypotheses in spiders. Evolution 52:403-414

Burslem DFRP, Grubb PJ, Turner IM (1996) Responses to simulated drought and elevated nutrient supply among shade-tolerant tree seedlings of lowland tropical forest in Singapore. Biotropics 28:636-648

Cameron RAD, Cook LM, Hallows JD (1996) Land snails on Porto Santo – adaptive and non-adaptive radiation. Proc Roy Soc Lond Ser B 351:309-327

Carlquist S (1965) Island life. Natural History Press, New York

Carlquist S (1970) Hawaii: a natural history. Natural History Press, New York

Carlquist S (1974) Island biology. Natural History Press, New York

Carr GD, Robichaux RH, Witter MS, Kyhos DW (1989) Adaptive radiation of the Hawaiian silversword alliance (Compositae-Madiinae): a comparison with Hawaiian picture-winged *Drosophila.* In: Giddings LV, Kaneshiro KY, Anderson WW (Eds) Genetics, speciation, and the founder principle. Oxford Univ Press, New York, pp 79-97

Carson HL (1983) Chromosomal sequences and interisland colonizations in the Hawaiian *Drosophila.* Genetics 103:465-482

Chiba S (1999) Accelerated evolution of land snails *Mandarina* in the oceanic Bonin Islands: evidence from mitochondrial DNA sequences. Evolution 53:460-471

Clark A, Johnston IA (1996) Evolution and adaptive radiation of Antarctic fishes. Trends Ecol Evol 11:212-218

Clinebell HRR, Phillips OL, Gentry AH, Stark N, Zuuring H (1995) Prediction of neotropical tree and liana species richness from soil and climatic data. Biodivers Conserv 4:56-90

Condit R, Hubbell SP, LaFrankie JV, Sukumar R, Manokaran N, Foster RB, Ashton PS (1996) Species-area and species-individual relationships for tropical trees – a comparison of 3 50 haplots. J Ecol 84:549-562

Connell JH (1971) On the role of natural enemies in preventing competitive exclusion in some marine animals and in rain forest trees. In: den Boer PJ, Gradwell GR (Eds) Dynamics of populations. Centre for Agricultural Publishing and Documentation, Wageningen, The Netherlands, pp 298-313

Cook JA, Lessa EP (1998) Are rates of diversification in subterranean South American tuco-tucos (genus *Ctenomys,* Rodentia: Octodontidae) unusually high? Evolution 52:1521-1527

Foote M (1996) Ecological controls on the evolutionary recovery of post-Paleozoic crinoids. Science 274:1492-1495

Francisco-Ortega J, Crawford DJ, Santos-Guerra A., Jansen RK (1997) Origin and evolution of *Argyranthemum* (Asteraceae: Anthemideae) in Macaronesia. In: Givnish TJ, Sytsma KJ (Eds) Molecular evolution and adaptive radiation. Cambridge Univ Press, New York, pp 407-431

Fryer G, Iles TD (1972) The cichlid fishes of the great lakes of Africa. T. F. H. Publications, Neptune City, NJ

Funk VA, Wagner WL (1995) Biogeography of seven ancient Hawaiian plant lineages. In: Wagner WL, Funk VA (Eds) Hawaiian biogeography: evolution on a hot spot archipelago. Smithsonian Institution Press, Washington, D. C., pp 160-294

Gagné WC, Cuddihy LW (1990) Vegetation. In: Wagner WL, Herbst DR, Sohmer SH (Eds) Manual of the flowering plants of Hawai'i. Bishop Museum Publications, Honolulu, pp 45-116

Gentry AH (1982) Patterns of neotropical plant species diversity. Evol Biol 15:1-84
Gentry AH (1988) Changes in plant community diversity and floristic composition on environmental and geographical gradients. Ann Mo Bot Gard 75:1-34
Gillespie RG, Croom HB(1995) Comparison of speciation mechanisms in web-building and non-web-building groups within a lineage of spiders. In: Wagner WL, Funk VA (Eds) Hawaiian biogeography: evolution on a hot spot archipelago. Smithsonian Institution Press, Washington, D. C., pp 121-146
Gittenberger E (1991) What about non-adaptive radiation? Biol J Linnean Soc 43:263-272
Givnish TJ (1988) Adaptation to sun vs. shade: a whole-plant perspective. Austral J Plant Physiol 15:63-92
Givnish TJ (1995) Plant stems: biomechanical adaptation for energy capture and influence on species distributions. In: Gartner BL (Ed) Plant stems: physiology and functional morphology. Chapman and Hall, New York, pp 3-49
Givnish TJ (1997) Adaptive radiation and molecular systematics: aims and conceptual issues. In: Givnish TJ, Sytsma KJ (Eds) Molecular evolution and adaptive radiation. Cambridge Univ Press, New York, pp 1-54
Givnish TJ (1998) Adaptive plant evolution on islands: classical patterns, molecular data, new insights. In: Grant P (Ed), Evolution on islands. Oxford University Press, New York, pp 281-304
Givnish TJ (1999) On the causes of gradients in tropical tree diversity. J Ecol 87:193-210
Givnish TJ, Sytsma KJ (1997) Consistency, characters, and the likelihood of correct phylogenetic inference. Mol Phyl Evol 7:320-330
Givnish TJ, Sytsma KJ, Smith JF, Hahn WJ (1994) Thorn-like prickles and heterophylly in *Cyanea*: adaptations to extinct avian browsers on Hawaii? Proc Natl Acad Sci USA 91:2810-2814
Givnish TJ, Sytsma KJ, Smith JF, Hahn WJ (1995) Molecular evolution, adaptive radiation, and geographic speciation in *Cyanea* (Campanulaceae, Lobelioideae). In: Wagner WL, Funk VA (Eds) Hawaiian biogeography: evolution on a hot spot archipelago. Smithsonian Institution Press, Washington, D. C., pp 299-337
Givnish TJ, Sytsma KJ, Smith JF, Hahn WJ, Benzing DH, Burkhardt EM (1997) Molecular evolution and adaptive radiation in *Brocchinia* (Bromeliaceae: Pitcairnioideae) atop tepuis of the Guayana Shield. In: Givnish TJ, Sytsma KJ (Eds) Molecular evolution and adaptive radiation. Cambridge Univ Press, New York, pp 259-311
Givnish TJ, Evans TM, Pires JC, Sytsma KJ (1999) Polyphyly and convergent morphological evolution in Commelinales and Commelinidae: evidence from *rbcL* sequence data. Mol Phyl Evol (in press)
Goldblatt P, Manning JC, Bernhardt P (1995) Pollination biology of *Lapeirousia* subgenus *Lapeirousia* (Iridaceae) in southern Africa – floral divergence and adaptation for long-tongued fly pollination. Ann Mo Bot Gard 82:517-534
Gould SJ (1989) Wonderful life. Norton, New York
Grant PR (1986) Ecology and evolution of Darwin's finches. Princeton Univ Press, Princeton
Greenwood PH (1974) Cichlid fishes of Lake Victoria, East Africa: the biology and evolution of a species flock. Bull Brit Mus Nat Hist (Zool) Suppl 6:1-134
Greenwood PH (1978) A review of the pharyngeal apophysis and its significance in the classification of African cichlid fishes. Bull Brit Mus Nat Hist (Zool) 33:297-323
Greenwood PH (1984) African cichlids and evolutionary theories. In: Echelle AA, Kornfeld I (Eds) Evolution of fish species flocks. Univ of Maine Press, Orono, pp 141-154

Hapeman JR, Inoue K (1997) Plant-pollinator interactions and floral radiation in *Platanthera* (Orchidaceae). In: Givnish TJ, Sytsma KJ (Eds) Molecular evolution and adaptive radiation. Cambridge University Press, New York, pp 433-454

Jackman T, Losos JB, Larson A, de Queiroz K (1997) Phylogenetic studies of convergent adaptive radiationsa in Caribbean *Anolis* lizards. In: Givnish TJ, Sytsma KJ (Eds) Molecular evolution and adaptive radiation. Cambridge Univ Press, New York, pp 535-557

Janzen DH (1970) Herbivores and the number of tree species in tropical forests. Am Nat 104:501-528

Jernvall J, Hunter JP, Fortelius M (1996) Molar tooth diversity, disparity, and ecology in Cenozoic ungulate radiations. Science 274:1489-1492

Johnson TC, Scholz CA, Talbot MR, Kelts K, Ricketts RD, Nogobi G, Beuring K, Ssemanda L, Mcgill JW (1996) Late Pleistocene desiccation of Lake Victoria and rapid evolution of cichlid fishes. Science 273:1091-1093

Johnson SD, Linder HP, Steiner KE. (1998) Phylogeny and radiation of pollination systems in *Disa* (Orchidaceae). Am J Bot 85:402-411

Kirsch JAW, Lapointe F-J (1997) You aren't always what you eat: evolution of nectar-feeding among Old World fruitbats (Megachiroptera: Pteropodidae). In: Givnish TJ, Sytsma KJ (Eds) Molecular evolution and adaptive radiation. Cambridge Univ Press, New York, pp 313-330

Knox E. Downie SR, Palmer JD (1993) Chloroplast genome rearrangements and the evolution of giant lobelias from herbaceous ancestors. Mol Biol Evol 10:414-430

Kocher TD, Conroy JA, McKaye KR, Stauffer JR (1993) Similar morphologies of cichlid fish in Lakes Tanganyika and Malawi are due to convergence. Mol Phyl Evol 2:158-165

Lammers TG (1989) Revision of *Brighamia* (Campanulaceae: Lobelioideae), a caudiciform succulent endemic to the Hawaiian Islands. Syst Bot 14:133-138

Lammers TG (1990) Campanulaceae. In: Wagner WL, Herbst DR, Sohmer SH (Eds) Manual of the flowering plants of Hawai'i. Bishop Museum Publications, Honolulu, pp 420-489

Lammers TG (1992) Two new combinations in the endemic Hawaiian genus *Cyanea* (Campanulaceae: Lobelioideae). Novon 2:129-131

Lammers TG (1996) A new linear-leaved *Cyanea* (Campanulaceae:Lobelioideae) from Kaua'i, and the "rediscovery" of *Cyanea linearifolia.* Brittonia 48:237-240

Lammers TG, Freeman CE (1986). Ornithophily among the Hawaiian Lobelioideae (Campanulaceae): evidence from floral nectar sugar compositions. Am J Bot 73:1613-1619

Lammers TG, Givnish TJ, Sytsma KJ (1994) Merger of the endemic Hawaiian genera *Cyanea* and *Rollandia* (Campanulaceae: Lobelioideae). Novon 3:437-441

Losos JB, Jackman TR, Larson A, de Queiroz K, Rodriguez-Schettino L (1998) Contingency and determinism in replicated adaptive radiations of island lizards. Science 279:2115-2118

Mabberley DJ (1974) The pachycaul lobelias of Africa and St. Helena. Kew Bull 29:535-584

Mabberley DJ (1975) The giant lobelias: pachycauly, biogeography, ornithophily and continental drift. New Phytol 75:289-295

Mayr E (1970) Populations, species, and evolution. Belknap Press, Cambridge, MA

Meyer A, Kocher TD, Basasibwaki P, Wilson AC (1990) Monophyletic origin of Lake Victoria cichlid fishes suggested by mitochondrial DNA sequences. Nature 347:550-553

Meyer A, Kocher TD, Wilson AC (1991) African fishes. Nature 350:467-468

Meyer A, Montero C, Spreinat A (1994) Evolutionary history of the cichlid fish species flocks of the East African great lakes inferred from molecular phylogenetic data. Adv Limnol 44:409-425

Meyer A, Knowles L, Verheyen E (1996) Widespread geographic distribution of mitochondrial haplotypes in Lake Tanganyika rock-dwelling cichlid fishes. Mol Ecol 5:341-350

Nishida M (1991) Lake Tanganyika as an evolutionary reservoir of old lineages of East African cichlid fishes: inferences from allozyme data. Experientia 47:974-979

Patterson TB (1998) Phylogeny, biogeography, and evolutionary trends in the core Liliales and *Calochortus* (Calochortaceae): insights from DNA sequence data. Ph.D. dissertation, University of Wisconsin, Madison

Poulin E, Feral JP (1996) Why are there so many species of brooding Antarctic echinoids? Evolution 50:820-830

Price T, Gibbs HL, de Sousa L, Richman AD (1998). Different timing of the adaptive radiations of North American and Asian warblers. Proc Roy Soc Lond Ser B 265:1969-1975

Rainey PB, Travisano M (1998) Adaptive radiation in a heterogeneous environment. Nature 394:69-72

Reinthal PN, Meyer A (1997) Molecular phylogenetic tests of speciation models in Lake Malawi cichlid fishes. In: Givnish TJ, Sytsma KJ (Eds) Molecular evolution and adaptive radiation. Cambridge Univ Press, New York, pp 375-390

Rock JF (1919) A monographic study of the Hawaiian species of the tribe Lobelioideae, family Campanulaceae. Mem Bishop Mus 7:1-394

Sakai AK, Weller SG, Wagner WL, Soltis PS, Soltis DE (1997) Phylogenetic perspectives on the evolution of dioecy: adaptive radiation in the endemic Hawaiian genera *Schiedea* and *Alsinidendron* (Caryophyllaceae: Alsinoideae). In: Givnish TJ, Sytsma KJ (Eds) Molecular evolution and adaptive radiation. Cambridge Univ Press, New York, pp 455-473

Sang T, Crawford DJ, Kim S-C, Stuessy TF (1994) Radiation of the endemic genus *Dendroseris* (Asteraceae) on the Juan Fernandez Islands: evidence from sequences of the ITS regions of nuclear ribosomal DNA. Am J Bot 81:1494-1501

Sato A, O'Huigin C, Figueroa F, Grant PR, Grant BR, Tichy H, Klein J (1999) Phylogeny of Darwin's finches as revealed by mtDNA sequences. Proc Natl Acad Sci USA 96:5101-5106

Schluter D (1994) Experimental evidence that competition promotes divergence in adaptive radiation. Science 266:798-801

Seehausen O, van Alphen JJM (1998) The effect of male coloration on female mate choice in closely related Lake Victoria cichlids (Haplochromis nyererei complex). Behav Ecol Sociobiol 42:1-8

Seehausen O, van Alphen JJM, Witte F (1997) Cichlid fish diversity threatened by eutrophication that curbs sexual selection. Science 277:1808-1811

Seehausen O, Mayhew PJ, van Alphen JJM (1999) Evolution of colour patterns in East African cichlid fish. J Evol Biol 12:514-534

Shaw KL (1995) Biogeographic patterns of two independent Hawaiian cricket radiations (*Laupala* and *Prognathogryllus*). In: Wagner WL, Funk VA (Eds) Hawaiian biogeography: evolution on a hot spot archipelago. Smithsonian Institution Press, Washington, D. C., pp 39-56

Simpson GG (1953) The major features of evolution. Columbia Univ Press, New York

Smith AB, Littlewood DTJ (1997) Molecular and morphological evolution during the post-Palaeozoic: diversification of echinoids. In: Givnish TJ, Sytsma KJ (Eds) Molecular evolution and adaptive radiation. Cambridge Univ Press, New York, pp 559-583

Springer MS, Kirsch JAW, Case JA (1997) The chronicle of marsupial evolution. In: Givnish TJ, Sytsma KJ (Eds) Molecular evolution and adaptive radiation. Cambridge Univ Press, New York, pp 129-161

Stiassny MLJ, Meyer A (1999) Cichlids of the rift lakes. Sci Am 280:64-69

Terborgh J, Winter B (1980) Some causes of extinction. In: Soulé ME, Wilcox BA (Eds) Conservation biology: an evolutionary-ecological perspective. Sinauer Assoc., Sunderland, MA, pp 119-133

Turner PE, Souza V, Lenski RE (1996) Tests of ecological mechanisms promoting the stable coexistence of two bacterial genotypes. Ecology 77:2119-2129

Verheyen E, Ruben L, Snoeks M, Meyer A (1996) Mitochondrial phylogeography of rock-dwelling cichlid fishes reveals evolutionary influence of historical lake level fluctuations of Lake Tanganyika, Africa. Phil Trans Roy Soc Lond Ser B 351:797-805

Wagner WL, Herbst DR, Sohmer SH (Eds) (1990) Manual of the flowering plants of Hawai'i. Bishop Museum Publications, Honolulu

Wagner WL, Weller SG, Sakai AK (1995) Phylogeny and biogeography in *Schiedea* and *Alsinidendron* (Caryophyllaceae). In: Wagner WL, Funk VA (Eds) Hawaiian biogeography: evolution on a hot spot archipelago. Smithsonian Institution Press, Washington, D. C., pp 221-258

Whitmore, TC (1975) Tropical rain forests of the Far East. Oxford Univ Press, London

Wilf P, Labandeira CC (1999) Response of plant-insect associations to Paleocene-Eocene warming. Science 284:2153-2156

Wills C, Condit R, Foster RB, Hubbell SP (1997) Strong density-related and diversity-related effects help to maintain tree species diversity in a Neotropical forest. Proc Natl Acad Sci USA 94:1252-1257

Wimmer FE (1953) Campanulaceae-Lobelioideae, II Teil. In: Mansfield R (Ed) Das Pflanzenreich, Akademie-Verlag, Berlin, IV:276b:i-viii, 261-814

Wright SJ (1992) Seasonal drought, soil fertility and the species density of tropical forest plant communities. Trends Ecol Evol 7:260-263

6
Phylogenetic Analyses of Large Data Sets: Approaches Using the Angiosperms

DOUGLAS E. SOLTIS AND PAMELA S. SOLTIS

Department of Botany, Washington State University, Pullman, WA 99164-4238, USA

Abstract

Phylogeny is central to the understanding of biodiversity and evolutionary processes. However, elucidating phylogenetic relationships in many groups has remained problematic due to their sheer size. The feasibility of phylogenetic analyses of large data sets has been questioned on both theoretical and empirical grounds. Some have suggested that large data sets be broken into a series of smaller problems for phylogenetic analysis. However, recent empirical studies and critical developments in methods of data analysis indicate that large data sets are tractable. We have learned a great deal about the analysis of large data sets via the angiosperms, for which three large molecular data sets have been constructed (plastid *atpB* and *rbcL* and nuclear 18S rDNA). We discuss three approaches successfully applied in our analyses of these large data sets. Parsimony analyses of separate and combined data sets representing hundreds of taxa indicate that "bigger is better." That is, both empirical and simulation studies demonstrate that two solutions to dilemmas posed by large data sets is the addition of taxa as well as characters. Recent developments in software also greatly facilitate the parsimony analysis of large data sets. Applications such as NONA and the RATCHET can retrieve shorter trees than found by PAUP, and in much shorter run times. The recent development of "quick search" methods such as the fast bootstrap and fast jackknife are also of great utility in the analysis of large data sets. These methods are rapid and emphasize only those clades with strong support. All three of these approaches have recently been applied to a 567-taxon data set for angiosperms based on *atpB*, *rbcL*, and 18S rDNA sequences (a total of 4733 bp/taxon). Analyses of the combined three-gene data set have yielded

---

Key words. phylogeny, large data sets, angiosperms, parsimony jackknife, fast bootstrap

the best-resolved and best-supported topology to date for angiosperms, with virtually all major clades, as well as the spine of the tree, well supported. These developments indicate that the phylogenetic analysis of large data sets is not only feasible, but relatively straightforward.

1 Introduction

A firm understanding of phylogenetic relationships is obviously central to elucidating many questions involving evolution; a solid phylogenetic underpinning is also critical for a better understanding of biodiversity. Assessing phylogenetic relationships in many groups will require the compilation and phylogenetic analysis of data sets (DNA sequences and/or nonmolecular traits) for numerous taxa. This is particularly true, for example, in many large groups of organisms for which relationships have remained obscure, sometimes despite decades of study. Obvious examples include fungi, bacteria, insects, green plants, and large subclades within the green plants such as land plants, ferns, and angiosperms. However, the need for phylogenetic analysis of large data sets is not restricted to higher taxonomic groups, but may involve any portion of the taxonomic hierarchy, extending to the population level, or even to the level of "strains" of bacteria or viruses.

For convenience, we will arbitrarily define a large data set as one having over 150 placeholders. Although the phylogenetic analysis of large data sets often is central to understanding relationships within many groups, the feasibility of analysis of these data sets has been much debated (reviewed in Hillis 1996; Graybeal 1998; Soltis et al. 1998). For example, large data sets are problematic in parsimony searches because of the enormous number of trees possible for a large number of terminal taxa—the number of potential solutions increases logarithmically as taxa are added (Felsenstein 1978a). Hence, for just 20 taxa the number of possible rooted trees (8.87×10^{23}) slightly exceeds Avogadro's number—that is, roughly one mole of trees. For large data sets involving hundreds of taxa, the number of possible trees likely exceeds the number of atoms in the universe (Hillis, pers. comm.). A basic question, therefore, is how can we possibly examine a universe of possible trees, and be reasonably certain of the solution that we select?

Early simulation studies also suggested that the phylogenetic analysis of large data sets was impractical or impossible. For example, in some instances (those involving extreme branch-length heterogeneity), the correct reconstruction of phylogeny for only four taxa requires over 10,000 base pairs (bp) of sequence data (Hillis et al. 1994). Such problems and complexity with only four taxa prompted some to propose that large phylogenetic problems be broken into a series of smaller problems (e.g., Kim 1996; Mishler 1994; Rice et al. 1997), with one extreme view being to break large data sets into a large number of four-taxon problems (Graur et al. 1996).

Another problem posed by large data sets involves the assessment of support for individual clades. Two commonly used procedures for estimating branch support are the bootstrap (Felsenstein 1985) and Bremer support or decay index (Bremer 1988). Both are time-consuming, however, and impractical for large data sets given software presently available.

Early parsimony analyses of large data sets seemed to bear out, in part, the dire predictions of computational difficulty for large data sets (e.g., Felsenstein 1978; Graur et al. 1996; Hillis et al. 1994). For example, parsimony searches of a large data set of 500 *rbcL* sequences did not swap to completion (Chase et al. 1993), even after large investments of computer time. Rice et al. (1997) devoted "approximately 11.6 months of CPU time" using three Sun workstations in a reanalysis of the *rbcL* data set of Chase et al., yet none of their searches swapped to completion. Similarly, Soltis et al. (1997a) invested over two years of computer time in analyses of a 228-taxon data set of 18S rDNA sequences for angiosperms; again, no searches swapped to completion. Savolainen et al. (in press) encountered similar difficulties in phylogenetic analyses of a large data set of hundreds of *atpB* sequences representing the angiosperms.

Significantly, however, recent empirical and simulation studies suggest that large data sets are much more tractable than thought only a few years ago. Furthermore, much of what systematists have learned about "hands-on" analyses of large data sets has been garnered via the study of several large DNA sequence data sets compiled for angiosperms. Herein we attempt to review some of the approaches that can be used in the analysis of large data sets. We rely heavily on work recently completed using the three large DNA sequence data sets compiled for angiosperms to exemplify several of the possible solutions to the analytical problems posed by large data sets. We highlight here three general approaches (that are not mutually exclusive) to the analysis of large data sets encompassing hundreds of taxa: 1) the addition of taxa and characters; 2) the use of "fast" or "quick" searches such as the fast bootstrap and parsimony jackknife; and 3) the application of computer programs such as NONA and the RATCHET to facilitate faster searches of tree space. We also discuss two other approaches to the analysis of large data sets: 4) the supertree method, and 5) compartmentalization.

2 Background: Angiosperm Data Sets

Three large DNA sequence data sets have been assembled for angiosperms: 18S rDNA (1,855bp), *rbcL* (1,428bp), and *atpB* (1,450bp). From a historical perspective (reviewed in Soltis et al. 1997b; Chase and Albert 1998; Chase and Cox 1998), *rbcL* was the first gene to be sequenced widely in angiosperms (Chase et al. 1993), followed by 18S rDNA (Soltis et al. 1997a), and most recently *atpB* (Savolainen et al., in press). As noted, parsimony searches of these individual data sets did not swap to completion due to their large size. Nonetheless, the topologies obtained based on phylogenetic analysis of these individual genes were strikingly similar, prompting investigators to suggest that these genes were tracking the same organismal phylogeny and that analysis of large data sets might be a productive exercise (e.g., Chase and Albert 1998; Chase and Cox 1998; and Soltis et al. 1997b, 1998).

The initial data sets compiled for 18S rDNA, *rbcL*, and *atpB* did not "match" completely in terms of the exemplars used. However, via careful planning and collaboration, these three genes have now been sequenced for a nearly identical suite of taxa (often using the same DNA) for 560 angiosperms, as well as seven outgroups. This 567-taxon data set has been analyzed via several approaches: parsimony as implemented by PAUP*4.0, the ratchet approach of Nixon (unpubl.), and the parsimony jackknife method of Farris et al. (1996) (Fig. 1).

Several initial combinations of data sets contributed greatly to our understanding of how to solve large phylogenetic problems. These include the combination of 18S rDNA and *rbcL* sequences (Soltis et al. 1997b) and *rbcL* and *atpB* (Savolainen et al., in press), and 193 taxa for 18S rDNA, *atpB*, and *rbcL* (Soltis et al. 1998). Other noteworthy data sets and combined analyses include the *rbcL* and nonmolecular data sets compiled and analyzed by Nandi et al. (1998). Below we summarize some of the approaches used in the analysis of large data sets in general, with an emphasis on recent results for angiosperms.

3 Add Taxa and Add Characters

Both simulation and empirical studies indicate that adding taxa and increasing the number of characters (base pairs in this case) in phylogenetic analyses may not only increase the accuracy of the estimated trees, but also reduce the computational difficulty. These results are in contrast to earlier suggestions that large data sets may be extremely complex and difficult to analyze phylogenetically.

Several simulation studies have revealed the importance of adding taxa. Using the 228-taxon 18S rDNA data set for angiosperms and shortest trees obtained (Soltis et al. 1997a) as the basis for simulation studies, Hillis (1996) suggested that the phylogenetic analysis of large data sets may be more tractable than he and coworkers had previously suggested based on simulations involving only four taxa (e.g., Hillis et al. 1994). The more recent simulation studies of Graybeal (1998) further support this contention. These studies suggest that adding taxa, while perhaps counterintuitive, made phylogenetic analyses more straightforward, apparently because the addition of taxa breaks up long branches and disperses homoplasy.

Empirical studies further demonstrate the importance of adding both taxa and characters in the phylogenetic analysis of large data sets. Soltis et al. (1998) conducted heuristic parsimony searches and fast bootstrap analyses on separate and combined DNA data sets for 190 angiosperms and three outgroups; separate data sets of 18S rDNA, *rbcL* and *atpB* sequences were combined into a single matrix 4,733 bp in length. Analyses of the combined 193-taxon data set revealed great improvements in computer run times compared to the analyses of the separate data sets and the data sets combined in pairs. Six analyses of the combined three-gene data set were conducted, and in all cases TBR branch swapping was completed, on average in less than two days of computer run time. In contrast, the separate and pairwise analyses of the data sets did not swap to completion. These results agree with those of Chase et al. (1993), Rice et al. (1997), Soltis et al. (1997a) and Savolainen et al. (in press) for large data sets for the individual genes; none of these searches ever completed. Furthermore, whereas analyses of a 193-taxon data set for two genes (*rbcL* + 18S rDNA) for angiosperms did not swap to completion, analyses of somewhat larger (and highly similar data sets in terms of taxon composition) *rbcL* + 18S data sets for 228 and 267 taxa did swap to completion (Soltis et al. 1999). The individual data sets for *rbcL* and 18S rDNA alone for 228 and 267 taxa did not swap to completion. Thus, simply by adding taxa and sequence this large phylogenetic problem actually became more tractable.

The results summarized here illustrate a major advantage of combining sequences (or other characters) and adding taxa in studies of large data sets: the time needed for parsimony analysis actually decreases with the addition of characters and taxa. In addition to shorter run times, combined data sets for angiosperms show tremendous improvements in internal support for clades, as well as increased resolution (Soltis et al. 1998). The number of clades receiving fast bootstrap support of ≥ 50%

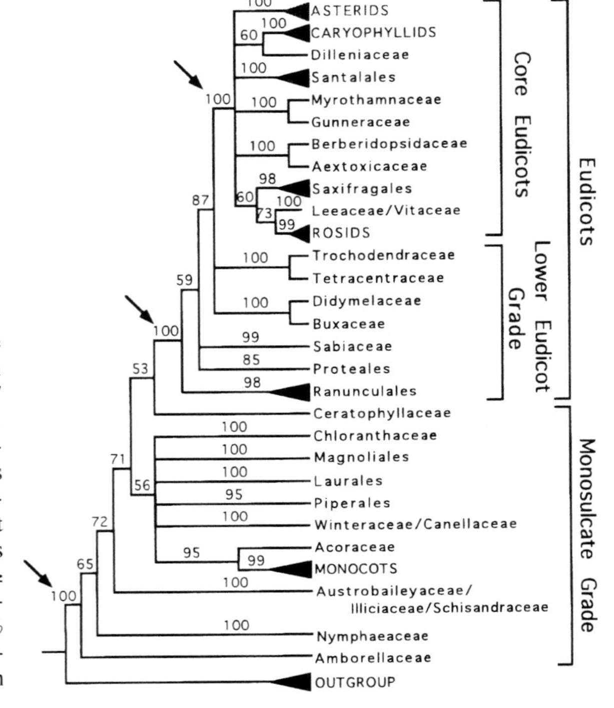

Fig. 1. Summary jackknife tree for angiosperms based on phylogenetic analysis of 567 taxa sequenced for *atpB*, *rbcL*, and 18S rDNA. Parsimony jackknife analysis was conducted by S. Farris. Values above branches represent jackknife values. Only nodes having jackknife support ≥ 50% were saved; nodes having support less than 50% were not saved and are depicted as polytomies (from Soltis et al., submitted)

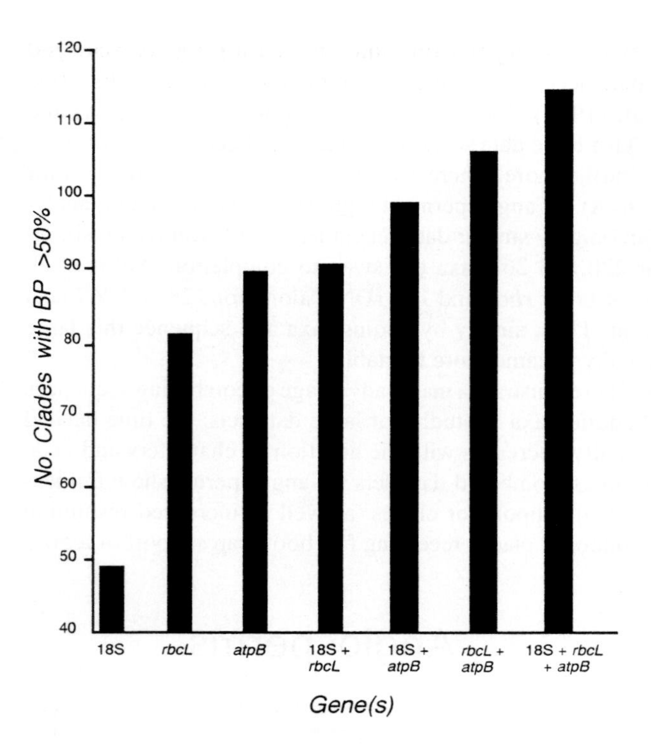

Fig. 2. Number of angiosperm clades having bootstrap percentage (BP) values ≥50% based on the phylogenetic analysis of separate and combined *atpB, rbcL*, and 18S rDNA sequence data sets for 190 angiosperms (modified from Soltis et al. 1998)

increased dramatically as data sets were combined compared to the separate data sets (Fig. 2). Most clades show a steady increase in bootstrap support as data sets are combined and additional characters are analyzed. Particularly noteworthy are those clades that are present in all of the shortest trees observed, but do not receive fast bootstrap support of ≥ 50% in the analyses of the separate data sets. As data sets are combined, the level of fast bootstrap support rises. Significantly, a number of large angiosperm clades do not exhibit fast bootstrap support of ≥ 50% until data sets are combined; examples include some of the largest, most critical clades of angiosperms, such as eudicots, monocots, Asteridae s.l., Asteridae s.s., and Caryophyllales).

As noted, the success realized with these "smaller" data sets of 193 taxa involving three genes served, in part, as the stimulus for the construction of a 567-taxon data set for angiosperms based on *rbcL* + 18S rDNA + *atpB*. Analyses of this combined data set have resulted in the best-resolved and best-supported topology yet retrieved for the angiosperms (Fig. 1; Soltis et al., submitted.). The internal support for clades and overall resolution for this three-gene tree for 567 taxa are much higher than obtained based on bootstrap or parsimony jackknife analyses of the 500-sequence *rbcL* data set for angiosperms and other seed plants (Chase and Albert 1998; Kallersjo et al., in press). Internal support and resolution are also much higher than achieved for large data sets of 18S rDNA alone (Soltis et al. 1997a) or for *atpB* and *atpB* plus *rbcL* (Savolainen et al., in press).

4 Quick Searches for Well-Supported Clades

A significant problem with the analysis of large data sets is the difficulty in assessing internal support for clades. As reviewed earlier, large data sets are not amenable to standard bootstrapping (Felsenstein 1985) and decay or Bremer support analysis (Bremer 1988). However, programs that conduct fast bootstrap and fast jackknife analyses are available and were designed with large data sets in mind. The fast jackknife (or parsimony jackknife) was recommended by Farris et al. (1996). A fast jackknife program is also available on PAUP*4.0, and the two methods yield similar values (Mort et al., in press).

Several studies illustrate the value of conducting relatively quick searches and "saving" only those clades above some threshhold of internal support. Below a minimal threshhold (e.g., a bootstrap or jackknife value of 50%), confidence in a clade is low or nonexistant. Hence, why conduct extensive, time-consuming parsimony searches looking for shorter and shorter trees when continued branch-swapping does not result in increased support for weakly-supported clades? The well-supported clades appear relatively quickly in the analysis of big data sets—continued branch-swapping involving poorly-supported branches does not suddenly result in strongly-supported clades. Lengthy parsimony analyses may, in fact, be a waste of time. This is exemplified well by the reanalysis of the 500-sequence *rbcL* matrix (Chase et al. 1993) by Rice et al. (1997). The initial searches of this large data set by Chase et al. (1993) did not swap to completion; furthermore, the authors realized trees shorter than those that they had found existed (reviewed in Chase and Albert 1998). Lengthy reanalysis of this data set (Rice et al. 1997) did find shorter trees than those reported by Chase et al. (1993); however, these searches did not result in a significantly different topology from that provided initially by Chase et al. (1993).

The parsimony jackknife method (Farris 1996; Farris et al. 1996) is well suited for the "quick" analysis of large data sets and has been applied to a data set of 2538 *rbcL* sequences (Kallersjo et al., in press). The approach uses jackknifing (resampling of characters without replacement and stepwise addition of taxa without branch-swapping). A consensus of the trees generated by the jackknife replications depicts only well-supported clades (usually with support of 50% or greater); thus, the approach provides a rapid assessment of branch support. A second version of the *Jac* program is now available (Farris et al., in prep); it has a faster tree-building algorithm, and also allows branch swapping to be conducted at each replication. With the second version, it is also possible to perform several random-addition sequences per replicate. The number of replications performed is typically 500 or 1000.

Several studies have employed the parsimony jackknife approach and illustrate its utility, as well as its speed. Soltis et al. (1997a) used the original *Jac* program without branch swapping in their analysis of 228 angiosperm 18S rDNA sequences. The parsimony jackknife analysis of this data set (based on 1000 replicates) was completed in only 949 seconds. An extremely promising example of the applicability of the parsimony jackknife is that of Kallersjo et al. (in press), who analyzed 2538 *rbcL* sequences representing a broad taxonomic range, from cyanobacteria to flowering plants. Kallersjo et al. used both the original and newer versions of the *Jac* program, with surprising results — numerous (a total of 1400) clades with *Jac* values ≥ 50% were retrieved, including major clades such as green plants, land plants, angiosperms, and eudicots. With regard to the two *Jac* programs Farris and coworkers have provided, Kallersjo et al. (in press) found that while branch-swapping improved both resolution and the support for some clades, most of the groups were recovered by the original program (which is faster and simpler). The original *Jac* program required 356 seconds per replicate; the new version with branch-swapping required approximately 1304.5 seconds per replicate. The initial analysis of the 2538-sequence matrix using the original *Jac* program with 1000 replicates required 99 hours on a 133 MHz Pentium computer. Because the newer *Jac* program has a stop-restart function (which was employed by Kallersjo et al., in press), the amount of time needed to complete the analysis using the new *Jac* program was not precisely known (Kallersjo et al., in press), but can be estimated to be roughly 360 hours for 1000 replicates. Soltis et al. (submitted) have applied the new *Jac* program (with the help of S. Farris) to the 567-taxon, three-gene angiosperm data set; 1000 replicates were completed in 60.63 hours. Application of the parsimony jackknife approach to this 567-taxon data set yielded a topology with few nodes lacking support ≥ 50%; even the spine of the tree is generally well supported (Fig. 1). In this case, one could legitimately ask if lengthy parsimony searches are even necessary.

5 Developments in Parsimony Analysis

The numerous improvements and new options available in PAUP* 4.0 have been a great asset to those interested in analyzing large data sets. One obvious improvement is the faster speed with which PAUP* 4.0 (Swofford 1998) conducts heuristic parsimony searches compared to version 3.1. Similarly, improvements in the maximum likelihood (ML) program as implemented in PAUP* 4.0 have made it possible to analyze much larger data sets than before using this approach. Although ML analyses are now possible with data sets up to 50-60 taxa, these data sets are not technically "large" as considered for this review (≥ 150 taxa). ML analysis is still not feasible with truly large data sets.

Algorithms for finding shortest trees such as Hennig 86, (Farris 1988), NONA (Goloboff 1993), and PAUP (Swofford 1993) were developed for working on what we consider here to be small data sets. Perhaps one of the most important developments in the analysis of large data sets is a new computer program that greatly enhances the ability to find shorter trees using parsimony. K. Nixon (unpubl.) has developed a new method called the RATCHET, which is applied using his program DADA along with P. Goloboff's program NONA.

When applied to the 567-taxon matrix of three genes for angiosperms (analyses conducted by K. Nixon), the RATCHET quickly found trees shorter than those recovered by PAUP* 4.0 (Swofford 1998). After several weeks, searches with PAUP* 4.0 found trees of 45,107 steps whereas the RATCHET found trees of length 45,101 steps (that is, six steps shorter than the trees found using PAUP*) in only a few hours, and subsequent analyses found trees of 45,100 steps. In addition to being fast, the RATCHET method makes it possible to consider very large numbers of trees from multiple islands, decreasing one's chances of being marooned on a single island of trees and improving the reliability of the consensus of the shortest trees. For example, with the 567-taxon, three-gene data set, PAUP* found approximately 1600 trees of length 45,107, all on one island over the course of several weeks, whereas the RATCHET recovered nearly 5000 trees of length 45,100 in approximately 24 hours.

The phylogenetic analysis of large data sets via parsimony has benefited greatly from recent developments in software; continued developments in this area are anticipated, making the analysis of large data sets ever more straightforward.

6 Other Approaches: Super Trees

Another approach that permits the construction of trees for large numbers of taxa is the "supertree" method (Sanderson et al. 1998). Using this approach, existing topologies that overlap in at least several taxa are "grafted" together (Sanderson et al. 1998). Although this approach may have some applications in the study of large numbers of taxa, it cannot, we believe, substitute for the actual analysis of the large number of taxa needed to infer phylogeny accurately across large groups. The supertree method has potential difficulties and shortcomings, some of which are reviewed by Wilkinson and Thorley (1998). Perhaps the biggest problem with the supertree approach is that it reconstructs a phylogeny across a diverse group, such as the angiosperms, via a piecemeal approach where only one or a few taxa may be shared among trees. To assess phylogeny across the angiosperms, a group for which the spine of the tree has been uncertain, as have the major clades, only a global approach would yield meaningful results.

There are, however, potential applications of a "tree grafting" approach when used in combination with the methods of analysis outlined in the sections above. That is, once a large data set has been analyzed phylogenetically and an overall topology obtained, it may be useful to graft existing clades that have been the subject of more thorough sampling onto the main tree. For example, using the three-gene topology for angiosperms (Soltis et al., submitted), it would seem appropriate to graft well-sampled and studied clades onto the overall tree to replace some of the placeholders used for larger, well-supported subclades. Saxifragales, Malvales, Sapindales, Brassicales (glucosinolate clade), Fabaceae, Cornales, and Hydrangeaceae are a few of the well-supported clades of flowering plants that have been examined with more taxa than employed by Soltis et al. (submitted). The existing topologies for these well-studied and well-supported clades could be "grafted" onto the three-gene tree, replacing the original placeholders. Donoghue et al. (1998) employed this basic approach in establishing a tree for Asteridae for use in analyses of floral evolution. Working from the tree for Asteridae s.l. provided by Olmstead et al. (1993) based on *rbcL* sequences, subclades better sampled by other investigators, such as Apiales and Cornales, were grafted onto the "parent" tree, in place of the initial exemplars.

7 Other Approaches: Compartmentalization

An alternative to analysis of an entire large data set involves the partitioning of taxa to permit analyses of subsets of the data, a general method of analysis referred to as "compartmentalization" (Mishler 1994). Known monophyletic groups (i.e., compartments) are represented by an inferred hypothetical ancestor in more inclusive analyses. The character states of this ancestor are based on those of all the taxa that compose the monophyletic group, rather than a single exemplar, and the ancestor will likely differ from all real taxa in the group. The position of each monophyletic group (as represented by the hypothetical ancestor) is free to move relative to other taxa and groups in the analysis. Mishler (1994) recommended a three-step procedure: (1) a global analysis of all taxa to identify well-supported groups (i.e., compartments); (2) local analyses within compartments; and (3) further global analysis, with compartments represented by hypothetical ancestors or with compartment topologies constrained as found in local analyses. Of course, the most difficult aspect of a compartmentalized analysis such as this is the recognition of groups sufficiently well supported to be considered compartments. To provide reliable estimates of clades in a large data set, the global analysis in step (1) will likely require extensive computation. A significant benefit of a compartmentalized analysis is improved homology assessment within compartments, whether the data are molecular or morphological. Consequently, inferred relationships within compartments are more reliable, with the use of additional characters, and homoplasy among compartments is reduced.

A compartmentalized analysis (Mishler et al. 1998) was applied to a data set of 100 18S rDNA sequences for land plants and green-algal outgroups (Soltis et al., in press). Unconstrained parsimony analyses of this data set recovered trees that portray clearly spurious relationships, and bootstrap support for most clades was low (Soltis et al., in press). Compartmentalization, recognizing eight clades as compartments (algae, hornworts, liverworts, mosses, lycophytes, ferns, conifers, and angiosperms), improved resolution among these groups, produced trees that are more consistent with other hypotheses of land-plant relationships (e.g., Kenrick and Crane 1997), and increased bootstrap support for the clades recovered (Mishler et al. 1998). Compartmentalization may be a particularly attractive option for analyses of large data sets that span broad phylogenetic distances, because it allows the use of clade-specific characters and homology assessments within compartments in addition to global characters that apply across compartments.

8 Recommendations

We recommend a combination of the approaches discussed here when working with large data sets at any taxonomic level. Quick search approaches, such as the parsimony jackknife or fast bootstrap can be applied to any large data set. These will quickly retrieve the well-supported clades and concomitantly illustrate those areas of the topology in which confidence is weak. Where critical portions of the topology are not well supported, investigators are advised to collect more data (additional characters) and perhaps add taxa, rather than conduct lengthy parsimony searches of a data set in the hope of finding shorter trees (e.g., Rice et al. 1997). Adding taxa can be particularly important in breaking up long branches (Felsenstein 1978b). Adding characters can increase resolution and internal support for clades, and shorten run times. Parsimony searches should be conducted taking advantage of advances in software, such as the RATCHET, which more rapidly recovers short trees, enabling the investigator to explore larger areas of tree space in a given time. Tree combination and grafting methods are not a substitute for the analysis of large data sets. In many cases, however, replacing the exemplars for a given with a topology for that clade that represents more thorough taxon and character sampling will be both useful and informative.

References

Bremer K (1988) The limits of amino acid sequence data in angiosperm phylogenetic reconstruction. Evolution 42:795-803

Chase MW, Albert VA (1998) A perspective on the contribution of plastid *rbcL* sequences to angiosperm phylogenetics. In: Soltis DE, Soltis PS, and Doyle, JJ (Eds) Molecular systematics of plants II. Chapman and Hall, New York, pp 488-507

Chase MW, Cox VA (1998) Gene sequences, collaboration and analysis of large data sets. Aust Syst Bot 11:215-229

Chase MW, Soltis DE, Olmstead RG, Morgan D, Les DH, Mishler BD, Duvall MR, Price RA, Hills HG, Qiu Y-L, Kron KA, Retig JH, Conti E, Palmer JD, Manhart JR, Sytsma KJ, Michaels HJ, Kress WJ, Karol KG, Clark WD, Hedren M, Gaut BS, Jansen RK, Kim K-J, Wimpee CF, Smith JF, Furnier GR, Strauss SH, Xiang Q-Y, Plunkett GM, Soltis PS, Swensen SM, Williams SE, Gadek PA, Quinn CJ, Eguiarte LE, Golenberg E, Learn GH, Jr., Graham SW, Barrett SC, Dayanandan S, Albert VA (1993) Phylogenetics of seed plants: An analysis of nucleotide sequences from the plastid gene *rbcL*. Ann Mo Bot Gard 80:628-580

Donoghue MJ, Ree RH, Baum DA (1998) Phylogeny and the evolution of flower symmetry in the Asteridae. Trends Plant Science 3:311-317

Farris JS (1988) Computer program and documentation Hennig 86. Port Jefferson, New York

Farris JS (1996) 'Jac. Version 4.4.' Swedish Museum of Natural History, Stockholm

Farris JS , Albert VA, Kallersjo AM, Lipscomb D, Kluge AG (1996) Parsimony jackknifing outperforms neighbor-joining. Cladistics 12:99-124

Felsenstein J (1978a) The number of evolutionary trees. Syst Zool 27:27-33

Felsenstein J (1978b) Cases in which parsimony or compatibility methods will be positively misleading. Syst Zool 27:401-410

Felsenstein J (1985) Confidence limits on phylogenies: an approach using the bootstrap. Evolution 39:783-791

Goloboff P (1993) Pee-Wee and NONA. Computer programs and documentation. New York

Graur DL, Duret L, Gouy M (1996) Phylogenetic position of the order Lagomorpha (rabbits, hares and allies). Nature 379:333-335

Graybeal A (1998) Is it better to add taxa or characters to a difficult phylogenetic problem? Syst Biol 47:9-17

Hillis DM (1996) Inferring complex phylogenies. Nature 383:130

Hillis DM, Huelsenbeck JP, Swofford DL (1994) Hobgoblin of phylogenetics? Nature 369:363-364

Kallersjo M, Farris JS, Chase MW, Bremer B, Fay MF, Humphries CJ, Petersen G, Seberg O, Bremer K (in press) Simultaneous parsimony jackknife analysis of 2538 *rbcL* DNA sequences reveals support for major clades of green plants, land plants, seed plants and flowering plants. Plant Syst Evol

Kenrick P, Crane PR (1997) The origin and early diversification of land plants. Smithsonian Institution Press, Washington, DC

Kim J (1996) General inconsistency conditions for maximum parsimony: Effects of branch length and increasing numbers of taxa. Syst Biol 45:363-374

Mishler BD (1994) Cladistic analysis of molecular and morphological data. Am J Phys Anthropo 94:143-156

Mishler BD, Soltis PS, Soltis DE (1998) Compartmentalization in phylogeny reconstruction: philosophy and practice. Technical Report, Symposium on Estimating Large Scale Phylogenies: Biological, Statistical, and Computational Problems. DIMACS, Princeton, New Jersey

Mort ME, Soltis PS, Soltis DE, Mabry ML (in press) A comparison of three methods for estimating internal support on phylogenetic trees. Syst Biol

Nandi OM, Chase MW, Endress PK (1998) A combined cladistic analysis of angiosperms using *rbc*L and non-molecular data sets. Ann Mo Bot Gard 85:137- 212.

Olmstead RG, Bremer B, Scott KM, Palmer JD (1993) A parsimony analysis of the Asteridae sensu lato based on *rbc*L sequences. Ann Mo Bot Gard 80:700-722

Rice KA, Donoghue MJ, Olmstead RG (1997) Analyzing large data sets: *rbcL* 500 revisited. Syst Biol 46:554-563

Sanderson MJ, Purvis A, Henze C (1998) Phylogenetic supertrees: assembling the trees of life. Trends Ecol Evol 13:105-109

Savolainen V, Chase MW, Morton CM, Hoot SB, Soltis DE, Bayer C, Fay MF, de Bruijn A, Sullivan S, Qiu Y-L (in press) Phylogenetics of flowering plants based upon a combined analysis of plastid *atp*B and *rbc*L gene sequences. Syst Biol

Soltis DE, Hibsch-Jetter C, Soltis PS, Chase MW, Farris JS (1997b) Molecular phylogenetic relationships among angiosperms: An overview based on *rbcL* and 18S rDNA sequences. In: Iwatsuki K, Raven PH (Eds) Evolution and diversification of land plants. Springer-Verlag, Tokyo, pp 157-178

Soltis DE, Soltis PS, Chase MW, Mort ME, Albach DC, Zanis M, Savolainen V, Hahn WH, Hoot SB, Fay MF, Axtell M, Swensen SM, Nixon KC, Farris JS (submitted) Angiosperm phylogeny inferred from a combined data set of 18S rDNA, *rbcL*, and *atpB* sequences. Bot J Linn Soc

Soltis DE, Soltis PS, Mort ME, Chase MW, Savolainen V, Hoot SB, Morton CM (1998) Inferring complex phylogenies using parsimony: an empirical approach using three large DNA data sets for angiosperms. Syst Biol 47: 32-42

Soltis DE, Soltis PS, Mort ME, Hibschjetter C, Zimmer EA, Morgan D (1999) Phylogenetic relationships of the enigmatic angiosperm family Podostemaceae. Mol Phyl Evol 11: 261-272

Soltis DE, Soltis PS, Nickrent DL, Johnson LA, Hahn WJ, Hoot SB, Sweere JA, Kuzoff RK, Kron KA, Chase MW, Swensen SM, Zimmer EA, Chow S-M, Gillespie LJ, Kress WJ, Sytsma KJ (1997a) Angiosperm phylogeny inferred from 18S ribosomal DNA sequences. Ann Mo Bot Gard 84:1-49

Soltis PS, Soltis DE, Wolf PG, Nickrent DL, Chaw S-M, Chapman RL (in press) The phylogeny of land plants inferred from 18S rDNA sequences: Pushing the limits of rDNA signal? Mol Biol Evol

Swofford DL (1993) PAUP: Phylogenetic analysis using parsimony, version 3.1. Illinois Natural History Survey, Champaign, Illinois

Swofford DL, (1998) PAUP*: Phylogenetic analysis using parsimony, version 4.0. Sinauer, Sunderland, Massachusetts

Wilkinson, M, Thorley, JL (1998) Reduced supertrees. Trends Ecol Evol 13:283

Part 2
Ecological Biodiversity

Part 2
Ecological Biodiversity

7
The Theory and Practice of the Science of Biodiversity: A Personal Assessment

Otto T. Solbrig

Department of Organismic and Evolutionary Biology, Harvard University, 22 Divinity Ave., Cambridge, MA 02138, USA

Abstract

Diversity is a property, not an entity in itself. It refers to the property of a set of objects of not being identical, of varying one from another in one or more characteristics. When applied to organisms, it refers to the universal attribute of all living things that each individual being is unique, that is, no two organisms are identical. The origin of this variability is to be found in the basic and fundamental property of the DNA molecule that the order of the bases does not affect the free energy of the molecule, in other words all combinations of the four bases that form the genetic code are chemically equally viable. This characteristic combined with natural selection allows the acquisition and accumulation of favorable mutations, and in the approximately 3 thousand million years that life has existed on Earth these processes have produced the enormous biological variation that we see today, which is at best only a very small percentage of all the variation that has existed in the past. Today this diversity of life is threatened by human activities, although the exact rate of species loss is difficult to ascertain. And species loss is only one aspect of the profound transformation of the terrestrial landscape that is presently taking place, brought about by the growth of human populations and their economic activities. This transformation extends from genes to ecosystems. The transformation of the world's habitats by people has many causes, and many effects. As long as there is doubt regarding the exact mechanism underlying this transformation, there will be differences of opinion concerning the impact, and people who always are reluctant to change their behavior will use these varying opinions to rationalize their actions.

Key words. agriculture, bio-ethics, diversity, evolution, landscape transformation

Only a rigorous, unbiased, and mechanistic science of biodiversity can help resolve these differences.

The Science of Biodiversity must include all the aspects of evolutionary and ecological theory concerned with the Origin and Maintenance of the diversity of living organisms. It also must include, the study of human behavior and their economic activities. Finally, in order to not only understand, but also influence human behavior so as to reduce the environmentally negative aspects of their economic activities, a new environmental ethics must be developed.

1 Introduction

In the early days of my scientific career only systematists, ecologists and a small band of conservationists were interested in studying how to maintain the rich and wonderful diversity of organisms. Yet the profound modification of many landscapes that have taken place in the last forty years, the actual and potential loss of species, especially the so called "emblematic" species, i.e. those used in heraldic emblems such as the Bald Eagle emblem of the United States, have brought the issue of the maintenance of biological diversity to the public's attention, and today even politicians feel obliged to mention loss of Biodiversity at least once in their speeches. Yet what is Biodiversity, and why is it important?

As I have mentioned repeatedly in my writings (Solbrig 1991, 1994), "diversity" is a property not an entity in itself. It refers to the attribute of a set of objects of not being identical, that is of varying one from another in one or more characteristics. When applied to organisms, it refers to the universal characteristic of all living things that each individual being is unique, that is, no two organisms are identical, with the possible exception of identical twins and clones.

The origin of this variability is to be found in the basic and fundamental property of the DNA molecule that the order of the bases does not affect the free energy of the molecule, in other words all combinations of the four bases that form the genetic code are chemically equally viable. This characteristic combined with natural selection allows the acquisition and accumulation of favorable mutations, and in the approximately 3–4 thousand million years that life has existed on Earth these processes have produced the enormous biological variation that we see today, which is at best only a very small percentage of all the variation that has existed in these thousands of million years.

Today this diversity of life is threatened by human activities, although the exact rate of species loss is difficult to ascertain. And species loss is only one aspect of the profound transformation of the terrestrial landscape that is presently taking place, brought about by the growth of human populations and their economic activities. This transformation extends from genes to ecosystems.

2 Landscape Transformation

Planet Earth is a blend of living and non-living elements interacting in innumerable ways that result in a variety of landscapes and ecosystems. This amalgam is very dynamic, both in time and in space, and is constantly changing, unfolding, and producing new combinations. The living elements — living by nature of their capacity to self-reproduce — come in a great variety of forms, shapes, and complexities, that are continuously changing and evolving, comprising the diversity of life.

Cosmic factors and local geological forces have sustained the process of ecosystem evolution. Organisms have in turn molded the process by modifying the characteristics of their surroundings. The most significant of these changes is probably the transformation by photosynthetic bacteria of the atmosphere of the Earth from a reducing to an oxidizing one, that reached present levels about a thousand million years ago. Humans, late arrivals to this series of events, have of late become important transformers of natural landscapes, although nothing they have done or are likely to do, other than an atomic holocaust, can match the transformation of the atmosphere by the early photosynthetic bacteria. Yet there are some parallels. Although transformation of the atmosphere from a reducing to an oxidizing one did not eliminate all anaerobic organisms, it reduced them significantly, both in numbers, biomass, and relative importance, and opened the way for an entire new biological world dominated by oxygen breeding, multicellular plants and animals. Today people are transforming the surface of the globe at such an accelerated pace that they have launched a new phase in the evolution of the planet. This new condition is characterized by the dominance of one species, our own, over all others, to an extent never before experienced on Earth. Because of our ability to transform landscapes, we, humans, are in a situation where we can choose to jeopardize our own existence as well as that of innumerable other species, or we can select to achieve a more harmonious coexistence with our biological and abiological surroundings. This presents us with an ethical and moral choice to which I return further on.

The transformation of the world's habitats by people has many causes, and many effects. My talk today will analyze some of these and the role that scientific knowledge can play in ameliorating the human impact on the environment.

3 The Process of Transformation

We transform our surroundings to suit them to our needs when we build structures such as houses, factories or roads, and especially when we replace natural vegetation and fauna with agro-ecosystems and pastures for our domestic animals; we further transform landscapes when we extract natural resources, be they renewable such as timber, or non-renewable such as oil or minerals. In short, the transformation of the Earth's landscapes is the result of our economic activities. And our manifold economic activities are the manifestation of the drive we share with every other species to obtain as many resources as possible to fulfill our need to survive, grow and reproduce.

The transformations of our surroundings have a direct impact on the functioning of ecosystems. Not only are native species populations reduced or eliminated, the abiotic components of the ecosystem are also profoundly affected and the flows of energy and materials modified. Drainage patterns are changed, resulting in new and sometimes very intense erosive processes, in silting of rivers and wetlands, in loss of soil nutrients and organic matter, and in the nutrient and water retention capacity of the soil. The loss of vegetation accelerates erosion. From an ecosystem perspective, the loss of species is only one aspect of the transformation of landscapes, and possibly not the most worrisome.

The transformations of natural landscapes is not the result of a grand conspiracy by particular individuals, or groups of individuals, industry, governments, or other institutions in the society. Neither the rich, nor the poor, the inhabitants of developed countries, or those living in developing nations, are especially interested in destroying or modifying the function of natural ecosystems. Much to the contrary, most everybody admires natural environments, and given a chance tries to recreate them in their homes and gardens. Yet we all are modifying the natural environment day by day, month by month, usually without realizing it. Each day, hundreds, thousands, if not millions of decisions are made by individuals, governments, businesses, and industries that affect our surroundings. Some decisions have only a small impact, and others a very great one. Most of these decisions are made by people to increase their wealth and improve their economic well being. A house here, a road there, a factory that will give employment, a business that cuts down only a part of a forest, an excess of fertilizer to insure a good crop, a little bit of some chemical poison dumped into a river, or into the air, a bigger car, an extra appliance, and so on, lead to the transformation and eventual degradation of the natural environment. While each decision will usually have only a small local effect, collectively they will have a tremendous effect, and because of the nature of the process nobody feels responsible for the global effect.

Because the process of habitat modification is normally highly dispersed, to understand it, the individual decisions and the motivations behind them need to be studied in detail. Aggregate data on how many hectares of land are lost to development, or guesses regarding the number of species lost, or calculations of how many tons of soil are lost to erosion each year, although useful as an indication of the overall human impact, cannot determine the causes of the process. Only by studying human decisions at the micro level can we ever understand the dynamics of the process of transformation and develop realistic policies to stop it. Some industry organizations are trying to do this (Hawken 1993) but ecologists have been very reluctant to study systems in flux due to human activity. Trained to control our variables, we prefer to study systems where the uncontrollable human impact is minimal, and our first inclination is to put a fence around our study sites to exclude disturbance by people. However, in order to understand human impact we must consider people as part of the system and include them in our studies just like any other variable. A science of biodiversity must incorporate the impact of humans on the landscape as well as their behavior *vis-à-vis* the environment, studied until now almost exclusively by anthropologists.

4 Impact of Human Activities on Ecosystem Function

Yet not everybody agrees that these transformations are serious or that they can have dire consequences for the survival of the human species. Global warming is denied by some, or its consequences are represented as benign; loss of biodiversity is considered inconsequential to human welfare, and those of us who worry about it are called alarmist; and chemical contamination is not taken seriously. As long as there is doubt regarding the impact of human activities there will be differences of opinion, and people who always are reluctant to change their behavior will use these varying opinions to rationalize their actions. A rigorous, unbiased, and mechanistic science of biodiversity can help resolve these differences.

Agriculture is one of the principal agents of landscape transformation and is a good example of the dilemmas faced by individuals and society in regard to ecosystem change. Let us therefore briefly analyze ecosystem function in relation to agriculture and assess the evidence regarding the significance of the changes taking place.

When our ancestors adopted agriculture they contracted a Faustian bargain. In exchange for a new and more reliable source of sustenance, not only would their toil increase, but they had to bear the consequences of the transformation of the land they plowed and the plants and animals they grew. Hunter-gatherers had appropriated over many previous millennia some of the products of nature — fruits, seeds, game — changing only minimally the characteristics of the ecosystem that produced these resources. However, agriculturalists and pastoralists purposefully increased the abundance of some species, and eliminated others. Furthermore, with time they also affected the characteristics of the crops and domestic animals they maintained to the point that they no longer can reproduce by themselves, or compete with the native vegetation or fauna. A case in point is the wheat plant. Hexaploid *Triticum aestivum,* the most widespread plant in the world, is not known in the wild and is incapable of reproducing by itself, and would disappear from this planet if farmers stopped planting it. Similar considerations apply to cultivated rice and maize, the plants that together with wheat constitute 50% of the food intake of humankind. Yet, while humans have assumed the responsibility of being the administrators of the evolution and reproduction of their domestic plants and animals, they have not assumed the same degree of responsibility in relation to the physical characteristics of the agro-ecosystems where these plants and animals live, resulting in enormous losses of topsoil.

Erosion is the most serious and costly of the negative effects of modern agriculture. Soil that is lost is difficult to recover, and restoring degraded soils is expensive (but not impossible). The second most serious problem — and in many irrigation areas more acute these days than erosion — is salinization. According to Norse et al. (1992) 2000 million hectares of land have been lost to erosion and salinization. It has been calculated that 35 % of the arable land of the world is affected by some degree of erosion (Norse et al. 1992). For the United States alone, the costs of erosion have been calculated at 900 million dollars/year (Pimentel et al. 1995) although not everybody agrees with these figures. In general, natural scientists tend to see the problem as very serious and economists tend to dismiss it as secondary. The difference lies in that natural scientists measure the changes that are occurring in the soil itself, while economists look at the production from these same soils. Production has not decreased in the same measure as erosion, and in mildly eroded soils it has actually increased due to the introduction of new varieties and increased use of fertilizers. The unresolved question is whether soil losses can be replaced for ever by technological improvements in agronomic technologies, or whether eventually a point is reached at which the soil loses its capacity to sustain agriculture. All indications (National Academy of Science 1989) are that the latter is true, and prudence calls for a careful approach. New minimum tillage and no tillage techniques can reduce soil erosion by at least an order of magnitude and should be adopted whenever possible. Every effort should be made to reduce erosion and salinization. The knowledge is there, so why are erosion and salinization increasing?

Table 1. Fertilizer use in Argentina 1983–1996

Year	Tons
1983	252 973
1984	310 148
1985	354 101
1986	272 489
1997	341 347
1988	357 754
1989	316 468
1990	303 379
1991	325 612
1992	516 155
1993	603 067
1994	922 121
1995	1 209 768
1996	1 700 000

Source: Salvador (1998).

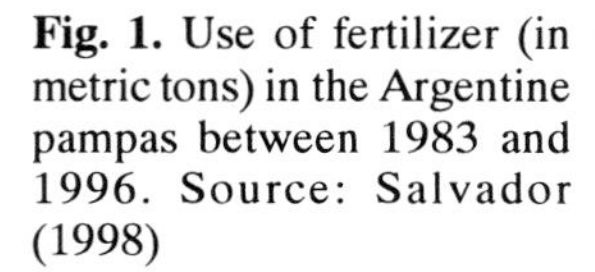
Fig. 1. Use of fertilizer (in metric tons) in the Argentine pampas between 1983 and 1996. Source: Salvador (1998)

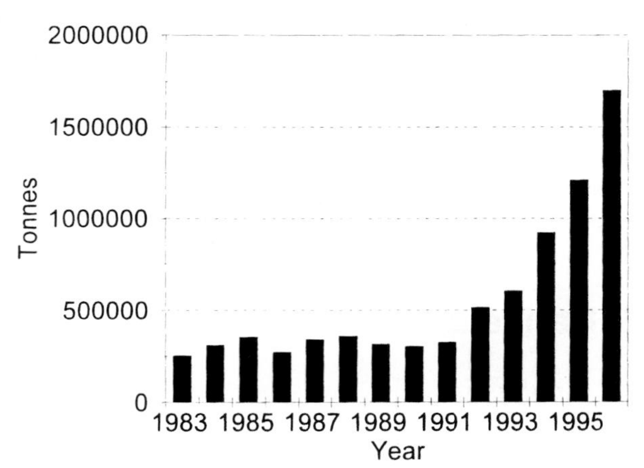

The answer has to be found in our economic behavior. Erosion and salinization (and other environmental problems) reduce productivity, but they also require investments of time and money to combat them. When the marginal costs of those investments are equal or less than the marginal income derived from the added productivity, then it pays to spend the time, money and effort to reduce the negative effects, otherwise it does not. An example from the Argentine pampa illustrates this fact.

In the period between 1970 and 1980, pampa farmers increased productivity largely as the result of better genetic varieties, mechanization and the use of chemical herbicides. Yet little or no fertilizers were used, except in high value horticultural crops resulting in loss of nutrients and reduction of soil quality (Casas 1998). Imported fertilizers were expensive, grain prices were low, and the government taxed agriculture (Fiorentino 1998). The marginal costs of increasing yields by using fertilizers and better farming techniques were not compensated by the increased productivity. Starting in 1990, the government reduced and then eliminated agricultural export taxes; international grain prices increased, and fertilizer prices were reduced as factories were established in the country. The result was a dramatic increase in fertilizer use (Table 1 and Fig. 1) and the adoption of no-tillage farming in 20% of the area.

Loss of organic matter and nutrient depletion reduce land productivity since they reduce the nutrient and moisture retention capacity of the soil. They result from continuous agriculture, especially when the same crop is planted year after year (King 1990). Alternating crops, especially with leguminosae, or even better, alternating agriculture and livestock raising, reduces nutrient depletion and loss of organic matter. No-tillage techniques in many cases also increase organic matter.

The high yielding varieties of our crops obtained through genetic manipulation are the basis of the increase in agricultural yields that have avoided widespread famines in spite of the tripling of the world population in the last fifty years. The drawback from the point of view of biodiversity conservation is their genetic uniformity. Traditional varieties were genetically diverse. As these are replaced by high yielding ones, genetic diversity is reduced dramatically, threatening the genetic base on which plant breeders depend to obtain newer and better varieties in the future. An effort to preserve genetic diversity is the establishment of a network of gene banks (National Academy of Science 1993). Yet, these gene banks only preserve a small part of the gene diversity. This is one of the most serious problems associated with the loss of biodiversity.

5 Native Ecosystems

For the most part people have given little thought to the conservation of that part of the Earth's surface not directly involved in the productive process, what has been called rustic landscapes (Halffter 1999). Natural ecosystems are as important as human-made ones in sustaining life, through their role in such processes as water and nutrient cycling, storing carbon, fixing atmospheric nitrogen and liberating oxygen. However, most of these landscapes have also been profoundly affected by human activities.

This problem has many facets. In the developed world it takes the form of urban sprawl, especially in the United States. In a belt surrounding large cities, housing developments and factories advance on natural landscapes and highly productive agricultural land destroying or profoundly affecting the function of the ecosystem. In other areas it takes the form of deforestation, or destruction of woody species in quest for firewood, or degradation of ecosystem by uncontrolled grazing. Whatever its form, the result is always the same: reduced primary productivity, reduced capacity of nutrient and water cycling, and loss of potential habitat for many species.

One of our innate beliefs derived from thousands, if not millions of years of human evolution, is that nature takes care of itself. When we pick an apple from a tree we never think of the consequences to the tree or the ecosystem. In the past, hunter-gatherer bands foraged in a given territory as long as the energy obtained from the food compensated for the effort involved in obtaining it. Once an area does not produce sufficient food to compensate for the energy needed to gather it, a hunter-gatherer band moves to another site. Natural ecosystems present us with the fruits, the seeds, the roots, and the game. "Nature" takes care of the upkeep of the system and the cost of production.

In a natural ecosystem, the impact of each species in the food chain, although not negligible, is part of the characteristics of the ecosystem, and the evolution of each species population is in part a response to the changes wrought by other species. As any given species increases its numbers at the cost of another it becomes a powerful selective force in the life of the secondary species. Sometimes that struggle leads to the disappearance of a prey species, which in turn affects the survival of the predator, but often an arms race ensues where each species increases its defenses, its competitive abilities, or its ability to penetrate the defenses of another one, as the case may be, resulting in a dynamic equilibrium.

The invention of tools, and much later the development of agriculture gave humans an almost absolute advantage over all species, those that are potential resources, those that are potential competitors, and those that are neither and therefore mostly neutral. Such clear advantage of one species over all others means that the effects of human interactions with other species are all in one direction, namely in favor of people, and against all other species. The resilience of the system that opposed an evolutionary reaction to every new adaptation that evolved is mostly gone, and unless people start taking this new situation into account, ecosystems can quickly become highly modified and degraded. This has occurred repeatedly in the history of humanity in the last 10 000 years. Examples are the salinization of irrigated land in Babylon; the desertification of once productive land in Cartage (present day Tunis); and soil erosion in Nepal following deforestation. The examples are very numerous, and all have the same cause: overutilization of resources without paying attention to the effects of that utilization.

If agricultural scientists have learned superbly how to manipulate plants and animals so as to transform them into highly efficient and productive food producing machines, scientists should also be capable of applying themselves to the careful management of the entire ecosystem, be it a human made agricultural field or pasture, or a natural ecosystem. And indeed many scientific advances have been made in our knowledge and ability regarding the management of agro-ecosystems, and forests. Yet never has there been a greater degradation of nature than what is being experienced today. Why this contradiction?

6 Need for an Environmental Ethics

From an evolutionary point of view, we perceive ourselves as sharing this planet with other species and being subjected to the same biological constraints. Our survival as living organisms depends on the same physical laws and ecological processes that govern every other species, present and past. Consequently we try to push ahead in the struggle for existence. But this understanding does not provide us with any particular directives as to how we should conduct our lives and how we should behave when confronted with the environmental changes with which we are faced today, because our role as moral agents is not deducible from our biological nature. On the contrary, the history of civilization is the development of normative behavioral rules that control human relations and that limit many of our innate biological impulses. Although countries still engage in mass killings when they go to war, individuals are no longer allowed to kill their rivals or seek private vengeance.

The characteristic of the human species that has given it an advantage in the struggle for existence over other species has been the development of rational thinking resulting from the large brain in relation to body weight. This and the acquisition of speech has been at the basis of our evolutionary success. Consequently we like to think that all our actions are the result of rational decisions. But this is true only in a minority of cases. Most of our behavior is innate and does not involve the forebrain.

Our behavior towards other species and towards the ecosystem in a world dominated by people is also not deducible from our biological nature or that of other species, or from knowledge regarding ecosystem function. This will require the development of new normative and evaluative judgments. Ecology can give us the pertinent factual understanding, but not the moral rules to determine our behavior *vis-à-vis* nature. These will have to be provided by the society at large taking biological constraints and human needs into account.

These rules are still being developed and there is as yet no general consensus as to what they should be. The guiding principle must be the concept of sustainability which maintains that we have an obligation to maintain for coming generations the productive capacity of the planet, including both areas developed for human use as well as the more natural environments.

Any rules of behavior are ultimately derived from some authority and require some mechanism of enforcement. The need for new rules relating to behavior *vis-à-vis* the environment is very clear but it is not at all certain what these rules should be, who should develop them, or how they should be enforced. There are a number of ideas floating around and international agencies such as IUBS through its Diversitas program or UNESCO's Man and the Biosphere program have formulated some general principles, but undoubtedly much more work is needed in this area. Although the Rio'92 Environmental Summit addressed a large number of issues including this one, no consensus has been reached.

7 Summary and Conclusions

Ecology has been concerned — and rightly so in my opinion — with the functioning of ecosystems independent of the activity of humans. This has allowed it to develop functional models of ecosystems notwithstanding the fact that these models do not take into account the complications brought about by human activities. Economy, on the other hand, has dealt with nature as a source of materials, but little else. This in turn has allowed it to develop useful but highly simplified functional models of human societies. Human behavior has also been studied independent of any special context. But the impact of humans on natural systems can no longer be ignored, neither can it be disregarded that, if abused by human societies, the capacity of natural systems to deliver natural products can be significantly reduced or even lost, as attested by numerous examples throughout the world.

A science of biodiversity must include all these aspects. It must be firmly anchored in biology. Biology has to provide a solid understanding of ecosystem function, and of human behavior. New economic thinking is needed that takes these biological constraints into account. Finally, a new morality that includes nature as well as humans must be adopted. This is, in my opinion, what the science of biodiversity must include.

Acknowledgments

I wish to thank Drs. Peter Raven and Edward O. Wilson for reading the manuscript and offering valuable comments.

References

Casas R (1998) Causas y evidencias de la degradacion de los suelos en la region Pampeana. In: Solbrig OT, Veinesman L (Eds) Hacia una agricultura productiva y sostenible en la pampa argentina. DRCLAS and CPIA, Buenos Aires, pp 99-129

Fiorentino R (1998) La politica agraria para la region pampeana entre 1940 y 1983. In: Solbrig OT, Veinesman L (Eds) Hacia una agricultura productiva y sostenible en la pampa argentina, DRCLAS and CPIA, Buenos Aires. pp 14-37

Halffter G (1999) Areas naturales protegidas y conservación de la biodiversidad: una perspectiva latinoamericana. In: Solbrig OT, Morello J, Matteucci S, Halffter G (Eds) Biodiversidad y Desarrollo: una visión latinoamericana. EUDEBA, Buenos Aires (in press)

Hawken P (1993) The ecology of commerce: a declaration of sustainability. Harper Collins, New York

King LD (1990) Soil nutrient management in the United States. In: Edwards CA, Lal R, Madden P, Miller RH, House G (Eds) Sustainable agricultural systems. St. Lucia Press, Delray Beach, FL, pp 89-106

National Academy of Sciences (USA) (1989). Alternative agriculture. National Academy Press, Washington, DC

National Academy of Sciences (1993). Managing global genetic resources. Agricultural crop issues and policies. National Academy Press, Washington, DC

Norse D, James C, Skinner J, Zhao Q (1992) Agriculture, land use and degradation In: Dodge C, Godman GT, la Riviere JWM, Marton-Lefevre J, Riordan TO, Pradeire F (Eds) An agenda of science for environment and development into the 21st century. Cambridge Univ Press, Cambridge, UK, pp 79-89

Pimentel D, Harvey C, Resosudarnmo P, Sinclair K, Kurz D, McNair M, Christ S, Shpritz L, Fitton L, Saffouri R, Blair R (1995). Environmental and economic of soil erosion and conservation benefits. Science 267:1117-1123

Salvador C (1998) La agroindustria argentina y la agricultura sustentable. In: Solbrig OT, Veinesman L (Eds) Hacia una agricultura productiva y sostenible en la pampa argentina. DRCLAS and CPIA, Buenos Aires, pp 189-201

Solbrig OT (1994) Biodiversity: an introduction. In: Solbrig OT, van Emden HM, van Oordt PGWJ (Eds) Biodiversity and global change. CAB and IUBS, Wallingford, UK, pp 13-20

Solbrig OT (1991) Biodiversity. Scientific issues and collaborative research proposals. Mab Digest No. 9, UNESCO, Paris, pp 1-77

8
Biodiversity and Ecosystem Processes: Theory, Achievements and Future Directions

JOHN H. LAWTON

NERC Centre for Population Biology, Imperial College, Silwood Park, Ascot SL5 7PY, UK

Abstract

Field and laboratory experiments with model plant communities show clear evidence of a reduction in ecosystem processes (productivity and nutrient cycling) with declining plant species richness within any one site. Across different sites, however, relationships can be more complex, so that mixing between- and within-site effects is confusing. A robust body of emerging theory underpins and explains these empirical results. I start by reviewing the theory, before briefly summarising the empirical data. The paper then describes preliminary new results from the European-wide BIODEPTH experiment to illustrate the key points. The challenges, now are: (i) To understand better the mechanism(s) underpinning these responses. (ii) To distinguish between the role of biodiversity in maintaining ecosystem processes under constant, or benign environmental conditions, and under extreme environmental perturbations (the 'insurance hypothesis'). (iii) To start to consider the role of animal and microbial diversity for ecosystem processes.

1 Introduction

Ecologists have made great strides over the last decade in understanding the relationship between biodiversity (here used in the narrow sense of species richness) and ecosystem processes (for example primary production, or nutrient cycling). In a characteristically pioneering book, Ehrlich and Ehrlich (1981) were among the

Key words. animals, between and within site effects, biomass, insurance hypothesis, microorganisms, niche complementarity, nutrient cycling, overyielding, plants, positive species interactions, primary production, redundancy, sampling hypothesis, soil fauna, species-richness

first ecological scientists to suggest that loss of species from ecosystems via local, regional or global extinctions may impair ecosystem processes. Their hypothesis lay essentially untested, despite its importance, for at least ten years. For example, in a book published in 1992 (arising from a symposium on Biodiversity and Global Change that took place at the Royal Netherlands Academy of Sciences, in Amsterdam in 1991, during the 24th Assembly of the International Union of Biological Sciences), the distinguished winner of the 1998 International Prize for Biology wrote: "...the fact is that we have very little quantitative evidence regarding the postulated negative consequences [of the loss of biodiversity]. We would like to see whether there is enough evidence to maintain that a certain level of diversity is essential for the function of biota and human society.... is there enough redundancy in the system that it can be argued that the system will continue to function [even with loss of species]?" (Solbrig 1992, page 15). Some six years later, we can provide detailed and definitive experimental answers to these questions.

I start by reviewing a rapidly emerging body of theory that now underpins Ehrlich and Ehrlich's insight. Then I briefly summarise the published empirical evidence addressing this theory, before presenting a brief summary of new results from a European-wide experiment called BIODEPTH. To date, much more attention has been paid, experimentally and theoretically, to single trophic-level systems, particularly plants. I finish by considering the roles of animal diversity in ecosystem processes, and touch briefly on microbial and soil-faunal diversity. Linking the diversity of micro-organisms and animals (meshed in webs of complex community interactions) to ecosystem processes is a theoretically challenging and unresolved problem.

2 Species Richness and Ecosystem Processes

2.1 Theoretical Relationships

2.1.1 Simple 'Pictorial' Possibilities

Theory and data have, until very recently, focussed on the relationship between plant species richness and ecosystem processes. Taking this state of affairs as our starting point, within any one site or location, we can imagine ecosystem processes responding in one of three ways to reductions in plant species richness (Vitousek and Hooper 1993; Lawton 1994; Naeem et al. 1995) (Fig. 1).

(i) The *redundant species hypothesis* suggests that there is a minimal diversity necessary for ecosystem functioning, but beyond this minimum, most species are redundant in their roles (Walker 1992; Lawton and Brown 1993); above this minimum, adding or deleting species has no detectable effect on the process or processes in question (Fig. 1a)

(ii) In marked contrast to (i), the *rivet hypothesis* postulates that all species are important (by analogy with the rivets holding an aeroplane together — Ehrlich

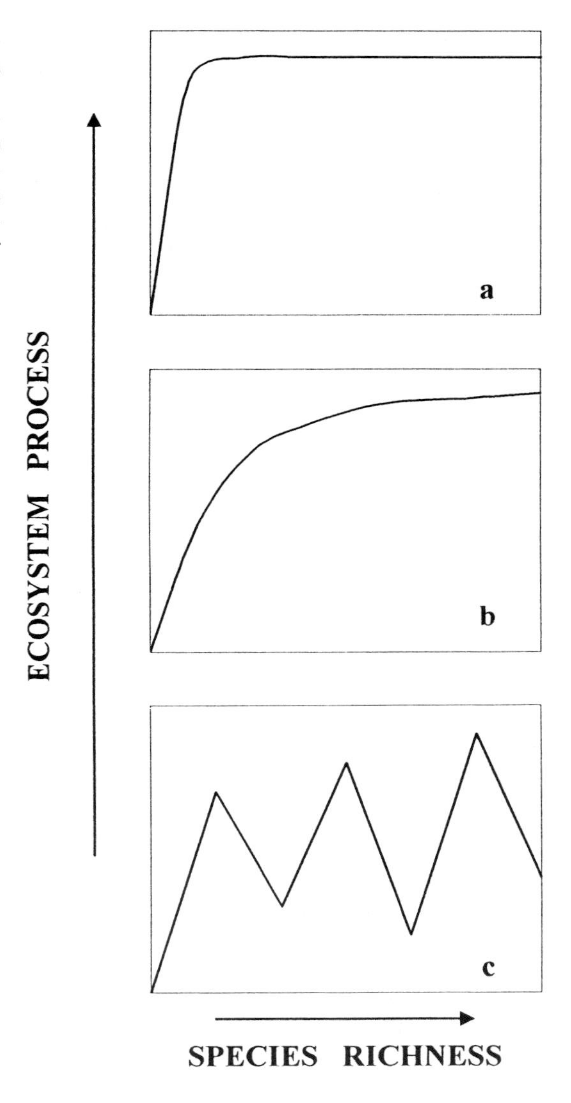

Fig. 1a-c. Hypothetical relationships between species richness and ecosystem processes (e.g. primary production, nutrient cycling) (from Naeem et al. 1995) within one locality. **a** The redundant species hypothesis. **b** The rivet hypothesis. **c** The idiosyncratic hypothesis. See text for details

and Ehrlich 1981), so that ecosystem processes are progressively more impaired as species are lost from the system (Fig. 1b).

(iii) The *idiosyncratic hypothesis* again postulates that species are important in ecosystem processes, but particular species identities matter more than species richness per se (Lawton 1994; Chapin et al. 1997; Tilman 1997). In consequence, ecosystem processes change erratically and unpredictably as species are lost from the system (Fig. 1c).

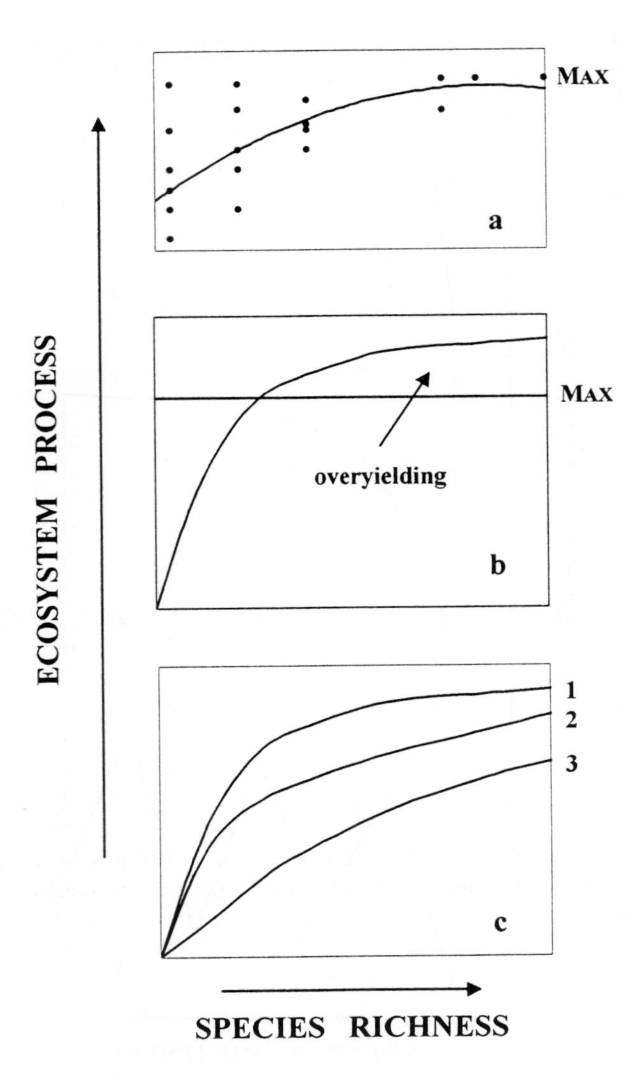

Fig. 2a-c. The general pattern illustrated in Fig. 1b can be generated by: **a** the sampling hypothesis; or **b** niche complementarity or positive species interactions. In **a**, the dots are hypothetical data (from left to right): individual species in monoculture, various two-species combinations etc. up to a single polyculture containing all species. Over yielding is impossible — the maximum productivity of mixtures (MAX) never exceeds that of the best monoculture. In **b** over yielding is possible. **c** In fluctuating environments with extreme perturbations, the insurance hypothesis predicts that more species are required to maintain ecosystem processes at a given level, the harsher, and/or the larger the number (1-3) of kinds of perturbations

Similar patterns could clearly also be generated by loss of functional groups of plant species, that is sets of species with similar growth-forms and ecologies. Functional groups can be broadly defined (e.g. legumes, grasses, and herbs), or more finely divided (e.g. early and late annuals, perennial grasses) (Hooper and Vitousek 1998).

2.1.2 More Formal Theory

Providing the species pool is reasonably large (thereby avoiding strong stochastic effects), the most likely theoretical relationship between loss of species and ecosystem processes is pattern (ii) (Fig. 1b), consistent with the rivet hypothesis. The fundamental underpinning mechanism is that species differ in their ecologies, that is, no two species have identical niches. But there are some important, and still poorly understood, subtleties underlying this simple statement, because there are at least three ways in which the different ecological requirements of different species can generate relationships broadly similar to Fig. 1b. They are, in order of increasing biological complexity:

(a) The *sampling hypothesis*. Species differ intrinsically in their potential maximum size, growth rate, and so on. Hence, as Huston (1997) points out, mixtures of many species are more likely to contain high-yielding species than monocultures or low-diversity mixtures. In consequence, random samples of species from a pool will, on average, show higher biomass, and higher productivity, as species richness increases (Tilman 1997) (Fig. 2a). In a sense, this is the biological null hypothesis for the relationship between biodiversity and ecosystem processes. It arises *because* species differ in their ecologies, and hence is due to 'hidden niche differences' between species. I refer to the niche differences in this model as 'hidden' because they are not explicit in the model. But clearly they are there, otherwise all species would behave identically. If all species are identical, there cannot be any relationship between species richness and ecosystem processes. Or, to rephrase this key point, I know of no theoretical mechanisms whereby a set of identical species could produce a relationship between species richness and ecosystem processes (Lawton et al. 1998).
(b) The *niche complementarity hypothesis*. Here the niche differences are explicit (Tilman et al. 1997a; Loreau 1998a). Niche-differences between species ensure that as species richness increases in ecological assemblages, so does the range of 'functional space' occupied by the assemblage (Carlander 1955; Tilman et al. 1997a). More diverse plant communities are more 'space-filling' above ground (Naeem et al. 1994, 1995), have a greater variety of rooting depths (Hooper 1998; Hooper and Vitousek 1997, 1998), a wider range of requirements for below-ground resources, and so on. Within any one habitat, and other things being equal, on average it is therefore almost inevitable that more species rich assemblages have greater productivity, because they are able to exploit a greater variety of limiting resources. Intercropping in agriculture and agroforestry exploits this phenomenon (Ewel et al. 1991; Swift and Anderson 1993). In the fisheries literature, Carlander (1955) recognised over 40 years ago that niche differences between species mean that species-rich fish assemblages sustain greater biomass than species-poor assemblages.

Although the sampling hypothesis and the niche complementarity hypothesis both rely on there being ecological differences between species, they differ in a subtle and important way. Under the simple sampling hypothesis, the maximum yield of a polyculture will not exceed that of the best-yielding monocultures. Under the niche complementarity hypothesis, it is possible for (but not inevitable that) species rich mixtures will out-perform the best performing monocultures (Fig. 2b). The phenomenon has been called 'overyielding' in the earlier botanical literature (Hector 1998), although detecting it statistically in experiments can be tricky (Loreau 1998b).

(c) The *positive species interactions hypothesis*. Species do not just compete with each other in communities. It is becoming increasingly clear that some plant species benefit from the presence of other species in the community. That is, some species interactions are ++, or +0 in Williamson's (1972) classification, for a variety of reasons (provision of shelter, other beneficial habitat modification, etc.) (e.g. Bertness and Leonard 1997). Now overyielding in mixed-species assemblages compared with monocultures is almost inevitable, and could be substantial (Fig. 2b).

The relative importance of models a, b and c in explaining empirical results (see Section 2.2) is currently unresolved, and presents a major challenge for future research.

2.1.3 Consequences

Several points follow from this theoretical framework. First, there is currently some debate in the literature about the relative importance of plant species richness vs. plant functional types, or plant species identity, in determining ecosystem processes (e.g. Grime 1997; Hooper and Vitousek 1997; Tilman et al. 1997b,c; Wardle et al. 1997a,b). A consideration of the underlying theory helps to clarify the debate. Because no two plant species within a functional group (however this is defined) have identical niches, species richness will influence ecosystem processes, by the arguments laid out above. However, if the niche differences between plant species are small, then the effects will be small. By these same arguments, deliberately selecting species to be as different as possible, by selecting different functional types is likely to have an even bigger effect on ecosystem processes. Because functional types are arbitrary divisions of continuous niche-space, deciding whether species richness or functional types has a bigger impact on ecosystem processes is to arbitrarily divide a continuum; the bigger the niche-differences between species in the assemblage, the bigger the effect we expect to observe on ecosystem processes.

Second, if there are no, or only very small differences between species influencing some ecosystem process of interest, then we are more likely than not to see data consistent with the redundant species hypothesis (Fig. 1a) — that is, statistically, we may be unable to detect any effect of species richness on the ecosystem process in question, except perhaps in very depauperate communities. An example might be leaf-litter decomposition. If litter quality is similar across a set of species, plant species richness and litter decomposition rates may be unrelated (e.g. Wardle et al. 1997c).

Third, if only a few species (or functional groups) are involved, and/or if there are major changes in dominants with diversity, the effects of species' identities (the idiosyncratic hypothesis) will be paramount. These ideas, particularly that species identities matter, are discussed in Lawton (1994), Naeem et al. (1996), Chapin et al. (1997), and Hooper and Vitousek (1997, 1998).

In sum, it is now possible, within a relatively simple conceptual framework, to embrace all the likely outcomes linking changes in ecosystem processes to changes in plant species richness within a community (Fig. 1), and to suggest that the most likely outcome is the relationship shown in Fig. 1b and Fig. 2: that is, a plot of biomass, or primary production etc. against increasing plant species richness will typically be a negatively accelerating, rising curve.

2.1.4 The Insurance Hypothesis

I have developed these arguments implicitly assuming a relatively benign world, with communities exposed to average environmental conditions. How might species richness affect ecosystem processes under extreme events? The theoretical answer again rests on niche differences between species. By exactly the same arguments developed in Section 2.1.2, niche differences between species 'spread risk', so that more species rich assemblages ought to be more resilient to extreme events (drought, late frost etc.) than species poor assemblages (Walker 1992; Lawton and Brown 1993). The effect has been variously called the *insurance hypothesis*, or the *portfolio effect* (from the analogous process of risk spreading by investing in a range of stocks and shares) and has strong theoretical support (Naeem 1998; Tilman et al. 1998).

A key point of the insurance hypothesis is that species that appear to be 'redundant' under more benign conditions (Fig. 1a, and the 'plateau' at higher species richness in Fig. 1b) may play an essential role in maintaining ecosystem processes under extreme events. In other words, more species are required to maintain ecosystem processes at a given level in a variable environment than in a constant one (Fig. 2c). Determining how different kinds of environmental disturbances interact with one-another, and with species richness, to influence ecosystem processes is an unresolved theoretical and experimental problem.

2.1.5 Within System and Between System Comparisons

There is a crucial difference between the impacts of changes in biodiversity on ecosystem processes within an ecosystem (*within-system effects*), and comparisons between different ecosystems (*between-system effects*) (Fig. 3). So far, I have been talking only about within-system effects, where the theoretical problem being addressed is about loss of species within one community, at one locality. Several experiments have now been carried out to test this aspect of theory (Section 2.2) (e.g. Naeem et al. 1994, 1995; Tilman and Downing 1994; Hooper and Vitousek 1997, 1998; McGrady-Steed et al. 1997; Naaem and Li 1997; Tilman et al. 1996, 1997b).

An alternative approach is to resort to correlative field studies on natural vegetation (e.g. Wardle et al. 1997a,b), using different sites to examine the relationship between biodiversity and ecosystem processes — that is to make between-system comparisons. This approach tests a fundamentally different problem, because it simultaneously examines changes in species richness *and* differences be-

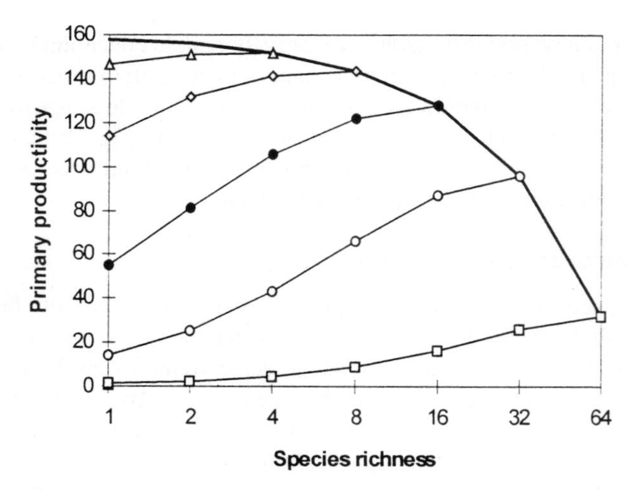

Fig. 3. The relationship between species richness and ecosystem processes need not be the same within a site and across sites. Here, ecosystem processes decline with maximum richness across 5 sites (single thick curve). Within-site effects are illustrated by the five thin curves, each site being given a different symbol (triangle, diamond etc). Within each site there is a positive diversity-process relationship, as in Fig. 1b and Fig. 2 (notice the log scale on the x-axis). Results from a model by Loreau (1998a). Between-site differences are generated by different nutrient transport rates (k) at each site; within-site differences are generated by loss of species, for fixed k, as in the BIODEPTH experiment (M. Loreau, pers. comm.)

tween environments (Tilman et al. 1997c). Models (Loreau 1998a) and empirical data (e.g. McNaughton 1993; Bulla 1996) show that unless environmental conditions are extremely similar, across-habitat or across-locality comparisons of the relationship between species richness and ecosystem processes are unlikely to look anything like Fig. 1b or Fig. 2, because between-site differences totally obscure within-habitat effects of species richness on ecosystem processes (Fig. 3). It is certainly important to know the relative magnitude of within-habitat species-richness effects, and across-habitat environmental effects on ecosystem processes, but it is not helpful to confuse the two (Grime 1997; Wardle et al. 1997b).

2.2 Empirical Data

2.2.1 Published Studies

Empirical support for the theoretical predictions is now reasonably strong: (i) for negatively accelerating increases in ecosystem function with increasing species richness or functional group diversity; (ii) for reasonably well understood exceptions to this general pattern; (iii) for the insurance hypothesis; and (iv) for differences between these intra-system effects and inter-system effects. Space precludes a detailed analysis and summary. The main results are as follows.

Relationships consistent with Fig. 1b, and the models discussed in Section 2.1.2, between plant species richness (or the variety of plant functional types) and ecosystem processes within a site have now been demonstrated in major field experiments at Cedar Creek (Tilman et al. 1996, 1997b). A similar experiment in California (Hooper and Vitousek 1997, 1998) varied functional group richness from 1–4. Although some combinations of plant functional groups were complementary, group composition was more important than the number of functional groups, consistent with the idiosyncratic hypothesis (Fig. 1c). Hooper and Vitousek did not vary plant species richness, and used only 2 or 3 species per functional group. Manipulating a small number of species and functional groups in this way is statistically likely to yield idiosyncratic responses, for the reasons explained in Section 2.1.3.

The first experiment to directly test the relationship between species richness and ecosystem processes in ecological communities used a controlled environment facility (the Ecotron at Silwood Park) (Naeem et al. 1994, 1995; Lawton et al. 1998), and manipulated both plant and animal diversity. In the Ecotron experiment, only primary production (above-ground biomass) increased with species richness. The sampling effect contributes strongly to this relationship (Naeem et al. 1995), but more subtle processes may also be involved (Naeem et al. 1996). Other ecosystem processes were *not* consistent with the rivet hypothesis (presumably for the reasons laid out in Section 2.1.3), but either showed no effect of richness (litter decomposition) or conformed to the idiosyncratic hypothesis (there were significant differences between treatments in nutrient cycling, but not simply related to species richness).

Most recently, again using experimentally synthesised plant communities in laboratory and field microcosms, van der Heijden et al. (1998) have moved the problem to another trophic level, and examined the influence of mycorrhizal fungal diversity on ecosystem processes. Once again, exactly as in Fig. 1b, plant biomass rises curvilinearly with increasing species richness of arbuscular mycorrhizal fungi (AMF); soil phosphorus concentrations do the opposite, and decline with increasing fungal richness. The authors explicitly explain their results in terms of niche complementarity: "increasing AMF biodiversity resulted in more efficient exploitation of soil phosphorus and to better use of the resources available in the system" (page 70).

Species richness has also been shown to buffer and stabilise ecosystem processes, exactly as predicted by the insurance hypothesis (Section 2.1.4), in plant communities at Cedar Creek (Tilman and Downing 1994; Tilman et al. 1998) and the famous Park Grass plots at Rothamsted (Dodd et al. 1994), in zooplankton assemblages (Frost et al. 1995) and in pioneering, three-trophic level experimental communities made up of bacteria and protists (McGrady-Steed et al. 1997; Naeem and Li 1997).

Finally, as predicted in Section 2.1.5, correlations between ecosystem processes and plant species richness based on across-site comparisons differ from the within-system results. On a series of Baltic islands, ecosystem processes declined with increasing plant diversity (Wardle et al. 1997a) (e.g. Fig. 3), because of over-riding inter-island differences in fire regimes and post-fire histories.

2.2.2 The BIODEPTH Experiment

Recently, I and colleagues throughout Europe have created the BIODEPTH experiment (Biodiversity and Ecosystem Processes in Terrestrial Herbaceous Communities). There are eight field sites (Umeå in Sweden, Sheffield and Silwood in the UK, Cork in Ireland, Bayreuth in Germany, Zurich in Switzerland, Lisbon in Portugal, and Lesbos in Greece). At all the field sites, we set up replicated model ecosystems with five levels of plant species richness, from monocultures to polycultures with a maximum richness per m^2 characteristic of natural vegetation for each site. We used the same basic protocol everywhere, and weeded the plots to maintain treatments (i.e. this is essentially Tilman's Cedar Creek experiment replicated across eight different sites). We now have three years data on the effects of species richness on ecosystem processes from BIODEPTH.

We asked whether the same within-site relationship between plant species richness and ecosystem processes would be found at every site, despite the very wide range of environments, soil-types and herbaceous plant communities embraced by the experiment. Our expectation was that we would find the same relationship (because there are no theoretical grounds for believing otherwise), and that it would again basically be like Fig. 1b, differing only in the positions of the curves from individual sites. And (as by now should be clear) we expected to find a totally different relationship plotting data across-sites.

The results of the experiment are still being worked up. Preliminary conclusions (A. Hector, B. Schmidt, J.H. Lawton and the BIODEPTH consortium, unpublished) are that both our major predictions are supported. If we fit one statistical model to the within-site responses from the second year's data, we find that for all eight sites, plant biomass increases linearly with the log of plant species richness (i.e. biomass rises curvilinearly with untransformed plant richness, as in Fig. 1b). There is no significant site × treatment interaction, meaning that all sites fit the same model, but sites differ from one another in overall productivity, because of inherent differences in soil fertility, the weather, and the length of the growing season. As well as within-site effects of plant species richness, there are also effects of plant functional group, particularly the presence of legumes, but only at some sites. Finally, taking the richest experimental communities from each site, and plotting a between-site graph of biomass vs plant species richness reveals no clear pattern. Between- and within-site effects, are, as predicted, totally different.

We do not yet know the mechanism(s) underpinning the strong and consistent within-site effects, but preliminary analyses suggest that the results cannot be explained solely by the sampling hypothesis, suggesting that complementarity or positive interactions play a role. Notice that all sites can be fitted to one statistical model, showing a linear increase in biomass with the log of plant species richness. In other words, ecosystem processes do not reach an asymptote; in this experiment, increasing plant species richness always leads to an increase in ecosystem performance, albeit at a decelerating rate as richness increases.

I emphasise that these are preliminary conclusions. Nevertheless, the BIODEPTH results point to a strong and consistent emerging picture of the way in which changes in plant diversity affect ecosystem processes within and across ecosystems. Moreover, it is a picture that appears to be entirely consistent with an equally rapidly developing body of theory.

3 Micro-organisms, Animals and Effects in Complex Food Webs

Although several studies discussed above have involved organisms other than plants, and/or more than one trophic level (Carlander 1955; Naeem et al. 1994, 1995; McGrady-Steed et al. 1997; Naeem and Li 1997; van der Heijden et al. 1998), a major challenge is now to develop better theoretical and empirical understanding of relationships between biodiversity and ecosystem processes for organisms embedded in complex webs of interacting species within communities (e.g. Bengtsson 1998). Obvious problems to tackle are the importance of microbial diversity, and by extension the role of the diversity of soil biota (microbes, micro-fauna and macro-fauna) in soil ecosystem processes. We know virtually nothing about the importance of herbivore diversity, and so on.

Pioneering experiments on soil biota (e.g. Mikola and Setälä 1998) suggest that the problems are theoretically and empirically much more complicated when complex food-web interactions modulate the effects of diversity treatments, and authors currently disagree about levels of redundancy in species-rich soil food webs (e.g. Wardle et al. 1998; Swift et al. 1998). A detailed review of this work is beyond the scope of the present paper. But it is quite clear that it is here, in these more complex food webs, that the next main challenges lie, linking biodiversity to ecosystem processes.

Acknowledgements

I thank the organisers of the 14th International Symposium of the International Prize for Biology for giving me the opportunity to present this paper, which is dedicated to the recipient of the Prize, Professor O. Solbrig. The quotation from Professor Solbrig at the start of my paper, inspired me to do the first Ecotron experiment with Shahid Naeem and colleagues, and launched me on the study of biodiversity and ecosystem processes. Andy Hector made valuable comments on a first draft of the present manuscript.

References

Bengtsson J (1998) Which species? What kind of diversity? Which ecosystem function? Some problems in studies of relations between biodiversity and ecosystem function. Appl Soil Ecol 315:1-9

Bertness MD, Leonard GH (1997) The role of positive interactions in communities: lessons from intertidal habitats. Ecology 78:1976-1989

Bulla L (1996) Relationship between biotic diversity and primary productivity in savanna grasslands. In: Solbrig OT, Medina E, Silva JF (Eds) Biodiversity and savanna ecosystem processes. Springer-Verlag, Berlin, pp 97-120

Carlander KD (1955) The standing crop of fish in lakes. J Fish Res Bd Canada 12:543-570

Chapin FS III, Walker BH, Hobbs RJ, Hooper DU, Lawton JH, Sala OE, Tilman DE (1997) Biotic control over the functioning of ecosystems. Science 277:500-504

Dodd ME, Silvertown J, McConway K, Potts J, Crawley M (1994) Stability in the plant communities of the Park Grass Experiment: the relationships between species richness, soil pH and biomass variability. Phil Trans Roy Soc Lond, B 346:185-193

Ehrlich PR, Ehrlich AH (1981) Extinction. The causes and consequences of the disappearance of species. Random House, New York

Ewel JJ, Mazzarino MJ, Berish CW (1991) Tropical soil fertility changes under monocultures and successional communities of different species. Ecol App 3:289-302

Frost TM, Carpenter SR, Ives AR, Kratz TK (1995) Species compensation and complementarity in ecosystem function. In: Jones CG, Lawton JH (Eds) Linking species and ecosystems. Chapman and Hall, New York, pp 224-239

Grime JP (1997) Biodiversity and ecosystem function: the debate deepens. Science 277:1260-1261

Hector A (1998) The effect of diversity on productivity: detecting the role of species complementarity. Oikos 82:597-599

Hooper DU (1998) The role of complementarity and competition in ecosystem responses to variation in plant diversity. Ecology 79:704-719

Hooper DU, Vitousek PM (1997) The effects of plant composition and diversity on ecosystem processes. Science 277:1302-1305

Hooper DU, Vitousek PM (1998) Effects of plant composition and diversity on nutrient cycling. Ecol Monogr 68:121-149

Huston MA (1997) Hidden treatments in ecological experiments: re-evaluating the ecosystem function of biodiversity. Oecologia 110:449-460

Lawton JH (1994) What do species do in ecosystems? Oikos 71:367-374

Lawton JH, Brown VK (1993) Redundancy in ecosystems. In: Schulze E-D, Mooney HA (Eds) Biodiversity and ecosystem function. Springer-Verlag, Berlin, pp 255-270

Lawton JH, Naeem S, Thompson LJ, Hector A, Crawley MJ (1998) Biodiversity and ecosystem function: getting the Ecotron experiment in its correct context. Funct Ecol 12:848-852

Loreau M (1998a) Biodiversity and ecosystem functioning: a mechanistic model. Proc Natl Acad Sci USA 95:5632-5636

Loreau M (1998b) Separating sampling and other effects in biodiversity experiments. Oikos 82:600-602

McGrady-Steed J, Harris PM, Morin PJ (1997) Biodiversity regulates ecosystem predictability. Nature 390:162-165

McNaughton SJ (1993) Biodiversity and function of grazing ecosystems. In: Schulze E-D, Mooney HA (Eds) Biodiversity and ecosystem function. Springer-Verlag, Berlin, pp 361-383

Mikola J, Setälä H (1998) Relating species diversity to ecosystem functioning: mechanistic backgrounds and experimental approach with a decomposer food web. Oikos 83:180-194

Naeem S (1998) Species redundancy and ecosystem reliability. Conserv Biol 12:39-45

Naeem S, Li S (1997) Biodiversity enhances ecosystem reliability. Nature 390:507-509

Naeem S, Thompson LJ, Lawler SP, Lawton JH, Woodfin RM (1994) Declining biodiversity can alter the performance of ecosystems. Nature 368:734-737

Naeem S, Thompson LJ, Lawler SP, Lawton JH, Woodfin RM (1995) Empirical evidence that declining biodiversity may alter the performance of terrestrial ecosystems. Phil Trans Roy Soc Lond, B 347:249-262

Naeem S, Håkansson K, Lawton JH, Crawley MJ, Thompson LJ (1996) Biodiversity and plant productivity in a model assemblage of plant species. Oikos 76:259-264

Solbrig OT (1992) Biodiversity: an introduction. In: Solbrig OT, van Emden HM, van Oort PGWJ (Eds) Biodiversity and global change. The International Union of Biological Sciences, Paris, pp 13-20

Swift MJ, Anderson JM (1993) Biodiversity and ecosystem function in agricultural systems. In: Schulze E-D, Mooney HA (Eds) Biodiversity and ecosystem function. Springer-Verlag, Berlin, pp 15-41

Swift MJ, Andrén O, Brussaard L, Briones M, Couteaux M-M, Ekschmitt K, Kjoller A, Loiseau P, Smith P (1998) Global change, soil biodiversity, and nitrogen cycling in terrestrial ecosystems: three case studies. Global Change Biol 4:729-743

Tilman D (1997) Distinguishing between the effects of species diversity and species composition. Oikos 80:185

Tilman D, Downing JA (1994) Biodiversity and stability in grasslands. Nature 367:363-365

Tilman D, Wedin D, Knops J (1996) Productivity and sustainability influenced by biodiversity in grassland ecosystems. Nature 379:718-720

Tilman D, Lehman CL, Thomson KT (1997a) Plant diversity and ecosystem productivity: theoretical considerations. Proc Natl Acad Sci USA 94:1857-1861

Tilman D, Knops J, Wedin D, Reich P, Ritchie M, Siemann E (1997b) The influence of functional diversity and composition on ecosystem processes. Science 277:1300-1305

Tilman D, Naeem S, Knops J, Reich P, Siemann E, Wedin D, Ritchie M, Lawton JH (1997c) Biodiversity and ecosystem properties. Science 278:1866-1867

Tilman D, Lehman CL, Bristow CE (1998) Diversity-stability relationships: statistical inevitability or ecological consequence? Am Nat 151:277-282

van der Heijden MGA, Klironomos JN, Ursic M, Moutoglis P, Streitwolf-Engel R, Boller T, Wiemken A, Sanders IR (1998) Mycorrhizal fungal diversity determines plant biodiversity, ecosystem variability and productivity. Nature 396:69-72

Vitousek PM, Hooper DU (1993) Biological diversity and terrestrial ecosystem biogeochemistry. In: Schulze E-D, Mooney HA (Eds) Biodiversity and ecosystem function. Springer-Verlag, Berlin, pp 3-14

Walker BH (1992) Biodiversity and ecological redundancy. Conserv Biol 6:18-23

Wardle DA, Zackrisson O, Hörnberg G, Gallet C (1997a) The influence of island area on ecosystem properties. Science 277:1296-1299

Wardle DA, Zackrisson O, Hörnberg G, Gallet C (1997b) Response. Science 278:1867-1869

Wardle DA, Bonner KI, Nicholson KS (1997c) Biodiversity and plant litter: experimental evidence which does not support the view that enhanced species richness improves ecosystem function. Oikos 79:247-258

Wardle DA, Verhoef HA, Clarholm M (1998) Trophic relationships in the soil microfood-web: predicting the responses to a changing global environment. Global Change Biol 4:713-727

Williamson M (1972) The analysis of biological populations. Edward Arnold, London

9
Creeping 'Fruitless Falls': Reproductive Failure in Heterostylous Plants in Fragmented Landscapes

IZUMI WASHITANI

Institute of Biological Sciences, University of Tsukuba, Tsukuba 305-8572, Japan

Abstract

In recent years, increasing instances of 'fruitless falls', i.e., seed set failure of flowering plants, have been appreciated for both cultivated and wild plants. Species having a sophisticated entomophilous breeding system such as heterostyly are likely to be most vulnerable to the detrimental effects of pollinator loss, resulting in fruitless falls through compatible pollen limitation. Our studies on pollination and seed set in wild populations of heterostylous species, *Primula sieboldii*, *P. kisoana* and *Persicaria japonica,* and a few remaining populations of a highly endangered monomorphic species *Crepidiastrum ameristophyllum* suggested that fruitless falls have been already rather common among wild plants in present-day fragmented landscapes. Serious effects of pollinator loss have been recognized for a *Primula sieboldii* population in a small nature reserve in the floodplain of the Arakawa River, which is an 'insular habitat' surrounded by the urbanized area of Greater Tokyo. Fertility of the rare short-homostyle far surpasses those of normal heterostylous morphs, of which fertility is severely limited by pollinator availability. Model simulation predicts the possibility of a large loss of genetic variation within a few generations under the present strong fertility selection for the homostyle. For small populations or isolated individuals in fragmented habitats, reduced opportunity for mating, i.e., one typical form of the Allee effects, is another major cause of seed set failure. Among 20 populations of *P. sieboldii* investigated in southern Hokkaido, only a negligible number of seeds were set in smaller populations consisting of less than four genets, while in larger populations, population mean seed set depended strongly on pollinator availability which can be assessed by craw marks left on the flower petals by effective pollinator queen bumblebees. Seed set failure due to reduced

---

Key words. Allee effect, biodiversity, biological interaction, *Crepidiastrum ameristophyllum*, genetic variation, habitat fragmentation, heterostyly, isolation, pollination, seed set, small population, *Persicaria japonica, Primula kisoana*, *Primula sieboldii,* reproductive failure

opportunity for mating because of solitude was also demonstrated for isolated genets or smaller populations of other heterostylous species, *Primula kisoana* and *Persicaria japonica* in highly fragmented deciduous forests and moist tall grasslands, respectively. We also demonstrated the detrimental effects of isolation, i.e., absence of mating partners, in a highly endangered homomorphic species, *Crepidiastrum ameristophyllum* endemic to the Bonin Islands. Fruitless falls ascribed to the Allee effects are likely to be already ubiquitous among wild plants subjected to habitat fragmentation or other threats to biodiversity. Habitat and/or population restoration and 'pollinator therapy' management based on sound population and reproductive ecology are urgently required to strive against fruitless falls.

1 'Fruitless Falls': Signs of Biodiversity Impairment Appearing in Plant Reproduction

Maintaining the earth's biodiversity and integrated ecosystems is among the most important social aims of present-day human beings to guarantee intergenerational long-term sustainability (Cristensen et al. 1996). We should have keen powers to appreciate any signs or symptoms of impairment of biodiversity or ecosystem integrity. More than thirty years ago, Carson (1962) predicted a 'silent spring' devoid of bird chorus and buzzing bees. The prediction invoked great social impacts and has continuously drawn wide public attention to the problems of environmental pollution by synthetic chemicals until today. Another prediction she made in her famous book entitled "silent spring" was a 'fruitless fall' in which flowers fail in producing fruits and seeds because of pollinator absence (Carson 1962).

Fruitless fall has been creeping up more secretly than the silent spring, but recently increasing instances of 'fruitless falls' have been appreciated for both cultivated and wild plants (Buchmann and Nabhan 1996). According to Buchmann and Nabhan (1996), among 258 species in which limiting factors for fruit set under natural conditions were investigated in detail, 62 percent were shown to suffer limited fruit set from insufficient pollinator services.

As sessile organisms, plants have evolved a tremendous variety of mechanisms assuring reproduction in case of isolation or pollinator absence, which include various self-pollination and self-fertility mechanisms. Simultaneously, various physiological, morphological and phenological mechanisms to facilitate outbreeding and thus to prevent selfing have been evolved (Silvertown and Lovett Doust 1993), probably under strong selective pressures imposed by inbreeding depression (Barrett and Kohn 1991). Contradictory demands for reproductive assurance vs. those for outbreeding and their temporal and spatial variations as selective forces would be largely responsible for the highly diversified plant breeding systems we can see in the present day world. Reproductive resistance to isolation or pollinator absence may vary according to the breeding system indigenous to the species or population. Plants in which breeding systems tilt more to outbreeding may be less resistant to habitat fragmentation or sudden population decline.

Heterostyly, which has evolved repeatedly in approximately 25 angiosperm families is a genetic polymorphism in which plant populations are composed of two (distyly) or three (tristyly) morphs that differ reciprocally in the heights of stigmas and anthers in their hermaphrodite flowers (Ganders 1979; Barrett 1992). The floral polymorphism is usually linked with a diallelic sporophytic self-incompatibility system. Most heterostylous plants are entomophilous (or pollinated by humming birds) and grow clonally, so that the breeding system is thought to be adaptive to avoid geitonogamous self breeding. A sophisticated entomophilous breeding system like heterostyly is likely to be most vulnerable to the detrimental effects of pollinator absence or isolation.

Our studies on pollination and seed set of heterostylous plants, *Primula sieboldii* E. Morren, *P. kisoana* Miq. and *Persicaria japonica* (Meisn.) H. Gross, and a highly endangered homomorphic species, *Crepidiastrum ameristophyllum* (Koidz.) Nakai in their natural habitats suggested that fruitless fall has already been rather common among wild plants which have suffered population decline or genet isolation due to habitat fragmentation or other threats to biodiversity.

2 Reproductive Failure of *Primula sieboldii* Due to Pollinator Loss

Primula sieboldii is a perennial clonal herb that occurs in a range of moist habitats throughout Japan. Like many other *Primula* species, *P. sieboldii* is distylous (Richards 1986). Although the species was once very common, in recent years it has declined and is now listed in the Japanese plant red list (Environmental Agency 1997).

Serious effects of pollinator loss have been recognized for a *P. sieboldii* population of a small nature reserve (4 ha) in the floodplain of the Arakawa River. The nature reserve encloses a remnant moist tall grassland and is an 'insular habitat', thoroughly surrounded by the urban area of Greater Tokyo. The nature-reserve population of *P. sieboldii* contains approximately the same proportions of normal heterostylous morphs, long- and short-styled morphs, with a very low frequency of self-fertile short homostyle morphs, less than 1%. Although such a morph composition can be expected for a population in which the heterostylous breeding system is normally functioning (Charlesworth and Charlesworth 1979), it should not necessarily reflect the current state of reproduction of the population, since the genet longevity of this perennial is likely to exceed several decades.

In the nature reserve population of *P. sieboldii*, seed set is generally low and varies significantly among the morphs (Washitani et al. 1994a). Fertility of the rare short-homostyle far surpasses those of normal heterostylous morphs, of which fertility is severely limited by pollinator availability (Washitani et al. 1991). Artificial legitimate pollination greatly enhanced seed set in long-styled and short-styled morphs, suggesting that limited supply of compatible pollen was the main reason for the low seed set of the normal heterostylous morphs. We could not observe any insect visitation during a continuous monitoring period for a total of sixteen hours with 68 flowers (Washitani et al. 1991). The entire insect fauna of the reserve appears impoverished, but the absence of long-tongued bumble bees may be the main reason for the fruitless fall of the species, since the queens of long-tongued bumblebees, of which tongue length is nearly equal to corolla tube length of *P. sieboldii*, are known to be especially effective pollinators (Washitani et al. 1994b).

The strong fertility selection for the rare homostyle morph suggests the possibility of a large loss of genetic variation within a few generations. Simulation with a genetic population model predicts either a strong genetic bottleneck due to fertility selection or an overall recruitment failure in the case of the presence of very strong inbreeding depression (Washitani 1996).

3 Reproductive Failure Due to Solitude: Low Quantity and Quality of Progeny

Pollinator loss is not the sole reason for seed set failure of plant populations in fragmented landscapes. Reduced opportunity for mating due to isolation, i.e., a typical form of the Allee effects (Allee 1951), may be another major cause of fruitless falls for small populations or isolated genets of declining populations. Evidence for a case of the Allee effect was obtained from our study, in which seed set and various biological and other environmental factors potentially affecting reproductive success were compared among 20 *P. sieboldii* local populations in a fragmented landscape in the Hidaka region, southern Hokkaido (Matsumura and Washitani, unpublished). Although various biological agents including herbivores and pathogens affected the fertility of the plants (Washitani et al. 1996), seed set of the populations was strongly dependent on the population size and pollinator availability, which can be assessed by craw marks left on the flower petals by long-tongued bumblebee queens (Washitani et al. 1994b). Among the 20 populations investigated, negligible numbers of seeds were set in smaller populations consisting of less than four genets, while in larger populations, population mean seed set strongly depended on pollinator availability especially in years with generally low pollinator activities.

Results of our study in another deciduous forest habitat suggested that not only the quantity of the seeds, but also their quality, i.e., the fitness of the few seeds which were produced on isolated mother plants of *P. sieboldii* was also significantly lower than those of the seeds from less isolated mother plants having potential mating partner(s) in their vicinity (Otsuka, Goka and Washitani, unpublished). The study was performed in a typical habitat of the species in central Honshu, i.e., a forest dominated by *Quercus mongolica* Fischer ex Ledeb. ssp. *crispula* (Blume) Menitsky (35° 57' N, 138° 28' E, 1350 m a.s.l.), and seed set, germinability of the seeds and survival and growth of the germinated seedlings were examined for many mother plants growing along a stream. Seed set as well as germination of the seeds and early survival of the geminated seedlings were significantly lower and more variable among isolated mother plants having no opposite morph flower in the neighborhood within 5 m than for the mothers having potential mating partners in their neighborhoods. However, there was no significant difference in survival and relative growth rate of the established seedlings between the mother groups differing in degree of isolation. Moreover, the level of allozyme heterozygosity of the progeny surviving to this stage did not significantly differ between the isolated and less isolated mothers. Therefore, it is suggested that strong inbreeding depression manifests in the early life stages of the progeny, and purges less fit homozygotes by the stage of seedling establishment. It is suggested that isolated mothers can leave much fewer fit progeny than expected from the quantity of seeds produced.

4 Complete Reproductive Failure in the Highly Endangered Species *Primula kisoana*

We can find another remarkable example of reproductive failure due to solitude in *Primula kisoana* that is a rare local endemic with an extremely narrow geographic range in the northern part of Kanto district. The species is also distylous and closely associated with deciduous forest habitats on steep mountainous slopes. Phylogenetically, *P. kisoana* is very close to *P. sieboldii* and shares may floral traits, though its clonal growth mode with long underground stolons is much dissimilar to that of *P. sieboldii* with stunted rhizomes.

In recent decades, the habitats of this species have been greatly reduced and highly fragmented due to expanding cedar plantations. In addition, the original rareness tends to incur excessive exploitation of the species for commercial horticultural use. Therefore, the population has been continuously dwindling, and also might repeatedly experience strong population bottlenecks. Recent invasion of remaining habitats by wild boars, which have originated from those artificially released for sport shooting, are also deadly threats to the species through trampling and digging of the roots.

Each remaining patch of the species we can find today consists of a single genet of either long- or short-styled morph, and as a matter of course, no seeds are set despite occasional visitation of the flowers by long-tongued bumblebee queens.

In the flowering seasons of 1996–1997, we performed artificial legitimate pollination by using the pollen transported from the opposite morph flowers through several kilometers of trekking. The seeds set by the hand pollinated flowers, however, were less vigorous and could not germinate even after laboratory treatments which are known to be favorable for the germination of many herbs including *P. sieboldii*, i.e., moist chilling and temperature alternation (Washitani and Kabaya 1988; Washitani 1987; Washitani and Masuda 1990). After some trial-and-error, we could obtain seedlings through dormancy breaking with Giberellic acid. Planting the carefully raised seedlings in apparent safe-sites (*in sensu* Harper 1977) in the historical range of the species, we are attempting to restore local self-sustainable populations including both long- and short-styled morphs.

5 Pollination Failure in a Common Heterostylous Species, *Persicaria japonica,* Due to Isolation

Seed set failure due to pollination limitation was also demonstrated for isolated genets (Nishihiro and Washitani 1998b) and smaller populations (Nishihiro et al. 1998) of *Persicaria japonica*, which is a heterostylous perennial herb common to various moist or wet habitats throughout Japan. The species has dish-shaped corolla, and is supposed to be a generalist concerning pollinator insects, since many insects including syrphids, flies, bees, and butterflies can pollinate the flower.

Our study (Nishihiro and Washitani 1998a) demonstrated that the species has several features advantageous for studying pollination success separately from post-pollination processes: 1) compatibility of a given deposited pollen grain can be relatively precisely judged from the pollen size, which markedly differs between the floral morphs, 2) a flower can set a seed by the receipt of only a single pollen grain from the opposite morph, and 3) there is little possibility of self-pollen interference with legitimate fertilization. Therefore, female reproductive success at the pollination stage can be evaluated by presence or absence of pollen grains from the opposite morph on the stigma.

The species is common in moist tall grasslands dominated by *Phragmites australis* W. Clayton in floodplains and lakeshores of lowland Japan. However, recently these habitats of the species have been subjected to extensive fragmentation due to reclamation and other construction works. Moist tall grasslands on the shore of Lake Kasumigaura, the second largest lake in Japan, are not exceptional, and at present there remains a highly fragmented landscape of the grasslands with small and/or low-density populations of *P. japonica*.

In the remaining patches of fragmented grasslands on the lakeshore, we compared pollination success and seed set between genets which markedly differed in relative distance from a potential mating partner of the opposite morph (Nishihiro and Washitani 1998b). Artificial bank construction left small isolated patches of *P. japonica* containing only a single genet, while patches consisting of several different genets (congregating genets) remain near the shoreline. Pollinator insects are abundant even in the fragmented grasslands and 28 insect species including hymenopterous, dipterous, and lepidopterous species frequently visited the pollinator-generalist flowers even in isolated patches. However, the pollination success varied significantly according to the degree of isolation and was considerably higher in the congregating genets than in the single genets. Seed set also differed significantly between them, and higher seed set was recorded for the congregating genets. These results suggested high susceptibility of pollination success and thus female fertility to isolation or solitude.

We compared seed set among 25 local populations of *Persicaria japonica* growing in remnant grassland patches of various sizes scattered along 25 km of shoreline of the lake (Nishihiro et al. 1998). The populations greatly differed in the area of grassland patches inhabited (120–18000 m^2), genet number (1–53), and morph

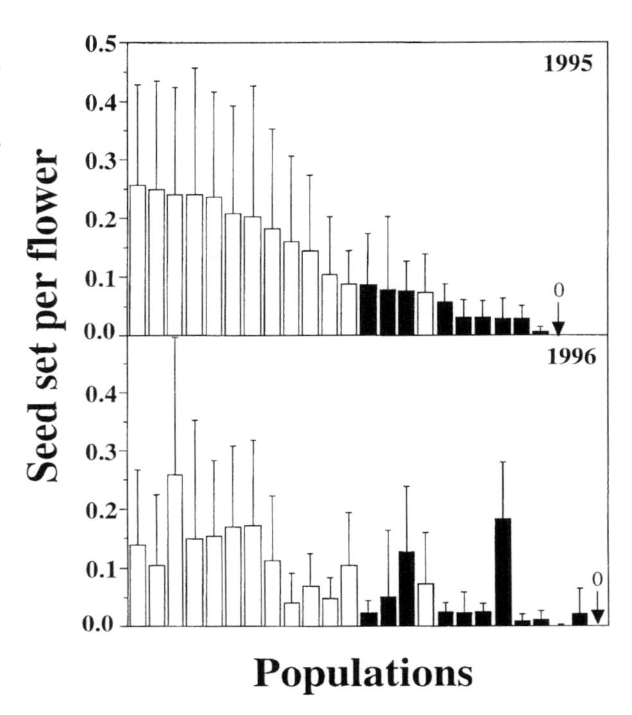

Fig. 1. Seed set (mean ± SD) of local populations of *Persicaria japonica* growing in fragmented moist tall grasslands on the shore of Lake Kasumigaura. Closed bars indicate populations with a small clone number (<3) or higher morph bias (>0.5), and open bars indicate populations with the other traits. Redrawn from Nishihiro et al. (1998)

bias index (| L-S | / (L+S) , where L and S are respectively the numbers of genets of long- and short-styled morphs; 0–1). Pollinator insects were generally abundant, and even the populations of isolated genets were frequently visited by pollinator insects such as *Eristalomyia tenax* L., *Campsomeris annulata* Fabricius and *Lucilia illustris* Meiga. However, seed set varied greatly among local populations but was rather consistent within populations between study years. Mean seed set of the local population was correlated with genet number ($r=0.59$, $p<0.01$ for 1995; $r=0.54$, $p<0.01$ for 1996) or index for morph bias ($r=-0.64$, $p<0.001$ for 1995 $r=-0.70$, $p<0.001$ for 1996). Most small populations found in small grassland fragments showed consistently low seed set in both years (Fig. 1). Artificial pollination with mixed pollen from three opposite morph genets significantly increased the seed set of 3 small populations each consisting of a single genet (Fig. 2). Therefore, it is suggested that lack of compatible pollen supply due to solitude is largely responsible for the reproduction failure of the isolated genets.

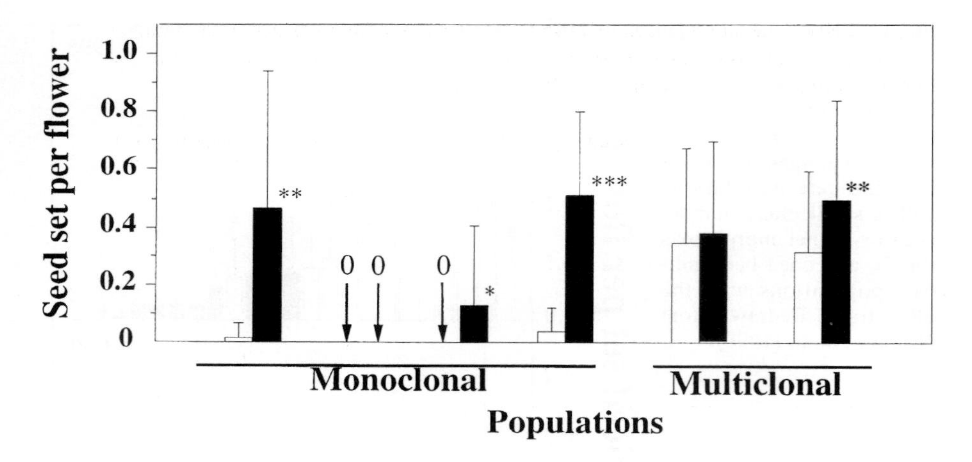

Fig. 2. Seed set (mean ± SD) without (open bars) and with (closed bars) hand pollination by mixed pollen from 3 opposite morph pollen sources in the genets and local populations of *Persicaria japonica*, which is a common species growing in moist tall grasslands. Redrawn from Nishihiro et al. (1998)

6 Allee Effects Revealed in Homomorphic Plants on Oceanic Islands

Differing from dioecious or heterostylous species having only a few mating groups which may be particularly sensitive to plant density and the special distribution of the individual of different mating types (Barrett and Thomson 1982; Feinsinger et al. 1991; House 1992, 1993; Wyatt and Hellwig 1979), a plant population with a homomorphic incompatibility system is less likely to experience reduced fecundity due to incompatibility since a large number of mating types is often present within the population (Laurence and O'Donnell 1981). However, if subjected to a strong bottleneck (Demauro 1993; Imrie et al. 1972), lack of compatible pollen may also cause reproductive failure in homomorphic species.

We found an example of strong Allee effects at the pollination stage of a homomorphic plant, *Crepidiastrum ameristophyllum*, which is a highly endangered shrub endemic to the Bonin Islands, the sole oceanic islands in Japan. Only two populations of considerable size, one large (1200 plants) and one small (31 plants), remain except for a few isolated plants. Seed set was negligible, seedling establishment was not observed in the small population (Table 1) and an isolated plant disappeared in 1996–1998 (Goto and Washitani, unpublished). Only in the large population were considerable numbers of seeds produced and seedling establishment was ascertained. We should suppose that the probability of the extinction of *Crepidiastrum ameristophyllum* is very high, since catastrophic damage by a typhoon could eliminate the sole self-sustaining population.

Table 1. Seed set (mean±SD) and established seedling density in the remaining populations of *Crepidiastrum ameristophyllum* on Hahajima island of the Bonin islands (Goto and Washitani, unpublished)

	Seed set per flower (%)	Seedling density (m^{-2})
Large population Shoot no.=1200	27.0±18.2	7
Small population Shoot no.=31	6.7±7.4	0
Isolated plant Shoot no.=1	0	0

7 How to Strive Against Creeping 'Fruitless Fall'

On the present-day globe, habitat fragmentation ranks among the most serious causes of biodiversity degradation (Wilcox and Murphy 1985; McNeely et al. 1990). Habitat fragmentation as well as other threats to biodiversity, such as over exploitation and biological invasion, results in small population size and also in isolated populations or genets that elevate the probability of extinction (Lande 1987; Menges 1991a; Holsinger and Gottlieb 1991) through various mechanisms represented schematically in Fig 3. Impairment of species interaction essential to reproduction may have a detrimental impact on an isolated plant population (Janzen 1974; Howe 1984). Especially, reduced pollinator services due to fauna degradation occur ubiquitously in present day landscapes throughout the world (Buchmann and Nabhan 1996). Absence of pollinators and solitude of isolated plants may impose several types of drawbacks on a plant population depending on the species' reproductive biology and population history: reproductive failure (Jennersten 1988; Bawa 1990; Washitani et al. 1994a; Groom 1998), decrease in effective population size through reducing gene flow, and breeding shifting to more selfing (Bawa 1990; Menges 1991b; Aizen and Feinsinger 1994; Washitani et al. 1994a). Thus an altered reproductive pattern in turn may cause loss of genetic diversity and/or decreasing progeny fitness due to inbreeding depression (Karron 1989; Barrett and Kohn 1991; Menges 1991b; Washitani 1996).

The results of our studies showing that seed production of both heterostylous and monomorphic plant species is limited by compatible pollen supply through reduced availability of either pollinators or mating partners suggest rather ubiquitous occurrence of 'fruitless falls' for wild plants with mating systems facilitating outbreeding. Plant species depending on specific pollinators, i.e., pollinator specialists, such as *Primula*, have a higher risk of pollination limitation due to pollinator loss compared to a pollinator generalist such as *Persicaria*. However, solitude is

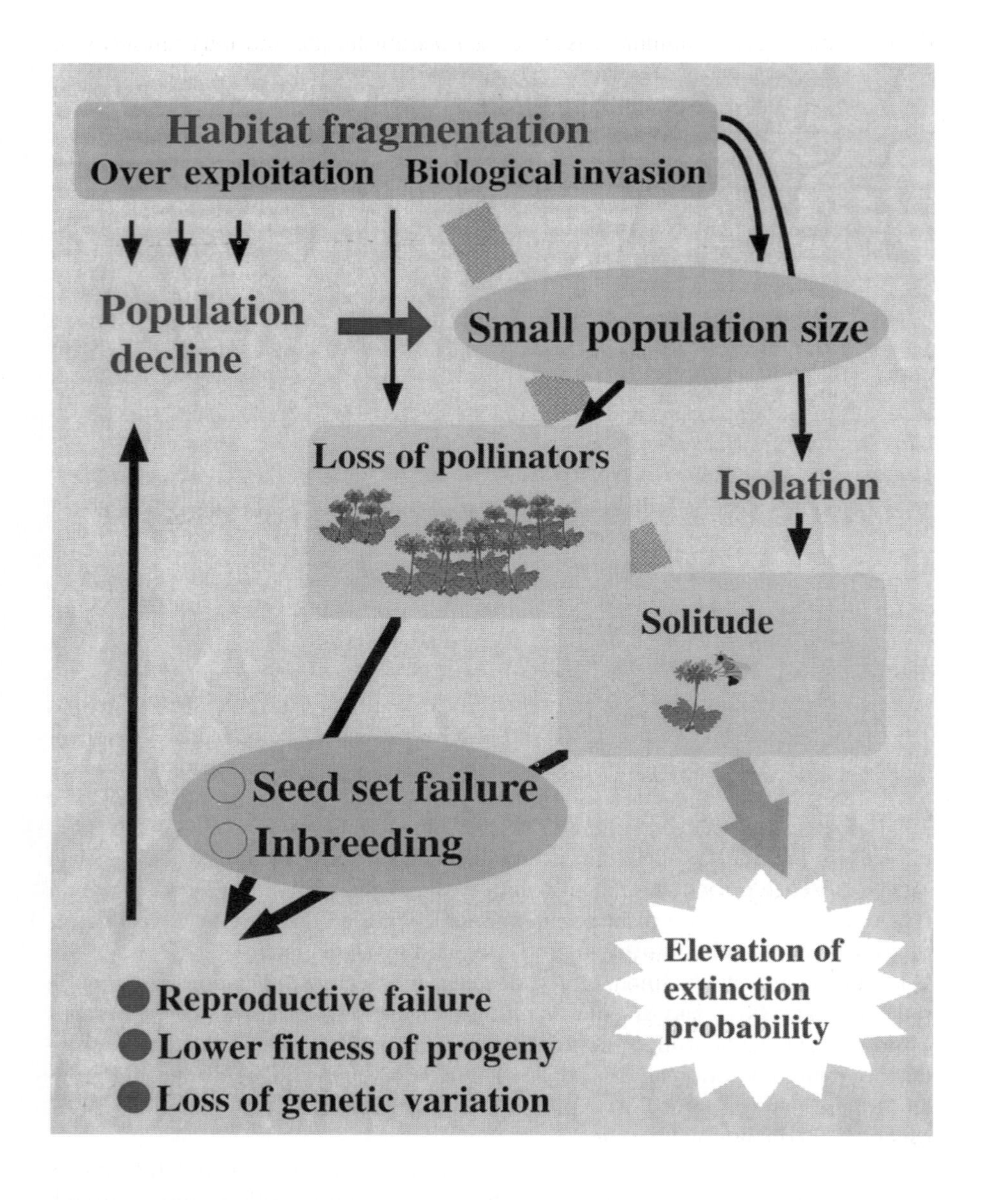

Fig. 3. Schematic representation of vortex for extinction for a flowering plant population brought about by habitat fragmentation and/or other threats to biodiversity. Reproductive failure, loss of genetic diversity and lower fitness of progeny, which are the consequences of loss of pollinator or mating partner associated with population decline or isolation, cause further decline and isolation of the population, thus elevating the probability of extinction

likely to be a more common reason for pollination failure and thus infertility of wild plants irrespective of whether the plant is pollinator specialist or generalist.

In highly fragmented anthropocentric landscapes, habitat and/or population restoration would be an effectual means for preventing species extinction. In order to restore plant populations and retain the present amount of genetic diversity, we should turn to the best plant ecology and population genetics currently available in designing the individual steps for the restoration practice: artificial crosses by hand pollination, seed collection, dormancy breaking and germination, and planting the obtained seedlings to appropriate safe-sites in natural or restored habitats. Soil seed banks may have great potential as plant materials for restoration of the vegetation of moist habitats. Monitoring the population status under several comparable management regimes (Pavlik et al. 1993) is desired to test the hypotheses on the factors causing population decline and to assess the self-sustainability of the restored population and appropriate management required for population persistence.

'Pollinator therapy', i.e., plant habitat management including reintroduction and re-establishment of suitable pollinator populations would be an indispensable measure in many cases for conservation of threatened populations suffering pollinator loss. Designing good pollinator therapy, however, requires scaling-up of pollination ecology to even higher levels of biological organization, i.e., interactions of plant and pollinator populations and community processes, for which we suffer an extreme poverty of information at present.

Restoration and management will provide not only effective measures for biodiversity conservation, but also good opportunities for population and community level experiments to enhance our understanding of 'a web of complex relations' (Darwin 1859), which is essential for wise, prudent use and management of our biodiversity.

Acknowledgments

The research described in this article was partly supported by Grants-in-Aid for Scientific Research, No. 08304040, 07304081 and 08454249, from the Ministry of Education, Science , Sports and Culture, and also by the Global Environment Research Fund (F-1) from the Environment Agency.

References

Allee WC (1951) The social life of animals. Beacon, Boston

Aizen MA, Feinsinger P (1994) Forest fragmentation, pollination, and plant reproduction in a Chaco dry forest, Argentina. Ecology 75:330-351

Barrett SCH (1992) Heterostylous genetic polymorphisms: model systems for evolutionary analysis. In: Barrett SCH (Ed) Evolution and function of heterostyly. Springer, Berlin, pp 1-29

Barrett SCH, Kohn JR (1991) Genetic and evolutionary consequences of small population size in plants: implications for conservation. In: Falk DA, Holsinger KE (Eds) Genetics and conservation of rare plants. Oxford Univ Press, New York, pp 3-30

Barrett SCH, Thomson JD (1982) Spatial pattern, floral sex ratios, and fecundity in dioecious *Aralia nudicaulis* (Araliaceae). Can J Bot 60:1662-1670

Bawa KS (1990) Plant-pollinator interactions in tropical rain forests. Annu Rev Ecol Syst 21:399-422

Buchmann SL, Nabahan CP (1996) The forgotten pollinators. Island Press, Washington

Carson R (1962) Silent spring. Houghton Mifflin Company, Boston

Charlesworth B, Charlesworth D (1979) A model for the evolution of distyly. Am Nat 114:467-498

Christensen NL, Bartuska AM, Brown JH, Carpenter S, D'Antonio C, Francis R, Franklin HR, MacMahon JA, Noss RF, Parsons D J, Peterson CH, Turner MG, Woodmansee RG (1996) The report of the Ecological Society of America Committee on the Scientific Basis for Ecosystem Management. Ecol Appl 6:665-691

Darwin C (1859) On the origin of species by means of natural selection. Murray, London

Demauro MM (1993) Relationship of breeding system to rarity in the lakeside daisy (*Hymenokys acaulis* var. *glabra*). Conserv Biol 7:542-550

Environmental Agency (1997) Japanese plant red list (in Japanese)

Feinsinger P, Tiebout HM, Young BE (1991) Do tropical bird-pollinated plants exhibit density-dependent interactions? Field experiments. Ecology 72:1953-1963

Ganders FR (1979) The biology of heterostyly. New Zeal J Bot 17:607-635

Groom MJ (1998) Allee effects limit population viability of an annual plant. Am Nat 151:487-496

Harper JL (1977) Population biology of plants. Academic Press, London

Holsinger KE, Gottlieb LD (1992) Conservation of rare and endangered plants: principles and prospects. In: Falk DA, Holsinger KE (Eds) Genetics and conservation of rare plants. Oxford Univ Press, New York, pp 195-208

House SM (1992) Population density and fruit set in three dioecious tree species in Australian tropical rain forest. J Ecol 80:57-69

House SM (1993) Pollination success in a population of dioecious rain forest trees. Oecologia 96:555-561

Howe HF (1984) Implications of seed dispersal by animals for tropical reserve management. Biol Conserv 30:261-281

Imrie BC, Kirkman CJ, Ross DR (1972) Computer simulation of a sporophytic self-incompatibility breeding system. Aust J Biol Sci 25:343-349

Janzen DH (1974) The deflowering of Central America. Nat Hist 83:48-53

Jennersten O (1988) Pollination in *Dianthus deltoides* (Caryophylaceae): effects of habitat fragmentation on visitation and seed set. Conserv Biol 2:359-366

Karron JD (1989) Breeding systems and levels of inbreeding depression in geographically restricted and widespread species of *Astragalus* (Fabaceae). Am J Bot 76:331-340

Lande R (1987) Extinction thresholds in demographic models of territorial populations. Oecologia (Berlin) 130:624-635

Laurence MJ, O'Donnell S (1981) The population genetics of the self-incompatibility polymorphism in *Papaver rhoeas*. III. The number and frequency of S alleles in two further natural populations (R102 and R104). Heredity 47:53-61

McNeeley JA, Miller KR, Reid WV, Mittermeier RA, Werner TB (1990) Conserving the world's biological diversity. IUCN, Gland, Switzerland; WRI, CI, SSF-US and the World Bank, Washington DC

Menges ES (1991a) The application of minimum viable population theory to plants. In: Falk DA, Holsinger KE (Eds) Genetics and conservation of rare plants. Oxford Univ Press, New York, pp 45-61

Menges ES (1991b) Seed germination percentage increases with population size in a fragmented prairie species. Conserv Biol 5:158-164

Nishihiro J, Washitani I (1998a) Patterns and consequences of self-pollen deposition on stigmas in heterostylous *Persicaria japonica* (Polygonaceae). Am J Bot 85:352-359

Nishihiro J, Washitani I (1998b) Effect of population spatial structure on pollination and seed set of a clonal distylous plant *Persicaria japonica* (Polygonaceae). J Plant Res 111 (in press)

Nishihiro J, Tomobe K, Washitani I (1998) Effects of habitat fragmentation on seed set in *Persicaria japonica* populations. Jap J Conserv Ecol 3:97-110 (in Japanese)

Pavlik BM, Nickrent DL, Howald AM (1993) The recovery of an endangered plant, I. Creating a new population of *Amsinckia grandiflora*. Conserv Biol 7:510-526

Richards JH (1986) Plant breeding systems. Allen & Unwin, London

Silvertown JW, Lovett Doust J (1993) Introduction to plant population biology. Blackwell, London

Washitani I (1987) A convenient screening test system and a model for thermal germination responses of wild plant seeds, behavior of model and real seeds in the system. Plant Cell Environment 10:587-598

Washitani I (1996) Predicted genetic consequences of strong fertility selection due to pollinator loss in an isolated population of *Primula sieboldii* an endangered heterostylous species. Conserv Biol 10:59-62

Washitani I, Kabaya H (1988) Germination responses to temperature responsible for the seedling emergence seasonality of *Primula sieboldii* E. Morren in its natural habitat. Ecol Res 3:9-20

Washitani I, Masuda M (1990) A comparative study of the germination characteristics of seeds from a moist tall grassland community. Functional Ecol 4:543-557

Washitani I, Namai H, Osawa R, Niwa M (1991) Species biology of *Primula sieboldii* for the conservation of its lowland-habitat population: I. Inter-clonal variations in the flowering phenology, pollen load and female fertility components. Plant Species Biol 6:27-37

Washitani I, Osawa R, Namai H, Niwa M (1994a) Patterns of female fertility in heterostylous *Primula sieboldii* under severe pollinator limitation. J Ecol 82:571-579

Washitani I, Kato M, Nishihiro J, Suzuki K (1994b) Importance of queen bumble bees as pollinators facilitating inter-morph crossing in *Primula sieboldii*. Plant Species Biol 9:169-176

Washitani I, Okayama Y, Sato K, Takahashi H, Ogushi T (1996) Spatial variation in female fertility related to interactions with flower consumers and pathogens in a forest metapopulation of *Primula sieboldii*. Res Pop Ecol 38:249-256

Wilcox BA, Murphy D (1985) Conservation strategy: the effects of fragmentation on extinction. Am Nat 125:879-887

Wyatt R, Hellwig RL (1979) Factors determining fruit set in heterostylous bluets, *Houstonia caerulea* (Rubiaceae). Syst Bot 4:103-114

10
Population Persistence and Community Diversity in a Naturally Patchy Landscape: Plants on Serpentine Soils

Susan Harrison

Department of Environmental Science and Policy, University of California, Davis, One Shields Avenue, Davis, CA 95616, USA

Abstract

Serpentine soils in California are patchily distributed at multiple spatial scales and support a distinctive flora. In a region of 10 × 30 km in Northern California, I compared plant species richness on 24 small (<5 ha) serpentine outcrops and 24 equally-spaced sites within large (> 1 km^2) outcrops. For serpentine endemic plants, alpha (local) diversity was lower but beta (differentiation) diversity was higher on small outcrops compared to sites within large outcrops. For alien plants, local diversity was higher on small outcrops than on sites within large outcrops. Experimental work examined mechanisms underlying these patterns. Two alien grasses (*Avena fatua* and *Bromus hordeaceus*) appeared to be more prevalent on small outcrops because of edge effects rather than because of habitat quality. One serpentine endemic herb (*Calystegia collina*) exhibited lower reproductive success on small outcrops, evidently because of a shortage of compatible pollen. Another endemic herb (*Helianthus exilis*) was absent from small outcrops because of the absence of seeps, a specialized habitat found within large serpentine outcrops. Direct evidence for local extinction and recolonization was found in five herbaceous species that inhabit serpentine seeps. These results illustrate the importance of large-scale landscape structure for population persistence and community diversity in plants.

Key words. diversity, species richness, fragmentation, patchy, landscape, serpentine, plants, pollination, edge effects, extinction, colonization, spatial, isolation, natural rarity, invasion

1 Introduction

Natural habitats range from relatively continuous to extremely patchy, and from relatively permanent to highly ephemeral. For the past several decades, ecologists have struggled to incorporate this natural spatial and temporal variation into our thinking about species interactions, population persistence and biological diversity. Meanwhile, human impacts continue to push natural systems in the direction of increasing patchiness and variability, and ecologists are increasingly called upon to identify strategies for mitigating these impacts. While we are blessed with a growing abundance of theory about how patchy habitats might affect ecological patterns and processes, we still lack a solid empirical foundation from which to evaluate these theories. This chapter describes an attempt to add one small piece to such a foundation, through a study of an ecological community that is naturally patchy and contains a number of rare taxa.

An impressive variety of theoretical perspectives may be brought to bear on how the patchiness of a habitat might affect ecological patterns and processes. MacArthur and Wilson's (1967) theory of island biogeography predicts that the smaller and more isolated a habitat, the lower an equilibrium level of diversity it will achieve. Metapopulation theory (Hanski 1997; Hanski and Simberloff 1997) predicts that species may become extinct regionally if patches of their habitat become fewer or more isolated. Models of the coexistence of predators and prey, or strong competitors, identify a more positive side to patchiness; a discontinuous environment can promote coexistence because temporarily vacant patches provide refuges for the victim species (reviewed in Harrison and Taylor 1997; Nee et al. 1997). Extending the latter idea to the "metacommunity" level, models by Caswell and Cohen (1991, 1993) show that local extinction, colonization and competition in a patchy environment can lead to high total species richness by promoting variation in species composition among patches (beta diversity, Whittaker 1960). However, Tilman et al. (1994) demonstrated that fragmentation can lead to the loss of the dominant species in a metacommunity, because these species are assumed to be the worst dispersers. A metacommunity model by Holt (1997) considers species that can either be patch specialists or patch-and-matrix generalists, and predicts that rare patch types will support a lower ratio of specialist to generalist species than will commoner types. This is closely related to the idea, basic to the field of landscape ecology, that diversity is shaped by the flow of organisms through mosaic habitats (Wiens 1997).

This brief and far from exhaustive review illustrates that ecological theory does not offer conservationists a simple, single message about patchiness and fragmentation. Moreover, in considering how such theories might apply to real species and communities, there are additional dimensions of biological realism that must be kept in mind. In plants, for example, long-distance movements are difficult to observe, and local extinction and recolonization are hard to determine because so many species have persistent seed banks. However, pollination represents a second avenue by which spatial isolation might affect the survival of plant populations (Rathcke and Jules 1993; Aizen and Feinsinger 1994a, 1994b; Groom 1998).

With my collaborators, I have examined patterns of plant diversity, individual species distributions, population persistence, and reproductive success in a landscape consisting of serpentine and nonserpentine soils in northern California. Outcrops of serpentine are associated with fault zones, and may range in size from a few m^2 to many km^2. They support a distinctive flora because their high Mg^{++}:Ca^{++} ratio excludes most species from the surrounding community, including the Mediterranean alien species that now dominate most of lowland California. Their flora includes a substantial number of strict serpentine endemics, many of which are listed as sensitive or rare taxa because of their narrow geographic distributions. The flora on serpentine also includes many non-endemic native species that are rare on nonserpentine because of competition from Mediterranean aliens (Kruckeberg 1984; Brooks 1987; Huenneke et al. 1990; Skinner and Pavlik 1994). For these reasons, the plant community on island-like outcrops of serpentine in my study region is both a good model system for studying patchiness, and of conservation interest in its own right.

Our work is being conducted in Lake, Napa, and Sonoma Counties, California, USA. The flora of this area is described by Barbour and Major (1977), Kruckeberg (1984) and Sawyer and Keeler-Wolf (1995). Most serpentine soils in this region are poorly developed, and seldom support grassland. Instead they support chaparral containing both serpentine endemics (e.g. *Quercus durata*, Fagaceae; *Arctostaphylos viscida*, Ericaceae; *Ceanothus jepsoni*, Rhamnaceae), and non-endemics (e.g. *Pinus sabiniana*, Pinaceae; *Adenostoma fasciculatum*, Rosaceae; *Heteromeles arbutifolia*, Rosaceae; *Umbellularia californica*, Lauraceae). Herbs are sparse, and occur mainly on rocky slopes interspersed with the chaparral. Non-serpentine soils in this region are mainly derived from sedimentary rocks; to the eastern (inland) side of the study area, the vegetation is predominantly blue oak (*Quercus douglassii*) woodland; toward the west, coastal mixed evergreen forest (*Quercus agrifolia*; *Pseudotsuga menziesii*, Pinaceae; *Arbutus menziesii*, Ericaceae).

2 Patterns of Diversity on Small and Large Serpentine Outcrops

In the initial phase of this study (Harrison 1997, 1999), I compared patterns of species richness (henceforth "diversity") of herbaceous plants in two settings: 24 small (mostly < 1 ha) and isolated (> 1 km from large outcrops) patches of serpentine, and 24 similarly spaced and identically sampled sites within very large (> 5 km^2) serpentine areas. Comparing patchy and continuous sites within a single type of habitat and community allowed me to analyze the effects of patchiness *per se* with respect to community structure, while avoiding the confounding effects that would arise in comparing two different communities with different evolutionary histories.

Using geologic maps, I located 24 small serpentine outcrops to which I was able to gain access. The patches chosen were found in 4 clusters of 5–7, within which outcrops were separated by 10–3200 m; distances among the 4 clusters ranged from 16–45 km. The four large outcrops chosen were approximately 6, 16, 30 and 55 km^2. Each large outcrop was matched with one of the four clusters of small ones, and within the large outcrop, a set of sampling sites was chosen with the same spatial configuration as that cluster of small outcrops. The plant community was sampled at all sites in April and May of 1996 and 1997, using three 5×50-m belt transects at each site.

From this data, I compared small and large outcrop sites with respect to Whittaker's (1960) components of diversity: local (alpha), regional (gamma), and among-site differentiation (beta diversity). Beta or differentiation diversity was measured by a metric proposed by Colwell and Coddington (1994): the total number of unshared species between two species lists divided by the number of species in the two lists. For each site, I calculated its (1) "total differentiation", the proportion of species unshared between that site and all other sites of its kind (small or large outcrop), and (2) "within-cluster differentiation", the proportion of species unshared between that site and all other sites in its cluster. I separately considered the woody flora and three nested sets of the herbaceous flora on serpentine: (1) all herb species, (2) herbs that are native versus alien to the study region, and (3) herbs that are strictly endemic to serpentine within the study region.

For serpentine-endemic herbs and woody species (most of which show high fidelity to serpentine), I found that small outcrops had roughly equal total diversity, lower local diversity, and higher differentiation diversity than sites within large outcrops (Table 1). The higher differentiation diversity on small outcrops prevailed at the "within-cluster" as well as the "total" level, supporting the interpretation that patchiness *per se* and not environmental gradients were responsible for the higher beta diversity among small outcrops. This means that for habitat specialists, communities on small patches consist of smaller samples drawn from the same total regional pool as communities on large continuous sites. This result is consistent with the idea that for habitat specialists, community structure is shaped by random colonization and local extinction.

For all herbaceous species together (85% of which are not endemic to serpentine), the only difference between the small outcrops and the sites within large continuous outcrops was higher local diversity on the small outcrops. Even this difference disappeared when only the native herbaceous species were considered (Table 1). These results indicate that, not surprisingly, generalist (non-endemic) species are far less affected by the spatial structure of serpentine patches than are the specialist (endemic) species. These results also indicate that small outcrops had a higher local diversity of non-native species; in fact, there were an average of 6.04 ± 3.88 (s.d.) alien species on small outcrops, versus 1.33 ± 1.76 on sites within large outcrops (MANOVA [multivariate analysis of variance], $p > 0.001$).

Environmental variables created additional gradients in diversity. For example, local diversity of all herbs decreased significantly with increasing elevation, and alien herb diversity decreased with increasing distance inland and decreasing levels of calcium. However, none of these variables explained away the differences in diversity between the small and large outcrops (Harrison 1997, 1999). Overall, the

Table 1. Components of diversity for plant species on small outcrops versus sites within large outcrops of serpentine. Standard deviations in parentheses. "Difference" is (small/large -1) × 100%. P-values are based on MANOVA with df = 3, 44

	Small	Large	Difference	p
a. Woody species				
Regional	21	18	17%	
Local	5.4 (2.5)	7.9 (2.3)	-31%	<0.01
Differentiation				
total	0.77 (0.15)	0.57 (0.12)	35%	<0.001
within-cluster	0.75 (0.21)	0.36 (0.18)	108%	<0.001
b. All herbs				
Regional	178	116	53%	
Local	38.8 (10.5)	30.8 (10.4)	26%	0.01
Differentiation				
total	0.83 (0.07)	0.79 (0.09)	5%	0.21
within-cluster	0.67 (0.11)	0.63 (0.15)	6%	0.34
c. Native herbs				
Regional	144	103	40%	
Local	32.8 (8.8)	29.4 (9.7)	12%	0.22
Differentiation				
total	0.81 (0.05)	0.79 (0.08)	4%	0.16
within-cluster	0.63 (0.11)	0.53 (0.23)	19%	0.08
d. Serpentine endemic herbs				
Regional	24	25	-4%	
Local	5.7 (3.2)	9.3 (3.2)	-39%	0.001
Differentiation				
total	0.78 (0.13)	0.62 (0.13)	26%	0.001
within-cluster	0.59 (0.21)	0.46 (0.17)	28%	0.02

results were strikingly consistent with the prediction by Holt (1997) that a patchy community should support a higher ratio of generalists to specialists than a continuous one. In the next phase of the study I attempted to examine more closely the mechanisms underlying this pattern.

3 Edge Effects and Alien Grasses

To better understand the greater prevalence of alien species on small outcrops than in the interiors of large ones, collaborators Kevin Rice and John Maron and I focused on the Mediterranean grasses *Avena fatua* and *Bromus hordeaceus*, because these two species showed strong patterns: *Avena* was found on 15 of 24 small patches and 0 of 24 sites within large ones, while *Bromus* was found on 20 of 24 small patches and 2 of 24 sites within large ones. Both species are extremely common in the nonserpentine matrix. We examined two non-exclusive hypotheses. First, the pattern might be a biological edge effect; small outcrops might receive a high influx of alien propagules from the nonserpentine matrix. Second, the pattern might be a physical edge effect; small outcrops might have more favorable soil properties than do the interiors of large outcrops.

We examined the edges of large outcrops to see whether, as the edge-effect idea predicts, these resembled small outcrops in terms of the prevalence of alien grasses. We found *Avena* at 3 of 24 large outcrop edges, intermediate between its prevalence in large outcrop interiors and small outcrops; neither difference was significant ($X^2_{(1)} = 3$, $0.05 < p < 0.10$). We found *Bromus* at 15 of 24 large patch edges, significantly more than large outcrop interiors ($X^2_{(1)} = 12.3$, $p < 0.001$) but not significantly different than on small outcrops ($X^2_{(1)} = 0.004$, $p > 0.10$). Average abundances of both species were significantly higher on nonserpentine than on serpentine, and on serpentine the abundance of *Bromus* decreased significantly with increasing distance from the edge ($r = -0.74$; df = 1,9; $p = 0.009$). These results suggest that as hypothesized, small patches are similar to the edges of large patches in terms of their prevalence of aliens (especially *Bromus*).

We performed a growth experiment to determine whether soil conditions were more favorable in small serpentine patches than the interiors of large ones, and also to test for ecotype formation. There were three soil treatments (small outcrop, large outcrop, and nonserpentine), and two seed source treatments (nonserpentine and serpentine) for each species. Neither *Avena* nor *Bromus* performed significantly better in any respect on the small outcrop soil than the large outcrop soil, thus supporting the idea of biological edge effects. We also found some evidence for ecotype formation; serpentine seeds outperformed nonserpentine seeds when grown on small-outcrop serpentine soil. Our results suggest the importance of a landscape-level approach to invasion ecology. Even serpentine soils, with their strong abiotic resistance to invasion, are more likely to be invaded when they are juxtaposed with other habitats that supply an abundance of alien propagules.

4 Reproductive Success in the Serpentine Morning Glory

To investigate how patchiness might affect the persistence of serpentine endemic species, collaborator Amy Wolf and I studied the serpentine morning glory *Calystegia collina* (Convolvulaceae), an outcrossing clonal perennial found only on serpentines in the north Coast Ranges of California (Wolf and Harrison, in press). We compared the reproductive success of *Calystegia* in 39 plant patches on 16 small outcrops and 7 large outcrops, using field measurements, pollen addition experiments, and transplant experiments to compare: (1) flower, fruit and seed production, (2) seed mass and seed germination, (3) pollinator abundance and visitation rate, (4) degree of pollen limitation, and (5) survival of transplants on small versus large outcrops.

Calystegia grows in discrete patches 2 m^2–300 m^2 in area , consisting of one to a few genetic individuals. Small outcrops (here, <5 ha) supported 1–2 plant patches, while large outcrops (here, >300 ha) supported many discrete patches. On large outcrops, marked ramets produced over twice as many flowers per ramet than on small outcrops, in both 1995 and 1996 (MANOVA; $p < 0.001$). Production of seed capsules was also higher on large outcrops; during 1995 and 1996 respectively, 25% and 27% of marked ramets produced seed capsules compared to 10.8% and 4.3% on small outcrops (MANOVA; $p < 0.005$). For the 25 plant patches in which seed capsules were produced (8 on small and 17 on large outcrops), there were no significant differences in seeds per capsule, seed mass, or proportions of seeds that germinated (MANOVA; $p > 0.5$).

Experimental augmentation with pollen from other plant patches led to significantly higher seed production compared with all other treatments; thus, *Calystegia* appeared to be self-incompatible and pollen-limited. On a per-flower basis, there was no difference in the number or diversity of pollinators visiting flowers on small versus large outcrops. However, production of seeds per flower increased significantly with the number of other plant patches within 100 m. For a given patch, there were 2.1 ± 0.5 other patches within 100 m on small outcrops, versus 5.9 ± 1.3 on large outcrops (MANOVA, $p < 0.01$). Inclusion of this variable removed the significant effect of large versus small outcrop on seed production. Thus, isolation from sources of high-quality (= non-self) pollen appeared to be one reason for lower reproductive success of *Calystegia* on small outcrops.

5 Microhabitats and the Serpentine Sunflower

With collaborators Amy Wolf and Paul Brodmann, I examined the distribution of another endemic, the serpentine sunflower *Helianthus exilis* (Asteraceae). In preliminary surveys in 1994, we found *H. exilis* on 13 of 15 large outcrops, in populations of 200–93 000 flowering individuals; we found no *H. exilis* on any of the 30 small outcrops. We undertook experiments and further surveys to explain this striking pattern of distribution (Wolf and Harrison 1999).

The answer proved to be simple; *Helianthus* grows in so-called serpentine seeps, where a spring-fed stream emerging from serpentine crosses a sandy or gravelly area that traps moisture and remains wet into midsummer. In our regional surveys, we never found seeps on small outcrops. While this explains *H. exilis*' absence from small outcrops, it does not explain any of the results of the first (diversity) part of this study, because seeps are sufficiently rare that none of my 24 randomly chosen sites on large outcrops included any seep habitat.

Plants that experimentally received extra pollen produced more seeds on average than naturally pollinated controls, suggesting that seed production in *Helianthus* is partly limited by pollen. However, we observed no significant relationships between the size or isolation of *Helianthus* populations and either the frequency of pollinator visits or the number of pollinator species (multiple regressions, p always > 0.20). Rates of attack by seed predators were not related to either population size or isolation (multiple regressions, p always > 0.20).

6 Local Extinction and Colonization in Five Serpentine Seep Plants

The studies reported above all hint at the importance of local extinction and recolonization in shaping the patterns of plant diversity on patchy serpentine outcrops, but they lack direct evidence for this proposition. However, for the specialized microhabitat just described, there is direct evidence on population turnover. Serpentine seeps are typically isolated by hundreds to thousands of meters from one another, and support a specialized flora; thus they form patches within large patches of serpentine. In the study region there are 5 serpentine-seep specialist plants, all of which are considered uncommon to rare by the California Native Plant Society (Skinner and Pavlik 1995). These are *Helianthus exilis* (Asteraceae), *Senecio clevelandii* (Asteraceae), *Astragalus clevelandii* (Fabaceae), *Delphinium uliginosum* (Ranunculaceae), and *Mimulus nudatus* (Scrophulariaceae). With collaborators John Maron and Gary Huxel, I asked how isolation affected the chances of extinction and recolonization in these species.

In 1981-82, the five species' distributions were surveyed in a 4200-hectare area as part of an environmental assessment for a mine (D'Appolonia 1982). This survey found a total of 218 populations on 87 seeps. In 1997 and 1998, we resurveyed all these localities; 17 had been destroyed by construction, while at another 14 sites we were unable to find any seep habitat. On the remaining 56 seeps, there were 32 presences in 1981-82 followed by absences in 1997-98, i.e. local extinctions, and 100 presences in 1981-82 followed by presences in 1997-98, or non-extinctions. This subset of the data was used in logistic regressions to determine the correlates of local extinction. There were also 64 absences in 1981-82 followed by presences in 1997-98, or putative colonizations, and 79 absences in 1981-82 followed by absences in 1997-98, or non-colonizations. This subset of the data was used in logistic regressions to examine the correlates of colonization.

For each seep that was found in both sets of years, and for each of the five species, we measured the distances to the nearest three conspecific populations found in 1997-98 and calculated their harmonic mean as our measure of population isolation. We measured the downstream length of each seep on topographic maps as an index of seep size. We also measured the distance from each seep to the nearest major human-caused disturbance (e.g. new roads and mining activities).

Chances of local extinction increased with increasing isolation (Fig. 1) and decreased with increasing distance from disturbance (isolation, $t = 2.19$, $p = 0.029$; distance from disturbance, $t = -1.82$, $p = 0.068$; overall model, -2 log likelihood = 7.68, 2 df, $p = 0.021$). Chances of colonization were affected by species (*Mimulus,*

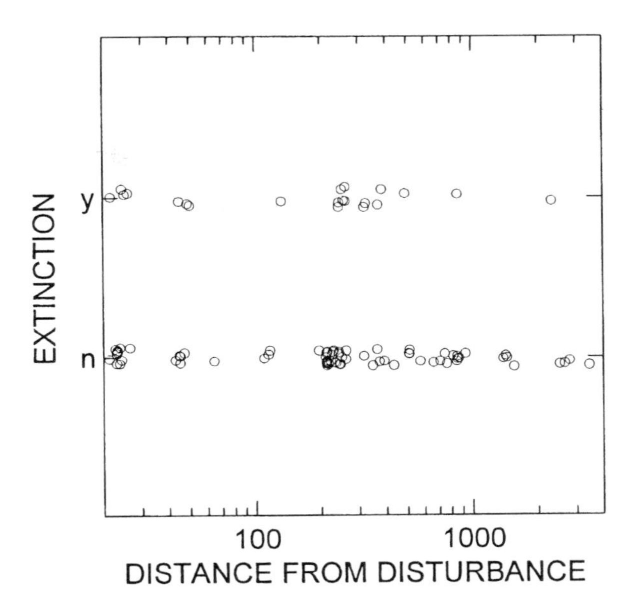

Fig. 1. Isolation, defined as the harmonic mean distances from the three nearest conspecific populations extant in 1997-98, for populations of serpentine seep plants that were found in 1981-82 and were extinct (y) versus not extinct (n) in 1997-98

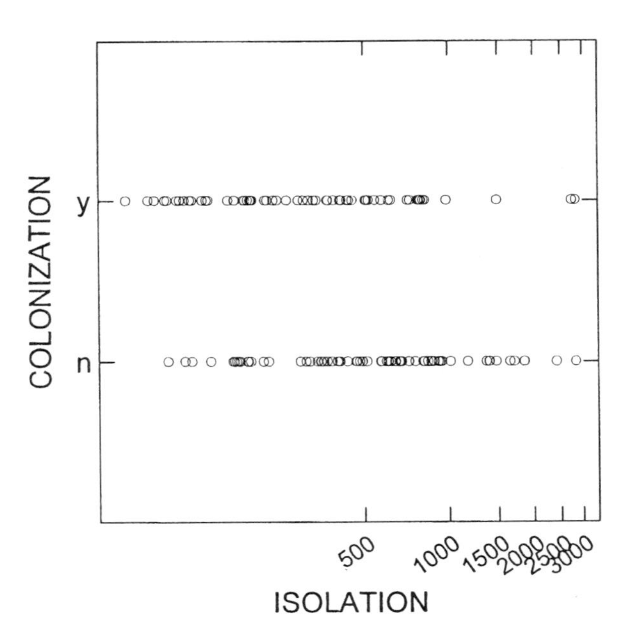

Fig. 2. Isolation, defined as in Fig. 1, for sites where populations of serpentine seep plants were absent in 1981-82 and became colonized (y) or not colonized (n) by 1997-98

t = 2.16, p = 0.030; *Helianthus*, t = 2.65, p = 0.008; indicating these two species had higher rates of colonization than the others), and showed a marginally significant decrease with increasing isolation (t = -1.91, p = 0.057; Fig. 2); the overall model was again significant (2 log likelihood = 20.97, 6 df, p = 0.002).

We also censused population densities for three of the seep species at 50 sites across a 20 × 40 km region, in 1997 and 1998. Between these years, *Helianthus exilis* decreased strongly, *Delphinium uliginosum* increased moderately, and *Mimulus nudatus* increased strongly at virtually every site in the region. Population fluctuations within each species were so completely synchronous at a regional scale that neither population density, nor change in density from 1997 to 1998, was significantly spatially autocorrelated using standard analyses (Cliff and Ord 1981; Isaaks and Srivastava 1989).

This study demonstrates that metapopulation dynamics may be relevant to the persistence of rare plants in patchy environments (Menges 1990). Like many plants, the species we studied are subject to regional-scale fluctuations driven by weather, and probably rely in part on long-lived dormant seeds to survive in their harsh and fluctuating environment. These considerations would appear to argue against the importance of metapopulation processes. Nonetheless, we found a detectable role for spatial isolation in determining the likelihood of local extinction and recolonization (cf. Sjogren Gulve 1994), implying an important role for large-scale dispersal in population persistence.

7 Discussion

The goal of this study was to examine as many facets as possible of how natural habitat discontinuity affects population persistence and community structure in an empirical system. While the study was guided in a general way by theoretical ideas, the intention was not to choose a single theory or hypothesis and then study the species most likely to exemplify it. Instead the study attempted to weigh the relevance of a variety of spatial theories to this particular study system.

Results of the diversity study were relatively supportive of the island biogeographic prediction of lower local diversity on islands compared with mainlands (MacArthur and Wilson 1967). But these results also went beyond island biogeography by considering total (regional) and differentiation (beta), as well as local (alpha) diversity. However, the results did not agree well with competition-based metacommunity models predicting strong effects of patchiness on regional diversity (Caswell and Cohen 1991, 1993; Tilman et al. 1994). Instead they were strikingly consistent with the prediction (Holt 1997) of a higher ratio of habitat generalists to specialists in rare or patchy communities. Holt's prediction was based in part on the idea that patches are influenced by the influx of propagules from the surrounding matrix (Wiens 1997). The results on edge effects in alien grasses supported this idea. However, while most spatial models treat extinction as a random process, the studies of the two serpentine-endemic plants also illustrated the importance of deterministic factors, such as plant-pollinator interactions and habitat availability. Finally, the distance-dependent extinction and colonization found in the study of serpentine seep plants provided good agreement with metapopulation theory (Hanski 1997).

In terms of conservation, the results tend to support the wisdom of setting aside large unfragmented areas to conserve maximal diversity of habitat-specialist species. The small outcrops of serpentine are more interesting from a scientific viewpoint, as places to study edge effects, ecotype evolution, and other questions relating to habitat and community structure, than they are valuable as habitats for rare species. However, the ultimate message of this study is not that any single ecological theory or conservation tactic is universally right or wrong. Rather, the point is that we have a great deal of work yet to do in evaluating the rich variety of ideas provided by spatial ecology. When such theory has been examined by a substantial number of quantitative studies at suitably large spatial scales, it may become a solid basis for deriving conservation strategies.

Acknowledgements

This study was made both possible and enjoyable by my collaborators Paul Brodmann, Gary Huxel, John Maron, Kevin Rice, and Amy Wolf. I also thank the many colleagues who have provided helpful advice and discussions, and the many landowners who allowed us access to the study sites. This work was supported by NSF 94-24137.

References

Aizen MA, Feinsinger P (1994a) Forest fragmentation, pollination, and plant reproduction in a Chaco dry forest, Argentina. Ecology 75:330-351

Aizen MA, Feinsinger P (1994b) Habitat fragmentation, native insect pollinators, and feral honey bees in Argentine "Chaco Serrano" Ecol Appl 4:378-392

Barbour MG, Major J (1977) Terrestrial vegetation of California. Wiley-Interscience, New York (reprinted 1988 by California Native Plant Society, Sacramento, CA)

Brooks RR (1987) Serpentine and its vegetation: a multidisciplinary approach. Dioscorides Press, Portland, OR

Caswell H, Cohen JE (1991) Disturbance, interspecific interaction and diversity in metapopulations. In: Hanski I, Gilpin ME (Eds) Metapopulation dynamics: ecology, genetics and evolution. Academic Press, London, pp 193-218

Caswell, H. and J. E. Cohen 1993. Local and regional regulation of species-area relations: a patch-occupancy model. In Ricklefs RE, Schluter D (Eds) Species diversity in ecological communities: historical and geographical perspectives. Univ Chicago Press, Chicago, IL, pp 99-107

Cliff AD, Ord JK (1981) Spatial processes: models and applications. Pion Limited, London
Colwell RK, Coddington JA (1994) Estimating terrestrial biodiversity through extrapolation. Phil Trans Roy Soc London B 345:101-118
D'Appolonia Company (1982) McLaughlin project: proposed gold mine and mineral extraction facility, Homestake Mining Company. Environmental report. D'Appolonia Company, San Francisco, CA
Groom M J (1998) Allee effects limit population viability of an annual plant. Am Nat 151:487-496
Hanski I (1997) Metapopulation dynamics: from concepts and observations to predictive models. In: Hanski I, Gilpin ME (Eds) Metapopulation dynamics: ecology, genetics and evolution. Academic Press, London, pp 69-91
Hanski I, Simberloff D (1997) Metapopulation dynamics: brief history and conceptual domain. In: Hanski I, Gilpin ME (Eds) Metapopulation dynamics: ecology, genetics and evolution. Academic Press, London, pp 5-26
Harrison, S (1997) How natural habitat patchiness affects the distribution of diversity in Californian serpentine chaparral. Ecology 78:1898-1906
Harrison S (1999) Local and regional diversity in a patchy landscape: native, alien and endemic herbs on serpentine. Ecology (in press)
Harrison S, Taylor AD (1997) Empirical evidence for metapopulation dynamics. In: Hanski I, Gilpin ME (Eds) Metapopulation dynamics: ecology, genetics and evolution. Academic Press, London, pp 27-42
Holt RD (1997) From metapopulation dynamics to community structure: some consequences of environmental heterogeneity. In: Hanski I, Gilpin ME (Eds) Metapopulation dynamics: ecology, genetics and evolution. Academic Press, London, pp 149-165
Huenneke L, Hamburg S, Koide R, Mooney H, Vitousek P (1990) Effects of soil resources on plant invasion and community structure in Californian serpentine grassland. Ecology 71:478-491
Isaaks E H, Srivastava RM (1989) An introduction to applied geostatistics. Oxford Univ Press, New York
Kruckeberg AR (1984) California serpentines: flora, vegetation, geology, soils and management problems. Univ California Press, Berkeley, CA
MacArthur R, Wilson EO (1967) The theory of island biogeography. Princeton Univ Press, Princeton, NJ
Menges ES (1990) Population viability analysis for an endangered plant. Conserv Biol 4:52-62
Nee S, May RM, Hassell MP (1997) Two-species metapopulation models. In: Hanski I, Gilpin ME (Eds) Metapopulation dynamics: ecology, genetics and evolution. Academic Press, London, pp 123-148
Rathcke BJ, Jules ES (1993) Habitat fragmentation and plant-pollinator interactions. Curr Sci 65:273-277
Sawyer JO, Keeler-Wolf T (1995) A manual of California vegetation. California Native Plant Society, Sacramento, CA
Sjogren Gulve P (1994) Distribution and extinction patterns within a northern metapopulation of the pool frog *Rana lessonae*. Ecology 75:1357-1367
Skinner MW, Pavlik BM (1994) California Native Plant Society's inventory of rare and endangered plants of California. Special Publication No. 1, 5th edition, California Native Plant Society, Sacramento, CA

Tilman D, May RM, Lehman CL, Nowak MA (1994) Habitat destruction and the extinction debt. Nature 371:65-66

Whittaker RH (1960) Vegetation of the Siskiyou Mountains, Oregon and California. Ecol Monographs 30:279-338

Wiens J (1997) Metapopulation dynamics and landscape ecology. In: Hanski I, Gilpin ME (Eds) Metapopulation dynamics: ecology, genetics and evolution. Academic Press, London, pp 43-62

Wolf AT, Brodmann PA, Harrison S (1999) Distribution of the rare serpentine sunflower (*Helianthus exilis* Gray, Asteraceae): the roles of habitat availability, dispersal limitation and species interactions. Oikos 84:69-76

Wolf AT, Harrison S (1999) Natural habitat patchiness affects reproductive success of serpentine morning glory (*Calystegia collina*, Convolvulaceae) in northern California. Ecology (in press)

11
Patterns of Tree Species Diversity Among Tropical Rain Forests

PETER S. ASHTON[1] AND J.V. LAFRANKIE[2]

[1] Arnold Arboretum and Organismic and Evolutionary Biology, Harvard University, 22 Divinity Avenue, Cambridge, MA 02138, USA
[2] Center for Tropical Forest Science, Nanyang Technical University, National Institute of Education, 469 Bukit Timah Road, Singapore 1025, Republic of Singapore

Abstract

In this early report of a global program of tropical rain forest research, the causes of tree species diversity are examined through comparisons at global, regional and local scales. Species diversity in communities of sessile organisms has been attributed to many factors. Hypotheses which address two of these, Connell's on intermediate levels of disturbance and Tilman's on competition for heterogeneous soil resources, are reviewed and clarified in context of exceptionally rich communities. Our evidence suggests that many factors are important, but at varying scales in space and time. These differing scales will be discussed with reference to our new data. Unresolved impediments to the community drift hypothesis will be presented. In the summary, lessons for conservation will be identified.

1 Introduction

The rapidly increasing research on plant biodiversity has come to focus on two kinds of questions: Where and why is biodiversity concentrated geographically; and what are the relative roles of selection (i.e., ecological deterministic forces) versus historical (i.e., stochastic) forces including catastrophe and dispersal opportunities, in determining patterns of biodiversity? Biodiversity encompasses not only the numbers of different entities, but also the diversity in their levels of abundance. Fundamentally, biodiversity refers to the genetic diversity of ecosystems, but the

---

Key words. correlates of species richness, tropical lowland evergreen forest, tropical Asia, long-term tree demography research plots

term has been used to describe concepts as different as landscape heterogeneity, community structural complexity and species variability. Species diversity is often used as a convenient proxy for genetic diversity, and will be here.

Ricklefs (1990) has stated, "Because the diversity of plants as primary resources rather straightforwardly determines the diversity of animals, the most rigorous tests of general explanations for diversity lie in their application to plant communities." This trend is particularly so because plants marshal an extraordinary array of physical

Table 1. Partner institutions and lead scientists with the Center for Tropical Forest Science, Smithsonian Tropical Research Institute

Country (Site)	Partner institution and lead scientist
CAMEROON (Korup Forest)	BioResources and Development Conservation Programme: N. Songwe
COLUMBIA (La Planada)	Instituto de Investigación de Resursos Biologicos "Alexander von Humboldt": C. Samper
CONGO (Ituri)	Centre de Formation et de Recherche en Conservation Forestière: J.-R. Makana Wildlife Conservation Society: T. Hart
ECUADOR (Yasuni)	Pontifica Universidad Catolica de Ecuador: R. Valencia University of Aarhus, Denmark: H. Balslev
INDIA (Mudumalai)	Indian Institute of Science: R. Sukumar
MALAYSIA (Pasoh)	Forest Research Institute: N. Manokaran National Institute for Environmental Studies (Japan): T. Okuda
MALAYSIA (Lambir)	Sarawak Forest Department: S. Tan Osaka City University: T. Yamakura
PANAMA (Barro Colorado Island)	Smithsonian Tropical Research Institute: R. Condit
PUERTO RICO (Luquillo)	University of Puerto Rico: J. Thompson U.S. Forest Service
PHILIPPINES (Palanan)	Isabela State University: R. Araño
SRI LANKA (Sinharaja)	University of Peradeniya: C.V.S. Gunatilleke Sri Lanka Forest Department
THAILAND (Huai Kha Khaeng) (Khao Chong)	Royal Thai Forest Department: S. Bunyavejchewin National Institute for Environmental Studies (Japan): T. Okuda

and chemical defenses against herbivores, often leading to species-specific interdependencies (e.g. Coley 1983). Species diversity in plant communities has been variously attributed to environmental heterogeneity in space and in time, including temporally differential influences on fecundity among species in competitive balance; to periodic climatic catastrophe and the stochastic mediators of extinction rate; to rates of accumulation by speciation or immigration of competitively equivalent species; and to the favoring of rarer species by consumer pressure (modified after Leigh 1982). None of these factors is likely to be universally preeminent in the sustainment of species diversity. An examination of changes in their relative importance under a variety of physical conditions may assist in understanding why species diversity is concentrated in certain habitats, and how it is maintained.

In this chapter, we will describe patterns of species diversity in the richest of all biomes, lowland evergreen tropical rain forest, on global, regional and local scales by means of a single life form, trees. This is possible for the first time thanks to collaboration of scientists in many tropical countries with the Center for Tropical Forest Science (CTFS) of the Smithsonian Tropical Research Institute (STRI) (Table 1): A network of large tree demographic plots in rain forests, censused according to a uniform protocol and replicated internationally, is being laid out along gradients of rainfall seasonality, soil fertility, canopy disturbance and insularity (Ashton 1998a). The plots generally represent samples of 50 ha, in which trees equal to or greater than 1 cm diameter are censused. Although the network is global, only the Asian component is near completion and we will concentrate on that. This is a progress report: research is at an early stage in respect of several factors which may enhance diversity in tropical forests.

Regional and local patterns of rain forest tree species diversity are now apparent in Asia and globally (Table 2). There is a strong decline in tree species richness correlated with the increasing length of the dry season, but no close relationship with mean annual rainfall. This trend is nevertheless partially obscured by variation in species richness within the aseasonal wet tropics, the island forests of Sri Lanka being exceptionally low and forests in edaphically heterogenous northwest Borneo unusually high.

2 Causes of Diversity

2.1 Accumulation of Competitively Equivalent Species Through Island Biogeographic Processes

Rates of immigration are influenced by degree of isolation of a forest area, rates of extinction by its area; the immigration and extinction rates of an area balance to create an equilibrium of species richness (MacArthur and Wilson 1969). As most forests differ in area and isolation, the species richness of each individual forest should be unique. This prediction may be obscured by the long life cycles of trees, and their generally short dispersal distances in the tropics. There may therefore be

Table 2. Summary data from 10 long-term demography plots of the Center for Tropical Forest Science

Plot	Mean annual rainfall (mm)	Length of dry season (months)	Site hetero-geneity	No. tree species (50 ha)	Basal area (m^2, indivs ≥ 1 cm dbh/ha)	Stand structure (No. indivs, cm dbh/ha)				
						≥ 1	1-2	≥ 11	≥ 60	≥ 100
Mudumalai, INDIA[a]	c. 1200	5	Low	65	23.7	349	10.7	271	12.9	0.66
Huai Kha Khaeng, THAILAND	c. 1300	5	Moderate	266	32.0	2183	676	428	18	3.8
Barro Colorado Island, PANAMA	2600	4	Low	303	31.8	4581	1791	368	17.4	3.5
Ituri, ZAIRE	1700	3	Low	450[b]		7475				
Korup, CAMEROON	5500	3	Moderate	>450[b]		c. 7000				
Khao Chong, THAILAND	2400	2	Moderate	450[b]						
Pasoh, MALAYSIA	1850	0	Low	810	30.3	6411	2201	498	11	1.1
Lambir, MALAYSIA	2650	0	High	1225	45.0	6883	2841	558	26	3.3
Yasuni, ECUADOR	2500	0	Moderate	1250[b]		6550				
Sinharaja, SRI LANKA	5050	0	Moderate	215[b]	66.0	8299	3721	653	24	2.1

[a] Deciduous forest.

[b] Predicted from a smaller sample.

few instances where immigration and extinction rates have achieved their equilibrium.

Our data are beginning to show a surprising level of congruence at global level (Table 2). Forests in similar habitats in the aseasonal tropics of the Ecuadorian Amazon and Northwest Borneo, both areas which appear to have escaped major climatic change during the Pleistocene, appear to have comparable richness, while forests in Panama, Zaire, and Thailand show a gradient of species richness in relation to rainfall seasonality, irrespective of their apparent relative insularity. These early results imply that forests reach an equilibrium richness according to their habitat, irrespective of their island biogeography. The Sinharaja forest in Sri Lanka is a major exception and also forests in eastern Borneo which, by comparison with those in northwest Borneo in similar habitats, have markedly fewer species (Ashton 1984). These two areas of aseasonal rain forest, so different in extent, nevertheless share a history of climatic change during the last glaciation. Sri Lankan and East Bornean forests may therefore still have been in a phase of species accumulation prior to human influence.

Very small islands nevertheless, as Hubbell (1995) has shown, have substantially reduced richness and also steeper dominance–diversity curves than in adjacent mainland forests, even where they have only become isolated during the Holocene. Moreover, and consistent with the theory of island biogeography, the dominant species are unpredictable.

We infer therefore that island biogeography has little influence on tree species diversity at community scale in areas greater than a certain minimum which is at present unknown. This may in part be due to the very long period needed for tree species richness to reach equilibrium in forests of substantial area, but also because richness there eventually reaches an asymptote in relation to specific physical conditions.

2.2 Temporal Variability in Fecundity

Established juveniles have a competitive advantage over germinants in all plant communities (see, e.g., Brown and Whitmore 1992 for tropical forest). Coexistence of ecologically equivalent species will therefore be promoted if species respond to different flowering cues, including endogenous cues, such that opportunities for successful establishment occur in different years for different species (Chesson and Warner 1981). This effect will be enhanced if seed dispersal is restricted, as in the case of most rain forest species. In this case, our data run contrary to theory. In south-east Asia, high tree species richness is closely associated with aseasonality of rainfall and drought. Climate changes from aseasonal to seasonal with 2-3 dry months, but without change in mean annual rainfall, across a narrow boundary, the Kangar-Pattani Line, which crosses the Thai-Malaysian frontier (Whitmore 1984). Though there is little change in generic composition, species richness is approximately halved in the seasonal climate (Table 2). Counter to expectations, canopy dipterocarps mass flower and synchronously mast fruit at intervals of 4-7 years

(Ashton 1989) in the aseasonal south, but annually and with different species participating each year in the less rich seasonal regions to the north (Appanah 1985, Ashton et al. 1988). Mass flowering has recently been shown to be a community-wide phenomenon by two Kyoto graduate students, Kuniyasu Momose and Shoko Sakai, at the Lambir CTFS site (Sakai et al. 1997).

There appears to be little difference in means of dispersal between forest flora of aseasonal and seasonal Asia. We conclude that the influence of temporal differentiation on species accumulation is either minor, or is overcome by a more important influence. For example, mast fruiting satiates predators and reduces seed mortality among dipterocarps (Janzen 1974, demonstrated by Curran 1994), though comparisons of seed and seedling predation north and south of the Kangar-Pattani line have yet to be made. Lower seed predation might contribute to higher species richness.

2.3 Density-Dependent Mortality

In most studies, more than half of all species in rain forest communities have proven not to be habitat specific and, after all factors are taken into account, it appears that there is a high level of spatial overlap if not ecological complementarity among co-occurring species. A persisting mystery of the tropical forest therefore is why most tree species maintain high outcrossing rates and genetic variability, which must be sustained at great cost in the prevailing low density populations and low nutrient availability (although these costs have yet to be critically documented). If endogenous reasons, such as the need to sustain crossing-over as a means to repair chromosomes are insufficient reason, then the sustainment of high genetic variability implies that some form of selection is operating and that the conditions necessary for survival are changing over evolutionary time. These changes are likely to be biotic, particularly in the uniquely ancient equable climates of the hyper-diverse forest of the putative refugia. The remaining, and still unanswered possibility is that the many apparently ecologically complementary tree species populations in these habitats may each be kept at reduced levels by species-specific interactions with motile organisms which may limit their fecundity or increase their mortality.

Janzen (1970) and Connell (1971) suggested that seed and seedling predators may cause density dependent mortality owing to the greater ease with which they may discover prey which is close to the parent, or to other conspecifics of similar size to themselves. Evidence for species-specific predation has proven elusive but there is growing, albeit inconclusive evidence of density-dependent, species-specific pathogenicity (Gilbert et al. 1994). The strongest support has come from studies of individual species' seedling populations (Augsberger 1983, 1992; Clark and Clark 1984; Howe et al. 1985; Webb and Peart, in press).

Attempts to test for density-dependence among all species in community samples have given mixed results (Hubbell 1980, 1997a; Hubbell and Foster 1990; Wills et al. 1997; Wills and Condit, in press; Webb and Peart, in press). Hubbell (1997a) argued that, though present, density dependent mortality is too weak to explain the levels of species diversity that exist. Wills et al. (1997) using a different method of analysis but the same data set as Hubbell (1997a), has shown higher density dependence. It should be expected that, the more equable the environment, the greater the hegemony of biotically over abiotically mediated selection. For species in similar population densities, biotically induced density-dependent mortality should increase in intensity with decline in rainfall seasonality. Yet Wills and Condit (in press) found that density dependent mortality is substantially lower in the aseasonal climate at Pasoh, Malaysia, than at Barro Colorado Island, Panama with its three month dry season. Could it be that drought-related, or generalized browsing-induced juvenile tree mortality is density dependent and stronger in seasonal climates? Evidence for this is so far lacking.

The CTFS tree demography plots are large enough to describe the canonical distribution of species abundances (Preston 1962). They indicate that the differing species richness of tropical forests along the rainfall seasonality gradient is not so attributable to differing numbers of rare species (e.g., Ashton 1984, 1998b) as to differing numbers in the peak abundance classes (Fig. 1). The number of very abundant species also does not greatly vary. There is a shift in the peak of species abundances caused by the differences in stand density. Preston predicted that the canonical distribution of species abundances should be bell-shaped. The number of species in low population densities in samples from less seasonal climates is greater than expected; but although the number of rare species hardly changes with increased seasonality within the evergreen forests, that number increasingly conforms to Preston's prediction as the peak of the abundance curve shifts to the left. These results imply that both density dependence and factors influencing population extinction such as Allee effects differ little in intensity between these forests; and that the key to the maintenance of species richness must be sought among the species majority, those of moderate population density. In forests of the aseasonal tropics but of differing richness, however, numbers of rare species and those of peak abundance covary (Fig. 2).

In a regenerating temperate woodland, we have observed an enormous diversity of patterns of tree population mortality, ranging from apparently random but continuing mortality of the powerfully defended *Rhus toxidodendron* L. by a pathogen, to the clustered but periodic mortality among mature trees of several genera occasioned by gypsy moth (*Porthetria dispar*) infestation, neither of which patterns would easily be identified by current methods of monitoring. The case of *Shorea albida* Sym., a dipterocarp which is the sole species in the closed canopy through thousands of square kilometers of peat swamp forests in northwest Borneo, should be a warning. Sharply circumscribed areas as great as 10 km^2 were defoliated and killed during the sixties by an unidentified hairy caterpillar, but attacks later ceased. These forests may have survived as pure stands for several thousand years.

Density dependence may also be induced by limitations on fecundity resulting from low carrying capacity of species-specific dispersal agents (Ashton 1998b). Most tree species in rain forests appear to have polylectic dispersal syndromes, but there are exceptions. An interesting example is *Durio,* Bombacaceae, which con-

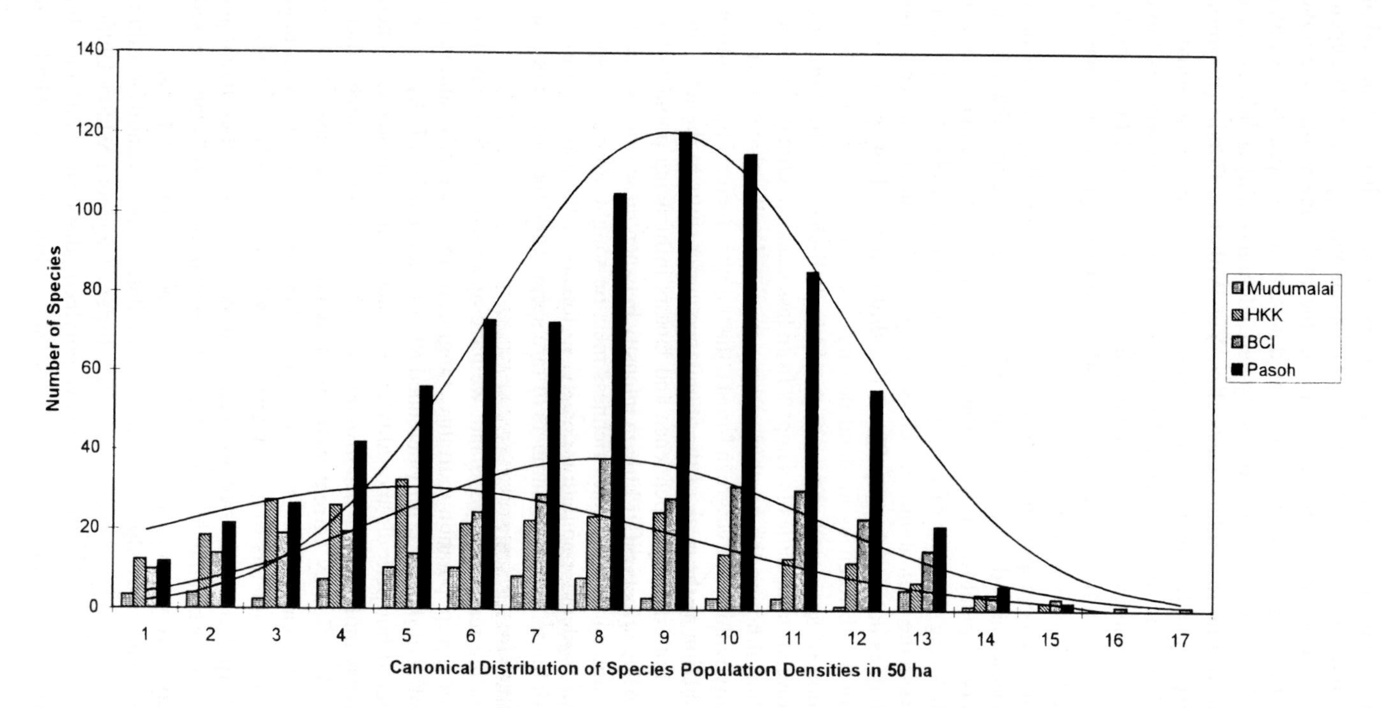

Fig. 1. Canonical distribution of tree species abundances in four plots along a regional rainfall seasonality gradient in tropical forest. Trees ≥ 1 cm dbh. Pasoh, peninsular Malaysia, aseasonal; Barro Colorado Island, Panama, 3 dry months; Huai Kha Khaeng, Thailand, 5 dry months; Mudumalai, India (deciduous), 5 dry months. Symmetric curves fitted using modal and higher species abundance classes

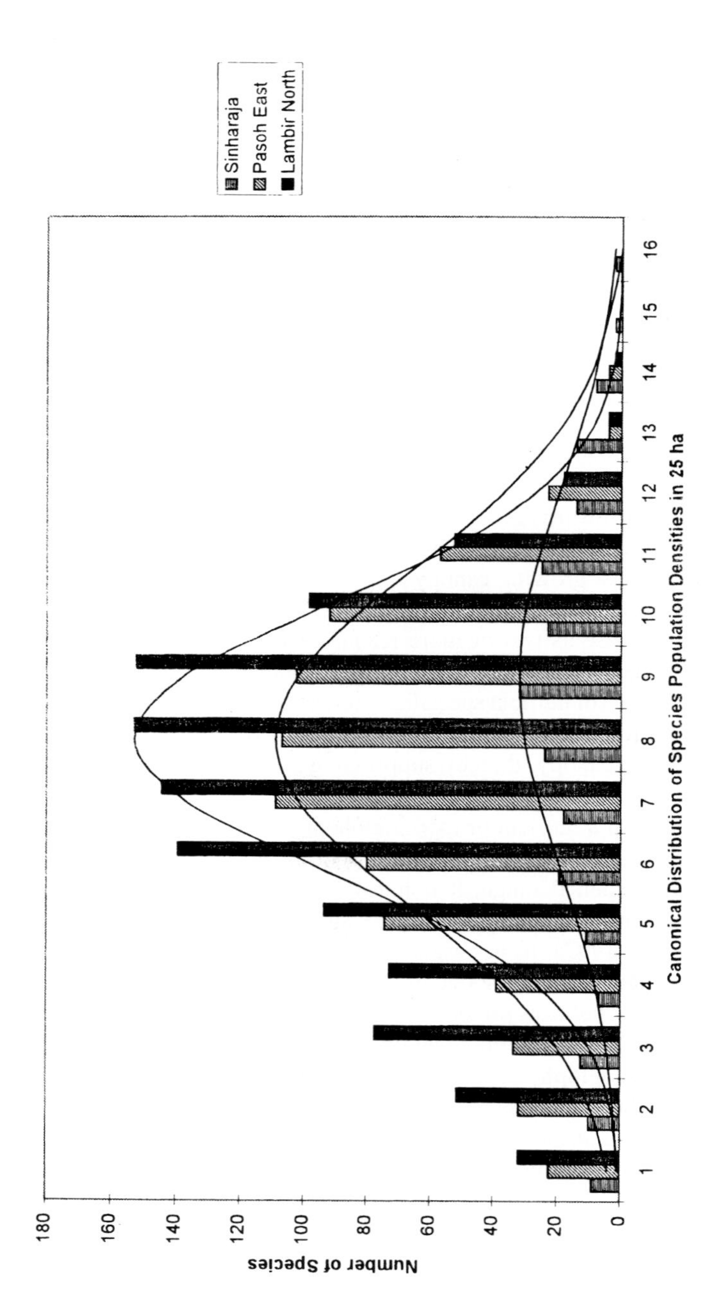

Fig. 2. Canonical distribution of tree species abundances in three plots in the aseasonal wet lowland tropical forests of Asia. Trees ≥ 1 cm dbh. Sinharaja, Sri Lanka, an island forest; Pasoh, peninsular Malaysia, on zonal udult ultisol soils; Lambir, northwest Borneo on udult and low nutrient humult ultisols. Symmetric curves fitted using modal and higher species abundance classes

sists of two subgenera, *Durio*, with relatively few, bat-pollinated flowers and large fruit, and *Boschia* with many, bee-pollinated flowers and small fruits. Species in subgenus *Durio* appear to be uniformly in low population density in the forest, whereas most *Boschia* are abundant in their preferred habitats (Ashton 1998b). *Durio* and other forest tree species visited by the nectarivorous bat *Eonycteris spelaea* Dobson mass flower seasonally or supra-annually. The carrying capacity of the bat will be limited by the reliability of the annual relay sequence of flowering trees.

The role of density dependence will not be resolved without much more, and longer term, study.

2.4 Environmental Heterogeneity in Space

It is well established that biodiverse tropical forest varies in species composition in relation to topography (Beccari 1904; Richards 1996; Ashton 1964; Austin et al. 1972; Hubbell and Foster 1990; Webb and Peart, submitted), and to soils (Richards 1996; Baillie et al. 1987; Ashton 1998b). Ashton (1977, in Terborgh 1994; Ashton 1998b) demonstrated a peaked distribution of species richness along a nutrient gradient. The peak occurs towards the low end of the nutrient range. Species richness at high nutrient concentrations was inversely correlated with relative dominance (density) of a few fast growing canopy species. These results are consistent with Tilman's (1982) resource competition hypothesis. Tilman predicted that plant species richness will increase as two or more resources increase in concentration from a low base, and also increasingly vary in space, because niche space and opportunities for coexistence of nutrient specialists increase; but above a given threshold, he argued, one or a few species with high growth rates will outcompete others for light and dominate the canopy, thereby suppressing species richness. This was the case here.

Yet within any sector of a nutrient or topographic gradient the more biodiverse rainforests carry many hundreds of tree species. Bearing in mind their longevity and frequently limited dispersal range, it is hardly conceivable that such a rich array can find and continue to occupy a complex spatial mosaic of multiple soil resources. Hubbell (1995, 1997a, b) has argued that these species assemblages are not in equilibrium but change over time due to random drift of species with limited dispersal capacities. This hypothesis predicts lack of consistent rank orders of abundances within community types, and the preeminent influence of the predictions of the theory of island biogeography. Only monitoring of marked individuals over many decades can test whether this is so, but it is more likely that increasing concentrations of nutrients at first merely increase the number of edaphically equivalent species that can co-occur. This conclusion is consistent with the very narrow range of maximum growth rates among species on the quite low nutrient soils which carry the richest forests (Ashton and Hall 1992).

2.5 Environmental Variability over Time

Connell (1978), on the evidence of benthic coastal communities, showed that a peak distribution of species richness can also occur along a gradient of canopy disturbance. In calm lagoons, and at depth, canopy closure by climax species leads to restriction of pioneers to scattered gaps which, as in an archipelago, experience low immigration and high extinction rates. Species richness is depressed therefore both by impoverishment of the pioneer community, and by a few dominant climax species which create a uniform and unfavorable light climate beneath their dense even canopy. Under the turbulent conditions of the wave-line, climax species cannot establish and the community is dominated by pioneers and is thus also impoverished; but under intermediate, moderately low disturbance regimes behind the reef edge, conditions favor a mixed community in which species diversity reaches its zenith. Each of these communities, in the classic case of coral reefs, is characterized by a distinct coral fauna.

Connell (1978) drew an analogy between coral reefs and tropical forests, but there are major differences. Canopy disturbance by wind, that which is most closely analogous to wave action, probably never differs in intensity on local spatial scales sufficiently consistently over long enough periods for communities of differing species richness *and composition* to evolve as a consequence. Examination by means of plots smaller than the scale of the community mosaic as a whole, of differences in species richness in relation to differences in disturbance intensity, precludes distinction between real differences and artefacts. Very large samples are required to capture the full array of pioneer species in a community where gaps are small and scattered in a matrix of the mature phase as is usual in tropical forest (Whitmore 1984), and even more so for climax species in severely disturbed forest. But both climax and pioneer species will be more easily captured by small plots in moderately disturbed forests, thereby artificially increasing the relative species richness of these communities.

The severe effects of typhoons (hurricanes) appear to influence forest structure, and possibly richness, in ways different from wave impact on a coral reef. Occasional typhoons cause catastrophic canopy damage and successional stages may patchily dominate the landscape for long periods (Wyatt-Smith 1954; Whitmore 1984; Ashton 1993). If such typhoons occur more frequently than the time required for the mature phase to reestablish and reproduce, extinction of climax species might be expected to result, but we have no example of such a case. By contrast, forests subject to frequent typhoons become adapted. The typhoons remove leaves and many twigs and the canopy is therefore diffuse. In extremely exposed southeast Taiwan, stunted closed-canopy forest on windward slopes is less rich in woody species than that in coves (Sun et al. 1998). Nevertheless, in Luzon heliophytes including some pioneers may persist in the understory (personal observation). A 16 ha tree demographic plot is under construction in such forest at Palanan on the windward eastern coast of Luzon, but detailed results are still awaited. We anticipate that here, in contrast to the coral reefs, species richness may be even greater in the most wind prone sites, where the diffuse canopy may shelter a richer subcanopy flora than beneath the denser canopy on lee slopes.

Intermediate disturbance regimes are associated with high species diversity in reef communities in part because of their heterogeneous light climate at the reef floor. Rain forest species richness and community composition, and also stand structure and perhaps gap size and frequency, may covary with soil and climate. The depression of species richness on mesic sites in forests of the calm, aseasonal tropics was already seen in part to be related to increasing canopy density (Ashton 1998b). The most impoverished rain forests here, the Heath forests, occur on podsols. They have low stature and a simple structure with even, albeit diffuse, canopy and small gap sizes (Brunig 1974; Gale 1997). The diffuse canopy structure is due to the small, generally upturned leaves. Forests with the greatest species richness have tall stature and a similarly diffuse albeit uneven canopy which may be sustained by patchy, often single tree mortality caused by lightning and occasional catastrophic drought (see e.g., Leighton and Wirawan 1986; also Gale 1997). The principles of the Connell hypothesis, therefore, obtain in forests of the aseasonal tropics, but canopy density, hence light intensity and heterogeneity, may here be determined as much by the influence of water stress and nutrients on leaf and twig growth, hence size, shape and density, as by disturbance.

Along the gradient of species richness which occurs from the aseasonal wet to the strongly seasonal dry tropics, the frequency and intensity of drought mediates the level of catastrophic mortality, and influence species' survival (Ashton 1993). Forest stature declines and the mature phase canopy becomes more even with increased rainfall seasonality (Richards 1996), but there is no obvious change in the frequency of canopy disturbance or the density of canopy trees, unless fire intervenes. The major gradient in mortality patterns, which could influence species richness, is in the understory; this accounts for declines in trees <10 cm dbh from >7500 to <100/ha (Table 1). Periodic intense drought following seedling establishment, and also fire and browsing, may each be influential. Their relative importance is as yet not known, and is a major objective of our research at Khao Chong with two, and Huai Kha Khaeng with five dry months, directed by Sarayudh Bunyavejchewin, Royal Thai Forest Department in collaboration with Toshinori Okuda, Japanese National Institute of Environmental Studies.

3 Conclusion

Our results already present important implications for conservation planning. Differences in rates of accumulation of species in rainforest tree communities do occur, but may reach a constant level beyond a quite small area. This implies that attrition of tree populations following forest fragmentation may be very slow and in large strict conservation areas may not occur at all over realistic time periods.

Variation in species richness correlated with climatically induced variation in phenology and fecundity occurs at a regional scale, but the relationship is contrary

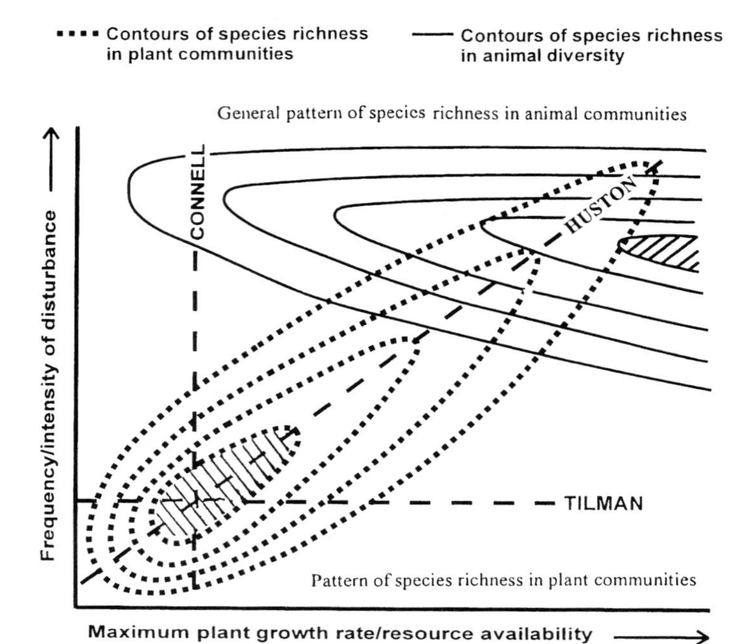

Fig. 3. Patterns of species richness in plant and animal communities. Contours indicate isoclines of richness, increasing to peaks in the cross-hatched areas. Conditions favoring plant and animal diversity do not coincide. Plant species richness follows peaked distributions in relation to frequency or intensity of canopy disturbance (Connell's hypothesis (1978)), or resource availability and maximum growth rates (Tilman's hypothesis (1982)), integrated in the hypothesis of Huston (1980); animals, including arthropods, increase in diversity and abundance in disturbed habitats, particularly where resource availability is high (Modified after Begon et al. 1996)

to current theory and is not yet understood. Variation in forest community composition and species richness may also be induced by variation in forest canopy heterogeneity over time at landscape or local scales. Then it covaries at least in part with variation mediated by environmental heterogeneity in space (Huston 1980). Rain forest communities are thereby often fragmented within the landscape in nature, again implying that forest fragmentation by man may not induce extinction levels predicted by the Theory of Island Biogeography (MacArthur and Wilson 1969). It appears, though, that patterns of tree species richness do not closely follow those for animal, notably vertebrate diversity which is centered where high disturbance and high nutrient resources, therefore abundance of high quality plant foods, are concentrated (M.R. Leighton, personal communication, Fig. 3).

Mediation of species richness by density-dependent factors acts locally, but is probably universal. Whether the influence of density-dependent mortality on species richness varies with habitat, especially climate, remains unknown.

The commonest, and some at least of the rarest species appear to be remarkably predictable in their relative and absolute abundance at community level (Ashton 1998b). But there are many also which are not. It is also curious that examples examined of the many closely related co-occurring species in diverse tropical forest appear to have distinct albeit overlapping life history traits, consistent with competitive coexistence (see, e.g., Davies et al. 1998). Such distinct and rather narrow characteristics do not of course, confirm niche specificity. On the other hand, would not survival be increased under conditions dominated by the drift model if species 'maintain their options' through retention of broad and complementary ecological amplitudes? Only simulation modeling tested by long-term monitoring of individuals and populations of putative competitors can resolve this question.

Acknowledgments

The authors gained from the analytical skills of Matthew Potts, Harvard University. We are particularly grateful for the rather complete data generated by S. Bunyavejchewin, R. Condit, C.V.S. Gunatilleke, N. Manokaran, R. Sukumar, and S. Tan.

References

Appanah S (1985) General flowering in the climax rain forest of South East Asia. J Trop Ecol 1:225-240

Ashton PS (1964) Ecological studies in the mixed dipterocarp forests of Brunei State. Oxford Forestry Memoirs 25. 197 pp

Ashton PS (1977) A contribution of rain forest research to evolutionary theory. Ann Missouri Bot Gard 64:694-705

Ashton PS (1984) Biosystematics of tropical forest plants: a problem of rare species. In: Grant WF (Ed) Plant biosystematics. Academic Press, Toronto, pp 497-518

Ashton PS (1989) Chapter 11: Dipterocarp reproductive biology. In: Leith H, Werger MJA (Eds) Tropical forest ecosystems: biogeographical and ecological studies. Ecosystems of the World Series, 14B. Elsevier, Amsterdam, pp 219-240.

Ashton PS (1993) The community ecology of Asian rain forests, in relation to catastrophic events. J Biosci 18: 501-514.

Ashton PS (1998a) A global network of plots for understanding tree species diversity in tropical forests. In: Dallmeier F, Comisky JA (Eds) Proceedings of the international symposium on measuring and monitoring forest biological diversity: the international network of biodiversity plots. Washington, D.C.: Smithsonian Institution- Man and the Biosphere Program, Paris, UNESCO and Parthenon Press, May 1995

Ashton PS (1998b) Chapter 18: Niche specificity among tropical trees: a question of scales. In: Newbery DM, Brown N, Prins HH (Eds) Dynamics of tropical communities. BES Symposium, Vol No. 37. Blackwell Scientific Publ, Oxford, pp 491-514

Ashton PS, Givnish TJ, Appanah S (1988) Staggered flowering in the Dipterocarpaceae: new insights into floral induction and the evolution of mast fruiting in the aseasonal tropics. Am Nat 132:44-66

Ashton PS, Hall P (1992) Comparisons of structure and dynamics among mixed dipterocarp forests of northwestern Borneo. J Ecol 80:459-481

Augspurger CK (1983) Offspring recruitment around tropical trees: changes in cohort distance with time. Oikos 20:189-196

Augspurger CK (1992) Experimental studies of seedling recruitment from contrasting seed distributions. Ecology 73:1270-1284.

Austin MP, Ashton PS, Grieg-Smith P (1972) The application of quantitative methods to vegetation survey, III. A re-examination of rain forest data from Brunei. J Ecol 60:309-324

Baillie IH, Ashton PS, Court MN, Anderson JAR, Fitzpatrick EA, Tunsley J (1987) Site characteristics and the distribution of tree species in mixed dipterocarp forest on tertiary sediments in central Sarawak, Malaysia. J Trop Ecol 3:201-220

Beccari O (1904) Wanderings in the great forests of Borneo. EH Gigliolo (Transl), Guillemard FHH (Ed). Constable, London

Begon M, Harper JL, Townsend CR (1996) Ecology. Individuals, populations and communities. Edition 3. Blackwell Science, Oxford

Brown ND, Whitmore TC (1992) Do dipterocarp seedlings really partition tropical rain forest gaps? Phil Trans Roy Soc B 335:369-378

Brunig EF (1974) Ecological studies in the kerangas forests of Sarawak and Brunei. Borneo Literature Bureau, Kuching, Sarawak, Malaysia

Chesson P, Warner RR (1981) Environmental variability promotes coexistence in lottery competitive systems. Am Nat 117:923-943

Clark DA, Clark DB (1984) Spacing dynamics of a tropical rainforest tree: evaluation of the Janzen-Connell model. Am Nat 124:769-788

Coley PD (1983) Herbivory and defensive characteristics of tree species in a lowland tropical forest. Ecol Monographs 53:209-233

Connell JH (1971) On the role of natural enemies in preventing competitive exclusion in some marine animals and in rain forest trees. In: den Boer PJ, Gradwell GR (Eds) Dynamics of populations. Proceedings of the advanced study institute in dynamics and numbers in populations, Oosterbeck. Wageningen, Center for Agricultural Publishing and Documentation, pp 298-310

Connell JH (1978) Diversity in tropical rain forests and coral reefs. Science 199:1302-1310

Curran LM (1994) The ecology and evolution of mast-fruiting in Bornean Dipterocarpaceae: a general ectomycorrhizal theory. Doctoral dissertation, Princeton University, Princeton, New Jersey

Davies SJ, Palmiotto PA, Ashton PS, Lee HS, LaFrankie JV (1998) Comparative ecology of 11 sympatric species of *Macaranga* in Borneo: tree distribution in relation to horizontal and vertical resource heterogeneity. J Ecol 86:662-673

Gale N (1997) Modes of tree death in four tropical forests. Doctoral dissertation, Aarhus University, Aarhus, Denmark

Gilbert GS, Hubbell SP, Foster RB (1994) Density and distance-to-adult effects of a canker disease of trees in a moist tropical forest. Oecologia 98:100-108

Howe HF, Schupp EW, Westley LC (1985) Early consequences of seed dispersal for a neotropical tree (*Virola surinamensis*). Ecology 66:781-791

Hubbell SP (1980) Seed predation and the coexistence of tree species in tropical forests. Oikos 35:214-229
Hubbell SP (1995) Towards a theory of biodiversity and biogeography on continuous landscapes. In: Carmichael GR, Folk GE, Schnoor J (Eds) Preparing for global change: a midwestern perspective. SPB Academic Publishing, Amsterdam, pp 173-201
Hubbell SP (1997a) The maintenance of diversity in a neotropical community: conceptual issues, current evidence, and challenges ahead. In: Dallmeier F, Comisky JA (Eds) Forest biodiversity: research: Monitoring and modeling. UNESCO and Parthenon Paris, Paris, pp 17-44
Hubbell SP (1997b) A unified theory of biogeography and relative species abundance and its application to tropical rain forests and coral reefs. Coral Reefs 16 supplement:59-521
Hubbell SP, Foster RB (1990) Structure, dynamics and equilibrium status of old-growth forest on Barro Colorado Island. In: Gentry AH (Ed) Four neotropical forests. Yale Univ Press, New Haven, pp 522-541
Hubbell SP, Condit R, Foster RB (1990) Presence and absence of density dependence in a neotropical tree community. Trans Roy Soc London (Ser B) 330:269-281
Huston M (1980) A general hypothesis of species diversity. Am Nat 113:81-101
Janzen DH (1970) Herbivores and the number of tree species in tropical forests. Am Nat 104: 501-528
Janzen DH (1974) Tropical blackwater rivers, animals and mast-fruiting by the Dipterocarpaceae. Biotropica 6: 69-103
Leigh EG (1982) Introduction: why are there so many trees. In: Leigh EG, Rand RAS, Windsor DM (Eds) The ecology of a tropical forest: seasonal rhythms and long-term changes. Smithsonian Institution Press, Washington, DC, pp 63-66
Leighton M, Wirawan N (1986) Catastrophic drought and fire in Borneo tropical rain forest associated with the 1982-1983 southern oscillation event. In: Prance GT (Ed) Tropical forests and the world atmosphere. Westview Press, Boulder, Colorado, pp 75-102
MacArthur RH, Wilson EO (1969) The theory of island biogeography. Princeton Univ Press, Princeton, New Jersey.
Preston, FW (1962) The canonical distribution of commonness and rarity. Ecology 43:185-215, 410-432
Richards PW (1996) The tropical rain forest: an ecological study. Cambridge Univ Press
Ricklefs RE (1990) Ecology. WH Freeman, New York
Sakai S, Momose K, Nagamitsu T, Harrison R, Yumoto T, Itino T, Kato M, Nagamasu H, Hamid AA, Inoue T (1997) An outline of plant reproductive phenology in one episode of general flowering cycle in 1992-1996 in Sarawak, Malaysia. In: Inoue T, Hamid AA (Eds) General flowering of tropical rainforests of Sarawak. Canopy Biology Program in Sarawak (BBPS) Series II. Center for Ecological Research, Kyoto University
Sun I-F, Hsieh C-F, Hubbell SP (1998) Structure and species composition of a sub-tropical rain forest in southern Taiwan on a wind-stress gradient. In: Dallmeier F, Comisky JA (Eds) Forest biodiversity: research, monitoring and modeling. UNESCO and Parthenon Paris, Paris, pp 563-590
Terborgh J. (1988) The big things that run the world: a sequel to EO Wilson. Conserv Biol 2:402-403
Terborgh J (1994) Diversity and the tropical rain forest. Scientific American Library, New York
Tilman D (1982) Resource competition and community structure. Princeton Univ Press, Princeton, New Jersey

Webb CO, Peart DR (in press 1998) Seedling density dependence promotes coexistence of Bornean rain forest trees. Ecology

Webb CO,. Peart DR (submitted) Habitat associations of trees and seedlings in a Bornean rain forest

Whitmore TC (1984) Tropical rain forests of the Far East. Clarendon Press, Oxford

Wills C, Condit RB, Hubbell SP (1997) Strong density- and diversity-related effects help to maintain tree species diversity in a neotropical forest. Proc Natl Acad Sci USA 94:1252-1257

Wills C, Condit RB (in press) Different relative roles of non-random processes and chance in the maintenance of diversity in two tropical forests. In: Losos E, Condit RB, LaFrankie JV (Eds) Forest diversity and dynamism: results from the global network of large-scale demographic plots. Smithsonian Institution, Washington, DC

Wyatt-Smith J (1954) Storm forest in Kelantan, Malay. Forester 17:5-11

12
Functional Differentiation and Positive Feedback Enhancing Plant Biodiversity

Takashi Kohyama[1,2], Eizi Suzuki[3], Shin-ichiro Aiba[3], and Tatsuyuki Seino[1]

[1] Graduate School of Environmental Earth Science, Hokkaido University, Sapporo 060-0810, Japan
[2] Harvard University Herbaria, Cambridge, MA 02138, USA
[3] Faculty of Science, Kagoshima University, Kagoshima 890-0065, Japan

Abstract

Ecosystems are under the control of negative feedback due to resource competition, and it is difficult to explain the coexistence of many species in an ecosystem. By contrast, evolutionary history suggests that positive feedback between living organisms and environments contributes to increasing biodiversity. This paper presents a conceptual framework to interface these feedbacks, taking forests and tree communities as an example. One of the prevailing global patterns of biodiversity is the latitudinal gradient of tree species diversity in forests. A tenfold difference exists in species diversity between tropical lowland forests and either tropical high-altitude forests or temperate forests. We examined tree census data from permanent plots across various forest types in eastern Asia. Tree species diversity increased exponentially along a geographic gradient while ecosystem measures such as biomass, biomass turnover rate and asymptotic canopy height increased only linearly. Examination of a size-structure-based dynamic model of tree populations suggests that these ecosystem measures multiplicatively contribute to the extreme species diversity in tropical lowland rain forests. The same model also shows that any singular species with higher resource-use efficiency replaces all coexisting species under the constraint of functional tradeoff. Such replacement brings about increasing efficiency of ecosystems in resource exploitation, and in turn presents a greater opportunity for species coexistence. The non-linear relationship between whole

Key words. asymptotic size, biomass, coexistence, ecosystem, feedback, forest, forest architecture hypothesis, functional differentiation, size-structure-based model, species diversity, recruitment capacity, resource gradient, tree community, tropical rain forests, turnover rate

ecosystem measures and species diversity develops through the process of positive feedback between the energetic efficiency of ecosystems and the functional differentiation among species, on evolutionary time scales.

1 Introduction

Ecologists and evolutionary biologists differ in their understanding of the mechanisms controlling the pattern of species assembly. One of the main concerns of ecologists is to describe the negative feedback that regulates populations, communities and ecosystems. Communities at the same trophic level are composed of species competing for an identical set of resources. In the view of this resource competition, competitive exclusion by one species is a consequence of the basic multiple-species competition model. Plenty of conditions have been proposed that relax competitive exclusion among species; however, this is still far from explaining the diversity of species in real communities.

By contrast, taxonomists, paleontologists, and evolutionary biologists share the view that the biota of the earth has been bringing about irreversible change in environments, and increasing in composite species diversity with time, beyond occasional interruption by mass-extinction events. This increase in organismic diversity is coupled with the exploitation of resources, and there acts a positive feedback between the improvement of resource-use efficiency and the promotion of biodiversity. However, ecological negative feedback by resource limitation observed in present ecosystems should have worked throughout the history of biota. This is thus an interesting question, i.e. how both negative and positive feedbacks explain the present assembly of species.

Forest ecosystems and forest tree communities are intriguing subjects of examination. Forests, developing in moist and warm climates, are the most complicated terrestrial ecosystems in terms of biological architecture, biomass density and species diversity. They are also the most stable systems due to cumulative growth habit and generation-overlapped populations of trees. Forest ecosystems show a prevailing trend of directional change in biomass, productivity and biodiversity from tropical moist climate to temperate and/or arid harsh climate. The International Biology Programme (IBP) in the 1970's intensively investigated biomass and productivity of forests and determined that a clear functional relationship between ecosystem attributes and climatic parameters existed (Lieth and Whittaker 1975; Kira and Shidei 1977). Communities of tree species primarily contribute to primary production and represent the complex and persistent architecture of forest ecosystems. Attributes of tree communities such as species diversity was, though, not explicitly related to ecosystem attributes during the IBP.

We are now accumulating data of repeated censuses of permanent plots from various types of forests to monitor the fate of tree individuals and to obtain demographic parameters of each component tree species. Based on these species parameters, we are able to analyze the results of competition using multi-species models to detect whether species differentiation is an assembly rule of tree communities. Employing a non-destructive technique of estimating ecosystem attributes, we can also analyze the relationship between ecosystem functions and tree community dynamics simultaneously at the same research plots.

This chapter examines data of forest plots over a climatic gradient in relation to a quantitative simulation model of the size-structure dynamics of forest trees, and presents a conceptual framework to interface ecological negative feedback and evolutionary positive feedback.

2 Geographic Pattern of Ecosystem Measures and Diversity

It is well known that tree species diversity of forest ecosystems declines with increasing altitude and latitude. Figures 1a and 1b show these relationships for 16 research plots in East Asia (see Appendix for plot description). Fisher's diversity index α employed here is almost proportional to the number of species on the scale of plots around 1 ha. The logarithm of the diversity index α is satisfactorily explained by the linear combination of altitude and latitude by ordinary multiple regression (Fig. 1c). The large-scale pattern of species diversity, e.g. at the grid resolution of a few degrees in latitude and longitude, has been analyzed in relation to climatic parameters. At least for forest trees, the pattern of species diversity is readily explained by available energy for communities in temperate regions across continents (Adams and Woodward 1989; Currie 1991). Our results from small-sized plots support these energetic explanations.

Using the data from the same forest plots, we can also estimate ecosystem at-

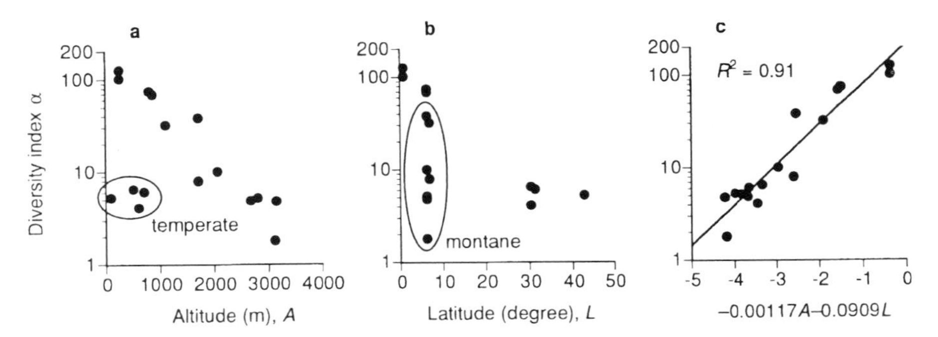

Fig. 1. The dependence of the Fisher's diversity index α of tree species on geographic location of research plots. (a) The relationship with altitude (temperate forests > 30 degree latitude circled), (b) that with latitude (montane forests > 1000 m altitude circled), (c) that with linear combination of altitude and latitude, from a multiple regression. Fisher's α is defined by $S = \alpha \ln(1 + N/\alpha)$, where N is the number of individuals, and S is the number of species

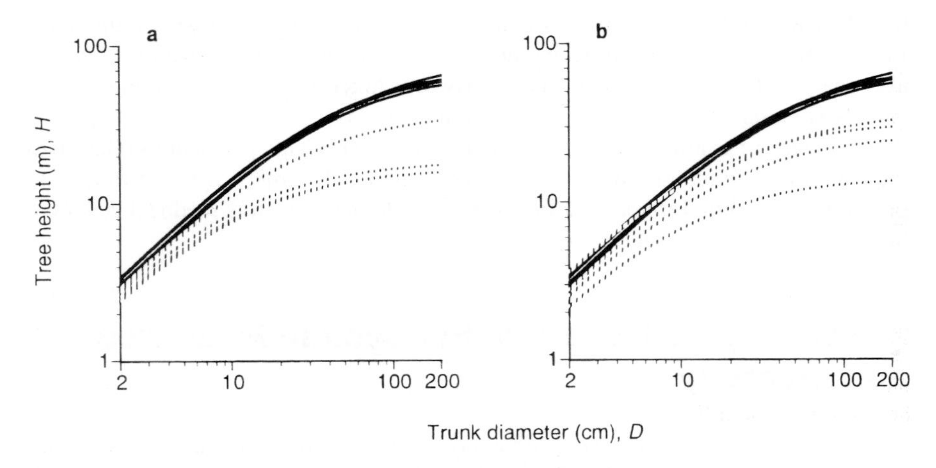

Fig. 2. Relationships between trunk diameter and tree height regressed by a reciprocal equation $1/H = 1/[AD] + 1/H^*$, where H is tree height (m), D is trunk diameter (cm), A is a coefficient of stand crowding (m/cm) and H^* is a coefficient of asymptotic height (m). (a) Low-altitude (< 1000 m) forests at different latitudes (full line for tropical < 10 degree latitude, dashed for temperate >30 degree). (b) Tropical (<10 degree latitude) forests with different altitudes (full line for lowland and foothill < 1000 m, dashed for montane > 1000 m; serpentine plots on Mt Kinabalu excluded)

tributes. Aboveground biomass is usually estimated from the census data of the distribution of trunk diameter D, the allometric relationship between D and tree height H, and the allometric regression between aboveground tree mass and D^2H (Ogawa and Kira 1977). The D-H relationship is readily regressed by a simple reciprocal model as shown in Fig. 2 (Ogawa and Kira 1977). Not only the asymptotic tree height, but also the height relative to diameter, which affect the possibility of canopy stratification in the upper forest profile, is different among forest types. Tropical lowland rain forests show remarkably large asymptotic height and increasing tree height with diameter in the larger size range, than either high-altitude tropical forests or high latitude forests (Fig. 2). This is related to the fact that the differentiation between canopy species and emergent species is only seen in tropical lowland rain forests.

Ecosystem measures for 16 permanent plots can also be expressed by the linear combination of an altitude-bound effect and a latitude-bound effect (Fig. 3). These attributes are linear-negatively related to altitude and latitude, unlike the exponential relation in case of species diversity (Fig. 1c). The contribution ratio between altitude and latitude, as the ratio of regression coefficients of multiple regression, is very similar between diversity *versus* altitude/latitude regression and ecosystem attributes *versus* altitude/latitude regressions: 1000-m increase in altitude corresponds to 13–15 degree (or 1400–1600 km) increase in latitude. The persistence of the contribution ratio suggests that the same determinant, possibly a cumulative thermal environment, controls both species diversity and ecosystem attributes

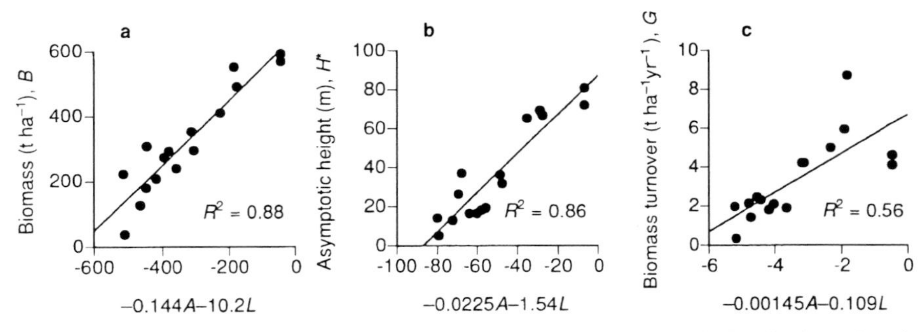

Fig. 3. Dependence of ecosystem attributes on altitude (A, m) and latitude (L, degree) estimated by multiple regressions. (a) Aboveground biomass; (b) asymptotic forest height; (c) turnover rate of aboveground biomass. Asymptotic height H^* is from trunk diameter-tree height regression in Fig. 2

(Ohsawa 1995). It is obvious from comparing Fig. 1 and Fig. 3 that the gradient in species diversity is not proportional to the gradient of ecosystem measures. There exists a ten-fold diversity in lowland tropical rain forests compared to high-altitude tropical forests or high-latitude moist forests, while this is not the case for ecosystem attributes. What then amplifies diversity in wet tropical lowland environments? The examination of a size-structure dynamic model offers some clue to solve this question.

3 Coexistence Enhanced by Biomass and Turnover

The role of local three-dimensional architecture of forests in promoting the coexistence of species has been examined by employing a size-structure-based model of forest tree dynamics (Kohyama 1993, 1996). Below we give an outline of the model and the consequences of simulation analyses.

The model employs a one-dimensional drift equation, so it is easier to trace the consequences of multi-species dynamics than otherwise more realistic individual-based models. Three demographic rates specific to species, namely recruitment rate, size growth rate and mortality, express the dynamics of the species population. All of the three rates are under the control of negative feedback through overall density, reflecting the competition for light resource. Here we examine a simple model where only recruitment and growth are subject to the control. Recruitment is suppressed by the total basal area, or the sum of squares of tree diameter of the stand, irrespective of species. The size growth rate is suppressed by the sum of basal area of trees irrespective of species larger than the size of target tree, reflecting the one-directional supply of light resource ('perfect one-sided competition').

The intensity of the suppression or the negative feedback employed here is identical across species, so that it is impossible for plural species to coexist without size

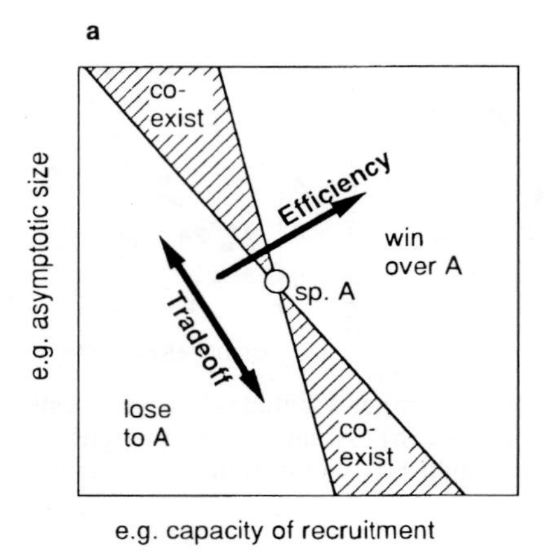

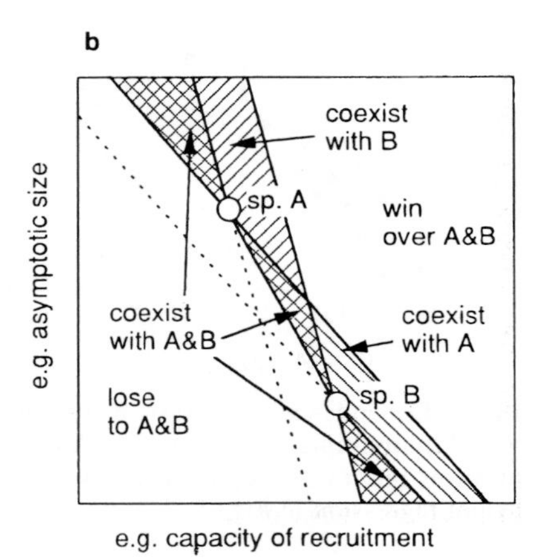

Fig. 4. Diagram showing the equilibrium condition of coexistence of the size-structured multi-species model on the space of two demographic parameters. The circle corresponds to a species with fixed parameter set; species with any parameter set within the hatched domains can coexist with the fixed species. (a) The result for a two species system, one (species A) is fixed. (b) The results for a three species system, where two (species A and B) are fixed. All other species parameters are set as equivalent over species. Parameterized versions are in Kohyama (1993, 1996)

structure. The size-structured model reveals that species can coexist when they adapt differentially along the light-resource gradient within a forest profile. Differential allocation among species is needed for stable coexistence: species with larger maximum size have a smaller *per-capita* recruitment rate, and *vice versa*. Additional tradeoffs such as differentiations between species with faster potential growth rate *versus* those with higher susceptibility to suppression can also create a stable coexistence in an extended model of gap-dynamic forest as a metapopulation of stands of differing stand-age after gap formation. Thus the three-dimensional architecture of the forest itself creates the resource heterogeneity and enables plural species to coexist ('forest architecture hypothesis', Kohyama 1993).

Examination of the parameter space (Kohyama 1993, 1996) shows that a species with a fixed set of parameters defines a bow-shaped domain of coexistence (Fig. 4a). Any species with parameter sets within that domain can coexist with the fixed species. Due to the bow shape, more dissimilar species have more flexibility in choosing parameters for coexistence. The system of two coexisting species with fixed sets of parameters further subdivides the coexistence domains for the third species, and so on (Fig. 4b). Therefore, the wider the domain of coexistence that any singular species defines, the more species are likely to coexist in the model results. The model shows that the domain of coexistence is increased if the potential rate of size growth is increased for all species together, and is decreased if the tree mortality and gap formation rate is increased (Kohyama 1996). Both increas-

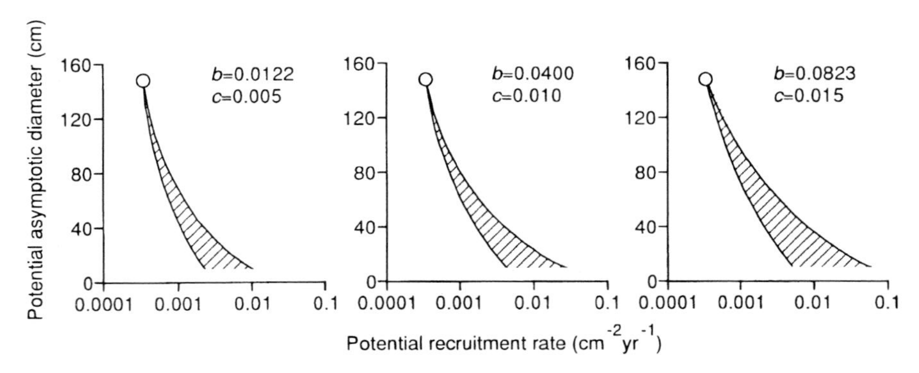

Fig. 5. Dependence of the coexistence domain of two-species systems on the turnover rate of populations. Both potential maximum relative growth rate of trunk diameter (b, cm cm^{-1} year^{-1}) and mortality (c, year^{-1}) for both species are adjusted to yield the same stationary basal area for the fixed tall species at 58.2 cm^2 m^{-2}. Any species within the hatched domain coexists stably with the fixed tall species shown by a circle. The model and other parameters are the same as in Table 2 of Kohyama (1993)

ing growth rate and decreasing mortality result in the increase of the equilibrium biomass (actually basal area in the model). Therefore, these results are easily explained in that the more developed architecture, or higher biomass density, the higher the resource heterogeneity, which provides greater opportunity for species coexistence.

We examined the effect of the vegetative turnover rate of trees independent of biomass density using this model. It is possible to keep biomass (actually basal area in the model) constant with different vegetative turnover rates (from 0.5% to 1.5% per year) through a balanced change of size growth rate and mortality. Results show that increasing turnover not only accelerates the dynamics toward equilibrium but also widens the domain of coexistence (Fig. 5). Therefore, the model suggests that biomass and its vegetative turnover rate differentially promote tree species diversity in forest ecosystems. However when we proportionally increase all vegetative turnover rates (*via* size growth and mortality) and reproductive turnover rate (*via* potential recruitment rate), such a proportional increase only results in faster convergence to the equilibrium independent of the condition of coexistence.

4 Significance of Functional Differentiation in Coexistence

From the permanent plot data, one can examine whether the theoretically suggested tradeoff relationship exists among co-occurring tree species. We found the constraint of tradeoff between asymptotic size and *per-capita* recruitment rate (re-

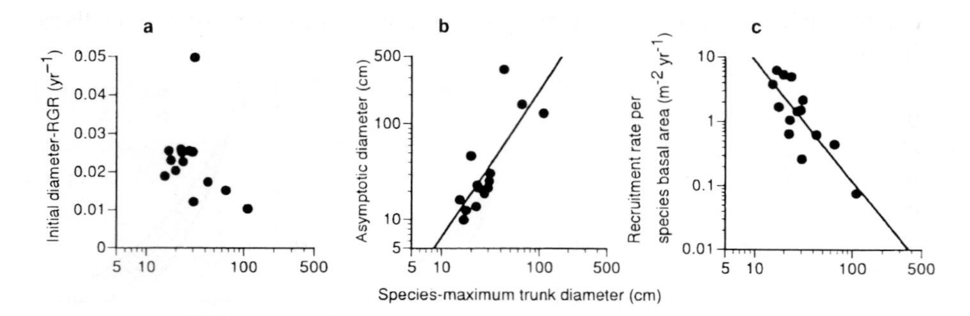

Fig. 6. Among-species difference in (a) initial relative growth rate (RGR) of trunk diameter, (b) the asymptotic diameter from RGR regression to diameter, and (c) the *per-capita* recruitment rate (i.e. per species basal area) to a census-defined minimum size of 3-cm diameter, plotted against the observed maximum diameter of each species. Data from the Koyohji plot of a primary warm-temperate rain forest on Yakushima Island, southern Japan, from Aiba and Kohyama (1996) and Kohyama and Takada (1998)

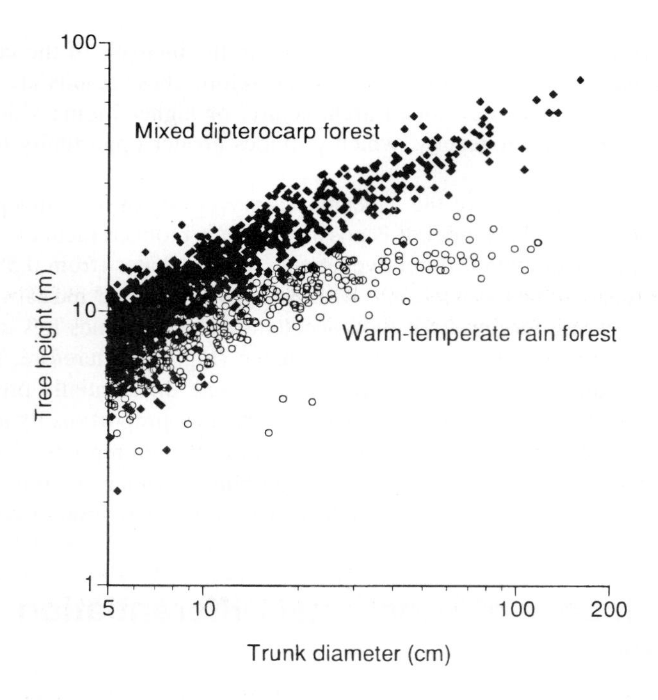

Fig. 7. The relationship between trunk diameter at breast height and maximum height of trees occurring in a mixed dipterocarp forest in West Kalimantan (Serimbu, plot-1, closed symbols) and in a warm-temperate rain forest in southern Japan (Koyohji, plot-1, open symbols)

cruitment rate per species basal area) both in a warm-temperate rain forest in southern Japan (Fig. 6) and in a mixed dipterocarp forest in west Kalimantan (T. Kohyama et al. in preparation, the pattern is very similar to Fig. 6). Thus the functional differentiation among species along a forest profile is likely to play an important role in the coexistence of species in these forests. The question then arises why there exists such a large difference in tree species diversity between these two forest types. It is worth paying attention to the fact that the above analysis was only done in terms of trunk diameter as a proxy for individual size.

Light-resource competition in a one-sided manner is a result of vertical crown stratification, so tree height is a better size dimension to use than trunk diameter. The relationship between diameter and height was quite different between these two forests (Fig. 7), even though the distribution in trunk diameter was similar. Trees of the warm-temperate forest approached their asymptotic height at a smaller diameter, above which differentiation among species in maximum size is fairly difficult. By contrast in the mixed dipterocarp forest, differentiation in terms of maximum height is still possible in larger size classes. Therefore, not only the biomass and the turnover rate of biomass, but also a measure of vertical differentiation among tree crowns such as the asymptotic height H^* of D-H allometry (Fig. 2) increase the possibility of coexistence. These attributes are co-related with each other (Fig. 3). Therefore, it is not necessary to assume any single factor amplifying species packing in wet tropical lowland, but these ecosystem attributes are likely to multiplicatively widen the resource axis for the coexistence of species (Fig. 8).

5 Positive Feedback Between Ecosystem Efficiency and Biodiversity

The results of the model competition between species (Fig. 4) suggest not only a condition of coexistence but also of exclusive replacement by superior species in terms of resource use efficiency. For instance, for the parameter space defined by reproductive capacity and asymptotic size (where other parameters are identical between species), species with larger asymptotic size in growth performance always wins over smaller species when the reproductive capacity is kept identical between the two species. When keeping the asymptotic size identical, species with higher reproductive capacity always wins. If there is only one species with superior performance beyond the tradeoff belt shared by many species, this species eventually excludes all other species. Therefore, succession or evolution is a series of replacement by superior species, which does not necessarily enhance species diversity.

However, when we examine the range of coexistence defined by the new, superior species in the monodominant ecosystem, it becomes clear that this species offers a wider domain of coexistence than the replaced less-efficient species does (Fig. 9). The pattern of widened domain of coexistence is also true in a series of replacements changing other parameters such as initial size growth rate. If there is

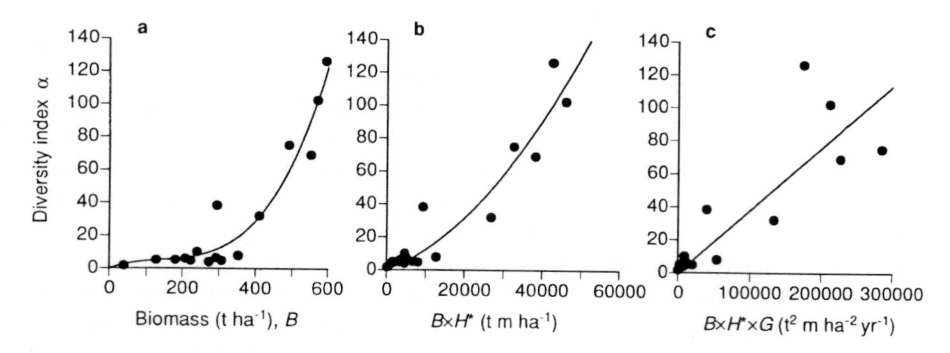

Fig. 8. The relationship between forest ecosystem attributes and tree species diversity. The diversity index α is plotted against (a) aboveground biomass B with cubic regression, (b) biomass multiplied by asymptotic height H^* with quadratic regression, and (c) the multiplication of biomass, asymptotic height and turnover rate of biomass G with linear regression

a large enough pool of similarly efficient species with sufficient functional differentiation, then the ecosystem will support more species. This course of improving reproductive capacity is comparable to such evolutionary events as the emergence of seed plants, and that of flowers in angiosperms with insect pollination.

This consequence of differential change of parameters for one species, which is related to succession and evolution, is similar to that of simultaneous change of parameters among competitors, which is related to climatic gradient. The replacement by superior species brings about higher resource-use or energetic efficiency at the level of a whole ecosystem. A more efficient ecosystem brings about larger heterogeneity of resources. Such a widened resource gradient can be subdivided by more species if all of them are efficient. In the time scale of succession, this cascade reaction triggered by the arrival of efficient competitors improves the energetic efficiency of ecosystems and increases biodiversity up to the maximum capacity under climatic environments. This provides a dynamic explanation of the coupling between climatic environments and ecosystem/biodiversity attributes.

On the evolutionary time scale, the emergence of a more efficient mutant in terms of resource use is coupled with the ecosystem-level improvement of resource exploitation. Such systems can potentially support more species. Therefore there is an evolutionary positive feedback between energetic efficiency of ecosystems and species diversity. The ecological mechanisms that regulate the performance of organisms through negative feedback under resource limitation play an essential role throughout evolutionary time scales.

The same process of positive feedback is likely to act at any trophic level. The energetic efficiency of autotrophs and heterotrophs can explain the diversity of trophic diversity, in terms of the number of trophic levels (Teramoto 1996). Therefore, the coupling of energetic efficiency and species diversity on an evolutionary time scale can be further amplified for the whole ecosystem level.

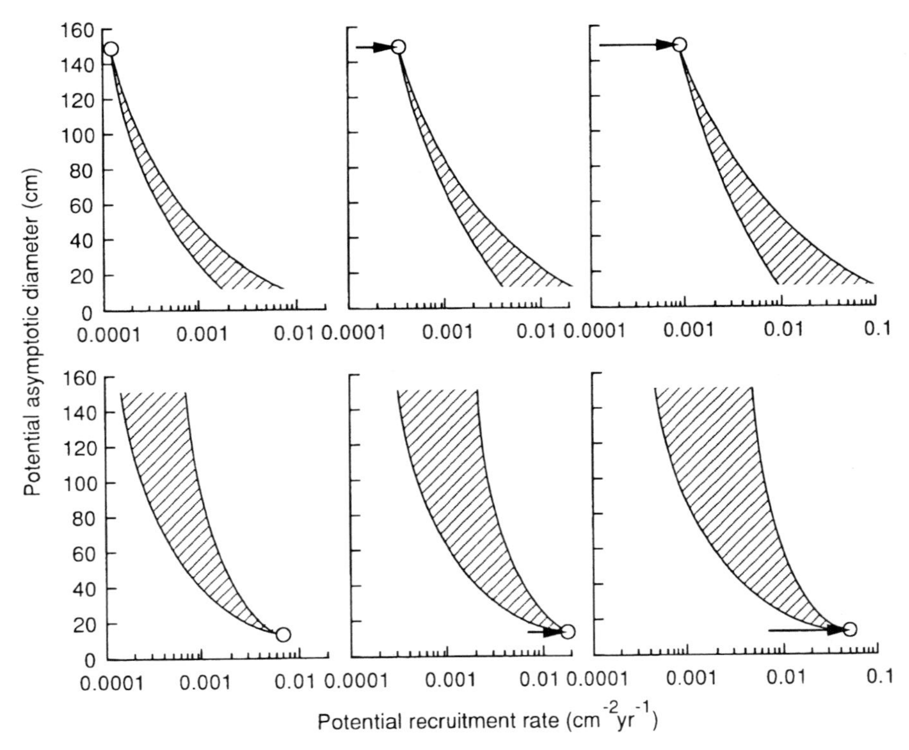

Fig. 9. Diagram of the equilibrium condition of coexistence of the size-structured multi-species model on the space of potential asymptotic size and *per-capita* potential recruitment rate. The two-species systems with one fixed at the open circle and another in the hatched domain coexist stably. From left to right, the three series of increasing recruitment rate for fixed species, shown by arrows, are presented for both fixed canopy species (upper) and fixed understory species (lower). The model and other parameters are the same as in Table 2 of Kohyama (1993)

Appendix

Data examined here are from these permanent plots: (1) lowland mixed dipterocarp forest on Gunung Berui, Serimbu, West Kalimantan (2 plots) by T. Kohyama et al., (2) montane forests in Gunung Halimun, West Java (2 plots) cf. Suzuki et al. (1997), (3) foothill to upper montane forests on Mount Kinabalu (8 plots) cf. Aiba and Kitayama (1998), (4) warm-temperate rain forest on Mount Inao, Osumi Peninsula, Southern Japan (1 plot) by E. Suzuki and S. Wakiyama, cf. Wakiyama (1998), (5) warm-temperate rain forest in Yakushima Island, Southern Japan (2 plots in Koyohji Basin) cf. Aiba and Kohyama (1996), and (6) cool-temperate mixed forest at Yufutsu, Tomakomai (1 plot), Hokkaido by Seino et al. Almost all plots are one hectare, while some (Yakushima and higher elevation Mount Kinabalu) are smaller in size. Measurements were made at least two times at intervals of a couple of years for trees ≥5 cm in trunk diameter. Aboveground biomass was estimated from trunk-diameter distribution, diameter-height allometry, and diameter-height *versus* tree mass allometry [the method of Nagano (1978) for all tropical and warm-temperate rain forests because of its robustness (cf. Aiba and Kitayama 1998); while for cool-temperate forest of Tomakomai, the method of Takahashi et al. (1998) was employed for deciduous broad-leaves, and the Research group on forest productivity of the four universities (1960) method for conifers]. The turnover rate of aboveground biomass used here was that of surviving trees (which should be balanced by the biomass loss by tree death, ignoring the small contribution of biomass increment by new recruits). This turnover rate is a component of aboveground net primary production rate together with non-mortal loss of leaves, branches and reproductive organs.

Acknowledgments

We thank Peter Ashton, Fakhri Bazzaz, Matthew Potts, Eric Macklin, Shin-Ichi Yamamoto, Masahiko Ohsawa, Masahiko Higashi, Yoh Iwasa, Masahiro Kato, and Kanehiro Kitayama for comments and suggestion at various stages of this study. The Bullard Fellowship of Harvard Forest to TK facilitated the analysis in this paper. This paper is a contribution to IGBP-GCTE-TEMA.

References

Adams IM, Woodward FI (1989) Patterns in tree species richness as a test of the glacial extinction hypothesis. Nature 339:699-701

Aiba S, Kitayama K (1999) Structure, composition and species diversity in an altitude-substrate matrix of rain forest tree communities on Mount Kinabalu, Borneo. Plant Ecol 140:139-157

Aiba S, Kohyama T (1996) Tree species stratification in relation to allometry and demography in a warm-temperate rain forest. J Ecol 84:207-218

Currie DJ (1991) Energy and large-scale patterns of animal- and plant-species richness. Am Nat 137:27-49

Kira T, Shidei T (Eds) (1977) Primary productivity of Japanese forests (JIBP Synthesis 16). Univ Tokyo Press, Tokyo

Kohyama T (1993) Size-structured tree populations in gap-dynamic forest – the forest architecture hypothesis for the stable coexistence of species. J Ecol 81:131-143

Kohyama T (1996) The role of architecture in enhancing plant species diversity. In: Abe T, Levin SA, Higashi M (Eds) Biodiversity: an ecological perspective. Springer-Verlag, New York, pp 21-33

Kohyama T, Takada T (1998) Recruitment rates in forest plots: gf estimates using growth rates and size distributions. J Ecol 86:633-639

Lieth H, Whittaker RH (Eds) (1975) Primary productivity of the biosphere. Springer-Verlag, New York

Nagano M (1978) Dynamics of stand development. In: Kira T, Ono Y, Hosokawa T (Eds) Biological production in a warm-temperate evergreen oak forest of Japan (JIBP Synthesis 18). Univ Tokyo Press, Tokyo, pp 21-32

Ogawa H, Kira T (1977) Methods of estimating forest biomass. In: Kira T, Shidei T (Eds) Primary productivity in forest ecosystems (JIBP Synthesis 16). Univ Tokyo Press, Tokyo, pp15-25

Ohsawa M (1995) Latitudinal comparison of altitudinal changes in forest structure, leaf-type, and species richness in humid monsoon Asia. Vegetatio 121:3-10

Research group on forest productivity of the four universities (1960) Studies on the productivity of the forest part I. Essential needle-leaved forests of Hokkaido. Kokusaku Pulp Ind Co, Tokyo

Suzuki E, Yoneda M, Simbolon H, Zainal F, Nishimura T, Kimura M (1997) Monitoring of vegetational change in permanent plots on Gn. Halimun National Park. Report of Cooperative Research of JICA and LIPI

Takahashi K, Yoshida K, Suzuki M, Seino T, Tani T, Tashiro N, Ishii T, Sugata S, Fujito E, Naniwa A, Kudo G, Hiura T, Kohyama T (1998) Stand biomass, net production and canopy structure in a secondary deciduous broad-leaved forest, northern Japan. Res Bull Hokkaido Univ Forests 55 (in press)

Teramoto, E (1996) Dynamical structure of energy trophic levels. Ecol Model 69:135-147

Wakiyama S (1998) Structure and dynamics of warm-temperate rain forests on Mount Inao. M Sc Thesis, Kagoshima Univ, Kagoshima

Part 3
Development and Evolution

13 Developmental Genetics and the Diversity of Animal Form: Hox Genes in Arthropods

MICHAEL AKAM

University Museum of Zoology, Department of Zoology, Downing Street, Cambridge, CB2 3EJ, England

Abstract

The wonderful diversity of animal forms is built upon a largely conserved set of genes and developmental mechanisms. The Hox genes provide one of the best studied examples of how such a conserved mechanism underlies a diversity of forms.

The Hox genes encode proteins that serve as molecular labels for the position of cells along the major body axis. The differential expression of these genes causes cells that would otherwise be equivalent to adopt different developmental fates in different regions of the body. This leads to the differentiation of segments in arthropods, and of regions along the axial skeleton of vertebrates. An analogous role is probably conserved in most other triploblastic bilaterian phyla.

The family of Hox genes diversified at an early stage in the radiation of the metazoa. The set of Hox genes present in each phylum provides phylogenetic characters that are useful for establishing the relationships between phyla. Characteristics of the 5' (*Abd-B* related) genes support the grouping of the metazoa into the same major lineages as the recent rRNA-based phylogeny proposed by Aguinaldo and colleagues.

The patterns of expression of the Hox genes provide markers to relate the body plans of distantly related animals. A good example is the comparison of head segmentation in chelicerate and mandibulate arthropods. Patterns of Hox gene expression refute a model of segment loss in chelicerates, and suggest that these very different arthropods retain a common set of uniquely defined head segments derived from their last common ancestor. No one gene can be taken to define "homology", but in cases such as this where comparisons of several genes provide a consis-

---

Key words. homeotic gene, arthropod, development, evolution, segmentation, tagmosis, Hox gene, chelicerate, *Drosophila*, protostome, phylogeny, crustacean, body plan, head, trunk

tent pattern, they may provide powerful discrimination between alternative hypotheses.

Comparisons of Hox gene expression in the arthropod trunk suggest a more flexible relationship between the processes of segment formation and Hox gene regulation. Changes in this relationship have led to changes in the pattern of tagmosis, which underlie the functional specialisation and evolutionary radiation of the arthropods.

1 Introduction

Diversity of form is one of the most striking characteristics of life on earth. This diversity depends on genetic instructions, acting through the mechanisms of development. Until recently we knew little of these mechanisms and even less about how changes in genes could generate diversity of form. Different phyla were considered to be so fundamentally different that comparisons between, for example, insects and vertebrates were thought unlikely to reveal many similarities in the molecular mechanisms of development.

Research in developmental genetics and genomics has radically changed that view. For a few model organisms it is now possible to outline how genes define the body plan. Moreover, it is clear that many, and perhaps most of the developmental mechanisms in metazoa depend on a "tool kit" of genes that arose before the major phyla of animals diversified. These conserved tools are used, sometimes in remarkably similar ways, to build the diversity of modern organisms.

Paradoxically, it is only because we can recognise these molecular similarities that we can now begin to study the mechanisms that generate the diversity of animal forms. We can use what we know from well studied model systems to inform comparative studies of other animals. In what follows, I explore this approach in the context of studies on the Hox genes of arthropods.

The Hox genes provided one of the first and still one of the most dramatic examples of universality in developmental mechanisms. Below, I introduce their role in development, largely as we understand it from studies in *Drosophila*. I then consider the antiquity and diversity of this gene family and, as a slight digression, mention some recent results using Hox gene sequences that address the relationships of the metazoan phyla. Finally, I return to the question of how studies of the Hox genes help us to understand the diversity of body plans among arthropods.

2 The Hox Genes

The Hox genes were first identified in *Drosophila* over 80 years ago, by a class of striking mutations that transform one part of the body into another (Bridges and Morgan 1923). Sporadic individuals showing similar phenotypes had previously been documented in a wide range of species by William Bateson, who coined the term "homeotic" for such transformations (Bateson 1894).

Drosophila contains two principal clusters of genes that give rise to such homeotic mutant phenotypes (Mahaffey and Kaufman 1987; Lawrence and Morata 1994). These genes derive from a single ancestral gene cluster that was split by a chromosome rearrangement relatively recently, during the radiation of the Drosophilids (Von Allmen et al. 1996). The homologous genes of other insects still comprise a single tightly linked cluster (Denell et al. 1996). Mutations in genes at one end of this cluster affect the head of the animal, while genes at the other end are needed for the normal development of the most posterior segments of the abdomen. The plan of the body from front to back is reflected in the sequence of the genes on the chromosome.

What these genes do became clear from genetic studies pioneered by E. B. Lewis at the California Institute of Technology, for which he was awarded the Nobel prize in 1995 (Lewis 1995). [Interestingly, some of Lewis' insights were informed by earlier studies, in Japan, of the homologous "E complex" genes in the silk moth during the 1940s and 50s, (Tazima 1964)]. Lewis showed that when the Hox genes were mutated, the normal diversity of segments was lost (Lewis 1963, 1978). When all genes in one half of the cluster were lost, all segments in the posterior thorax and abdomen of the fly developed like the wing bearing segment. It has since been shown in other insects that when the entire set of Hox genes are lost, each segment develops structures characteristic of the normal antennal segment (Beeman et al. 1993). Thus the Hox genes are required, not primarily for making segments, but for specifying the different pathways of development that each segment will follow.

We now know that the products of these genes are transcription factors — nuclear proteins that turn other genes on and off (Affolter et al. 1990). All of the genes in the Hox cluster encode related transcription factors with structurally similar, but biologically distinct, DNA binding domains called homeodomains. (Homeodomains have since been found in many other proteins, in addition to the Hox genes). Each of the Hox proteins is expressed in a restricted subset of segments, so dividing the body of the fly into a series of domains with different molecular labels (Akam 1987).

There are not enough Hox genes to give each segment a different label, though it is clear that in *Drosophila* each segment does have a unique morphology. There are several explanations for this (Lawrence and Morata 1994). Genes other than the Hox family provided a component of the address, particularly in the most anterior and posterior parts of the animal; genes are used in combination in some segments, and a single Hox gene can be expressed in different spatial and temporal patterns in different segments, affecting the type of segment that develops (Castelli-Gair and Akam 1995).

Genes of the Hox cluster are not unique to insects. Almost as soon as they were first cloned from *Drosophila*, homologous sequences were identified in vertebrates. These proved to encode closely related transcription factors of the same Hox family (McGinnis et al. 1984; Gehring 1987; McGinnis and Krumlauf 1992). Not only are the sequences conserved, but also three key characteristics of these genes. First, the Hox genes are clustered in vertebrates, as they are in insects. Second, they are expressed in restricted domains along the head/tail axis of the body. Third, when the function of these genes is disrupted, homeotic transformations are produced analogous to those seen in flies. For example, when the Hox gene A9 of the mouse is "knocked out", the most anterior of the lumbar vertebrae transforms into a thoracic vertebra (Fromental-Ramain et al. 1996).

These conserved characteristics suggest that the last common ancestor of insects and vertebrates possessed clustered Hox genes that were differentially expressed along the antero-posterior axis of the body, and were used to control the specific differentiation of different regions of the body, which may or may not already have been segmented (Akam 1989). Flies and humans use the same machinery to build their head-tail axis!

3 The Origins and Evolution of Hox Genes

Vertebrates and insects are representatives of two ancient lineages within the metazoa — the deuterostomes and the protostomes respectively. These lineages split near the base of the radiation of the metazoan phyla. How old then are the Hox gene clusters? The answer seems to be that the Hox clusters are a shared derived characteristic of the metazoa. They are by no means the only such characteristic, and here I disagree with the suggestion that these genes alone, as the Zootype, can be used to define the metazoa (Slack et al. 1993).

Hox genes belong to a larger family of transcription factors, the homeobox genes, which all encode transcription factors containing homeodomains. This protein domain is older than the metazoa — homeobox genes are known from plants and fungi as well as animals. Interestingly, in yeast and several other fungi the homeobox genes are specifically involved in controlling the choice between different cell fates, e.g. the mating types. The same seems to be true for the many classes of homeobox genes in animals. More than twenty classes of homeobox genes are recognised in animals, deriving from perhaps 4 or 5 prototypes that existed in the last common ancestor with plants and fungi (Bürglin 1995). This homeobox gene family appears to have radiated particularly dramatically in the animal kingdom, some time after the split of the first animals from the plants and fungi, but before the separation of the protostomes from the deuterostomes.

The precise sequence of gene duplications, and gene cluster duplications, that gave rise to the Hox gene family is not yet clear. At least one duplication of a proto-Hox cluster occurred, to give rise to Hox and para-Hox genes (Brooke et al. 1998). though whether this happened before or after the divergence of the cniadarians from the triploblasts is uncertain.

We can now say with some certainty that a cluster of at least 7 distinct Hox genes predated the radiation of the triploblast phyla. Hox genes have been sampled in nemerteans (non-segmented worms), annelids (segmented worms) and molluscs among the traditional protostomes; echinoderms, urochordates, cephalochordates and chordates among the deuterostomes. They have also been sampled in platyhelminths and nematodes among the supposed acoelomate phyla, and most recently, in brachiopods and priapulids. All of these phyla, with the exception of the nematodes, contain sets of Hox genes with multiple distinct gene classes, clearly more closely related to orthologous genes in other phyla than to other Hox genes in the same organism (Arenas-Mena et al. 1998; Snow and Buss 1994; Digregorio et al. 1995; Brooke et al. 1998; Kmita-Cunisse et al. 1998; de Rosa et al. 1999). These include orthologues for the five most "anterior" classes of Hox gene, together with at least one representative of the central genes, and one of the posterior (*Abd-B* related) genes.

These conserved similarities suggest that the Hox proteins diversified extensively in the stem lineage of the metazoa, but then their sequences became highly constrained, at least in the most critical parts of the molecule in and around the DNA binding domain, so that the unique characteristics of each subclass remain distinct in the multiple metazoan phyla.

Recent data emerging from a collaboration between the several labs cloning Hox genes suggests that, for some of the Hox gene classes, the process of sequence diversification continued after the basal radiation of the metazoan phyla. However, subsequent to this early radiation, the sequences have been conserved sufficiently well to retain a surprisingly strong phylogenetic signal linking groups of phyla (de Rosa et al. 1999). The posterior group (*Abd-B* related) Hox genes exhibit this pattern. They are the most variable between phyla, both in number and sequence. Deuterostomes have multiple, and quite diverse posterior genes, but within the protostomes, the sequences of these posterior genes fall into three clear classes. The genes of arthropods, priapulids and nematodes comprise one class, while annelids, molluscs and brachiopods share two classes of posterior gene, both present in each phylum. This distribution of genes provides support for the subdivision of the protostomes into two major lineages, the Ecdysozoa and the Lophotrochozoa, as proposed by Aguinaldo and colleagues (Aguinaldo et al. 1997). on the basis of ribosomal RNA sequences.

4 Hox Genes and Body Plans

The products of the Hox genes serve as molecular labels during development to define specific regions of the body. This suggests that patterns of Hox gene expression may be of use in comparisons between species to define underlying similarities in body plan that are obscured by morphological divergence. We must expect, of course, that patterns of gene expression will display both conserved ancestral features and novel, derived characteristics. In this regard, genes are not necessarily "better" than morphological characters (Abouheif et al. 1997). However, they do provide a new and rich data set, which is partially independent of morphology, and which may illuminate both underlying patterns of homology, and the processes that have led to the diversification of body plans.

I illustrate this below with reference to the diversity of arthropods. These studies are in their infancy, with few taxa sampled, and incomplete data available for even those few taxa. However, my preliminary conclusion from these studies is that patterning of the arthropod head and trunk have been evolving in rather different ways. In the head, the expression domains of Hox genes appear to have become fixed in one particular relation to segment formation before the major surviving clades of arthropods diverged, whereas in the trunk, changes in the relationship between segment formation and Hox gene expression have played, and probably continue to play, a major role in the diversification of arthropods.

4.1 The Arthropod Head

Segmentation of the head is one of the key characters used to define the four major subphyla of the arthropods (Brusca and Brusca 1990). Crustacea are defined by a head containing six segments — the eye bearing segment innervated from the first ganglion of the brain, followed by two pairs of segments bearing antennae, and three pairs of segments making up the jaws — a biting mandible, and two pairs of maxillae. Insects look superficially very different. They have only one pair of antennae, and the mouthparts may be highly modified. It has been one of the triumphs of comparative morphology to recognise that the diversity of insect mouthparts may be related to a common plan, and that this common plan is recognisably similar to that of Crustacea (Snodgrass 1931; Rogers and Kaufman 1997). The primary difference is that the appendage of the second antennal segment has been lost, though the segment which should carry it, the so-called intercalary segment, is clearly visible in the embryo. Both insects and crustaceans have a biting mandible as the first mouthpart segment, and so these arthropods are termed mandibulates.

Myriapods — millipedes, centipedes and their kin, have traditionally been viewed as closely allied to the insects within the mandibulate arthropods, but the phylogenetic position of the myriapods is now in doubt (Averof and Akam 1995b; Freidrich and Tautz 1995), and the similarity of their mouthparts to those in insects is questionable (Popadic et al. 1996). There are no data yet available on the expression of Hox genes in the head of myriapods, so I will leave them out of this discussion.

The organisation of anterior segments in chelicerates is indisputably very different from that of the mandibulates. These animals (spiders, scorpions, mites and horseshoe crabs) have no antennae, and only two pairs of specialised segments behind the eye — the chelicerae and the pedipalps (Brusca and Brusca 1990). The pedipalps are followed immediately by the walking legs, which are born on the same body region as the mouthparts, termed the prosoma.

The split between chelicerates and mandibulates has traditionally been seen as one of the most basal among living arthropods, and this is supported by molecular phylogenies, problems with the myriapods not withstanding. Therefore, if we are to have some idea of the ancestral morphology of the arthropods, and the evolution of appendage specialisation in each group, we need to be able to say how the structures of the mandibulate and chelicerate heads relate to one another. Unfortunately, arthropod biologists have not been able to agree about this for the last 100 years. There are two primary competing hypotheses. One simply counts segments backwards from the eye segment, homologising chelicerae with the insect antenna (crustacean first antenna), and so on. The other assumes that an ancestral antennal segment has been "deleted" in the chelicerate lineage, and equates the chelicerae with the second antennal segment of Crustacea (insect intercalary segment), and therefore the pedipalps with the mandible. The principle argument in favour of this second model relates to the innervation of the segments, and the position of segmental ganglia in relation to the oesophagus (Weygoldt 1985). In Mandibulates the mandibular ganglion is the first post-oral ganglion; in chelicerates it is the pedipalp ganglion.

There is a third possibility, not generally made explicit. This is that any attempt to homologise segments is meaningless, because at the time when these two lineages diverged, the post-antennal segments were not individuated, or differentiated from one another, and therefore deletion of a segment would have no meaning.

Two recent papers describing Hox gene expression patterns in chelicerates (Damen et al. 1998; Telford and Thomas 1998) go a long way towards proving that the question is meaningful, and that the first of the above hypotheses is correct.

We have seen that the individual anterior Hox genes themselves are older than the protostome/deuterostome split, and therefore more ancient than any conceivable divergence of chelicerates and mandibulates. We also know that in insects these genes serve as markers for specific head segments, and define at least some of the differences between them. Characteristically, the insect antennal segment expresses no Hox gene, and may be regarded as a ground state for a trunk segment, at least with respect to Hox gene function. The intercalary segment expresses the most anterior of the Hox genes, *labial*, and one segment behind this, the *Deformed* gene is active in the mandible (Rogers and Kaufman 1997). An anterior limit to the trunk is defined by a homeobox gene of a different class, *Orthodenticle*, which defines fore-brain territory, and specifically the ocular segment.

The Hox expression data relate to two different chelicerate groups, an Oribatid mite, *Archegozites*, and a spider, *Cupiennius*. As we would expect from earlier work with insects, the Chelicerate Hox genes are expressed in specific sets of segments with defined anterior boundaries. Strikingly, insects and chelicerates show a corresponding set of Hox gene boundaries, if it is assumed that no segment is missing (Fig. 1). The anterior limit of Hox gene expression is the pedipalp segment in chelicerates, suggesting that this is the homologue of the intercalary segment of insects; The *Deformed* genes (class 4) are expressed from the first leg bearing segment, and genes of class 5 (*Sex combs reduced*) from the third leg segment. There are differences between *Drosophila* and these chelicerates, particularly in the domains of the class 2 and 3 genes, but there are good reasons to believe that these genes have changed within the insects, and that the patterns in the chelicerates are ancestral. Other conserved details make it unlikely that these two corresponding patterns have been acquired independently in the two groups. For example, genes of class 1 and 4 appear to be expressed in register with the segment boundaries, whereas genes of class 5 respect anterior limits that lie out of register with seg-

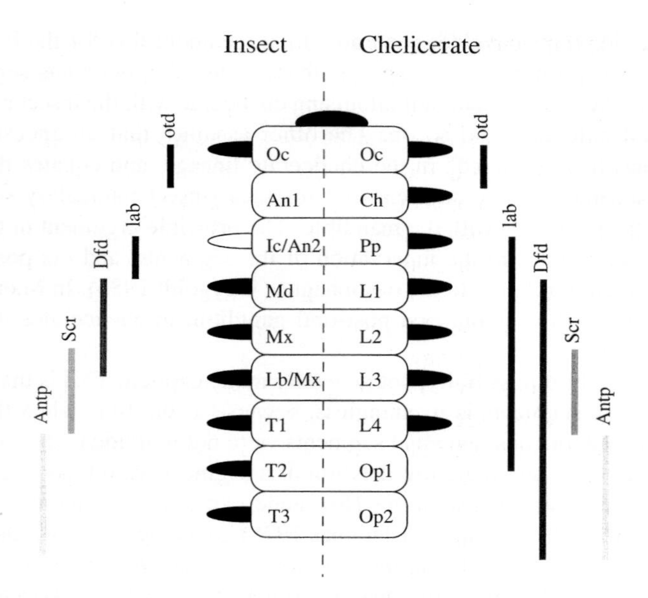

Fig. 1. Homologies between anterior segments of mandibulate and chelicerate arthropods, as suggested by comparisons of Hox gene expression domains in insects and arachnids. (Modified from Telford and Thomas, 1998, with additional data from Damen et al, 1998). Abbreviations — Segments: Oc, ocellar; An1, antennal (first antennal in crustaceans); Ic, intercalary (second antennal in crustaceans); Md, mandibular; Mx, maxillary; Lb, labial; T1,2,3, thoracic; Ch, cheliceral; Pp, pedipalpal; L1, 2, 3,4, Leg bearing segments in chelicerates; Op1,2, opisthosomal. Gene expression domains: otd, *orthodenticle*; lab, *labial/* Hox class 1; Dfd, *Deformed*/Hox class 4; Scr, *Sex combs reduced*/Hox class 5; Antp, *Antennapedia*. In chelicerates, genes of Hox class 2 and 3 are also expressed from an anterior limit in the cheliceral segment.

ments and reflect an early embryonic patterning boundary, the parasegment border (Damen et al. 1998).

These results strongly suggest that the heads of all these arthropods comprise a conserved array of uniquely defined segments, and that the patterning mechanisms which specify this array have been inherited, largely unchanged, from an ancestor that predates the radiation of the living arthropod groups. Hardly a vestige of this common inheritance is left in the functional morphology. Antennae are not obviously similar to chelicerae; mandibles and walking legs are as different as any two arthropod appendages. The meaning of the genetic address labels provided by these Hox genes has changed beyond recognition, but the underlying pattern of segment specification is conserved. Only in the eyes do we see an obvious conserved relation between the specification by a common gene (*Orthodenticle*) and the downstream fate of the segment — a fate that itself depends on other conserved genes [e.g., *Pax 6*, (Halder et al. 1995)].

4.2 The Arthropod Trunk

The situation in the arthropod trunk must be different. Segment number is not conserved between all arthropods, and even within closely related groups, segment numbers vary within well defined homologous regions of the body [e.g., in branchiopod crustaceans, and in geophilomorph centipedes (Schram 1986; Minelli and Bortoletto 1988)]. This alone suggests that there must be some flexibility in the relationship between the mechanisms of segment formation and those that regulate Hox genes, leading to segment identity. Here it makes no sense simply to count segments back from the head to establish "homologies". There need be no such thing as a one-to-one homology between segments in two different species. A more relevant comparison may be to ask how domains of Hox gene expression relate, not to individual segments, but to the "regiments" of similar segments long recognised as defining the pattern of tagmosis in the arthropod body. Our expectation is that boundaries of Hox gene expression will define positions along the body axis where segment morphologies change.

A clear correlation between tagmosis and Hox gene expression was revealed by our studies of Hox gene expression in the Branchiopod crustacean *Artemia* (Averof and Akam 1995a). *Artemia* is a crustacean with a relatively simple body plan, with clearly defined trunk tagmosis, and little segment variation within tagma. All the 11 segments of the thorax are of similar form, differing only in size. These uniform segments all express the same set of Hox genes, albeit at different levels. The distinct genital segments express a Hox gene of the more posterior *Abd-B* class, and the so-called abdomen of *Artemia*, lying posterior to the genitalia, expresses none of the Hox genes homologous to those in insects.

If we interpret these gene expression patterns as marking homology between body regions, we would conclude that the branchiopod thorax is homologous to the whole of the insect trunk anterior to the genital segments, while the branchiopod abdomen is either unrepresented in insects, or is represented only by the post-genital tenth abdominal segment (a segment which never bears appendages in insects). This was our interpretation when we published these data, and it remains my favoured hypothesis today. However, we will only know whether or not this interpretation makes sense when we have data on more genes, and have sampled more species.

Damen et al. (1998) have described Hox gene expression in the more posterior parts of the chelicerate body. In their spider, the transition from prosoma to opisthosoma ("abdomen") is marked by activation of the middle group Hox genes (Antennapedia, two distinct Ultrabithorax-like genes, and abdominal-A). In this regard Hox gene expression reflects tagmosis. Within the opisthosoma, each of these Hox genes is activated at a different segmental boundary. These authors suggest that this series of boundaries can be homologised with the boundaries observed in insects, as suggested above for the head segments, and so used to establish a one-to-one correspondence between insect and chelicerate segments throughout much of the trunk.

I find it hard to reconcile this interpretation with observations in crustacea and other arthropods (see below), which show no such precise correspondence in expression domains, and more importantly, show that even between quite closely related arthropods, the boundaries of Hox gene expression vary. I am more inclined to interpret the Chelicerate data as showing some degree of convergent similarity. It is important to remember that the anterior segments of the opisthosoma are highly differentiated in chelicerates, containing the male and female genitalia (Brusca and Brusca 1990). It is therefore hardly surprising that they are distinguished by distinct and precisely defined domains of Hox gene expression. Distinguishing between these interpretations will again require sampling a much wider range of taxa (and, I fear, resolving the phylogeny of the major arthropod groups).

The only relevant data so far available have been obtained by using a cross-reacting antibody (Kelsh et al. 1994) that detects one subclass of Hox proteins — the *Ubx* and *abd-A* gene products, in all tested arthropods, and in some closely related phyla. The availability of this antibody makes it possible to survey a range of species without having to clone genes from each. Using this antibody, Grenier and colleagues (1997) have shown that the Ubx/abd-A class Hox proteins are expressed throughout the trunk of centipedes, except in the first specialised poison claw segment. Our own work with a different species of centipede (*Lithobius forficatus*) suggests that these genes are initially expressed in all trunk segments, including the poison claw segment, and only later in embryogenesis become restricted to the leg bearing segments (Smith 1998). At first sight this pattern is directly comparable with that of the same genes in *Artemia*, defining a domain of the body lying immediately behind the head. However, the chelicerate data described above, and a very different pattern observed for these same genes in Onychophora (Grenier et al. 1997), suggests that it would be premature to posit this as an ancestral state for all arthropods.

Onychophora are probably the closest living relatives of the arthropods proper, and are generally assumed to share a common segmented ancestor. Behind the head, they have a uniform trunk with many similar lobopod bearing segments, which might therefore be expected to show a pattern of Hox gene expression ancestral to an unspecialised trunk segment. However, the cross-reacting Ubx/abd-A antibody stains only the last appendage-bearing segment and the terminal region of the Onychophoran trunk — a domain that does not obviously reflect any known segment specialisation. (In this case the relevant genes have been cloned, confirming that the epitope recognised by the antibody is conserved in the expected Hox gene products.) Unfortunately, we do not yet know which Hox genes are expressed in the more anterior segments.

On the basis of these observations, Grenier and colleagues suggest that "expression domains of Ubx/abd-A do not demarcate homologous body regions in these taxa [onychophorans, chelicerates, crustaceans], but instead, evolutionary changes in Ubx/abd-A deployment underlie the diversification of arthropod body plans that are evident in the Cambrian fossil record". This is perhaps correct, but too negative as an overall assessment of the significance of Hox expression domains. Demarcation of body regions within the arthropod trunk probably occurred independently within the most basal lineages, and for these lineages, we do not expect to be able to establish homologies between body regions. However, once regional specialisations were established, and linked to the expression of particular Hox (or other) genes, we can define homologies and ask specific questions concerning the relation of one body plan to another — for example, have the genes that define the post-genital abdomen of the branchiopod crustaceans lost any such role in insects and higher crustaceans, has their domain of function been collapsed to a single segment, or have they acquired new and unrecognisable downstream meanings, as seems to have occurred in the head.

4.3 Shifting Boundaries and Segment Specialisation Within Tagma

Comparisons between such distantly related organisms as insects, chelicerates and onychophorans provide little insight into the sequence of evolutionary intermediates that lead to these different body plans, or the functional consequences of those changes. For this purpose, comparisons between more closely related species are much more informative. Averof and Patel (1997) have used the cross-reacting Ubx/abd-A antibody in this way, to show that the diversification of segments within the crustacean thorax (pleon) has been accompanied by changes in the boundaries of expression of Hox genes. They have examined Crustacea in which one or more of the trunk segments have been recruited as auxiliary mouthparts, or maxillipeds.

The possession of a uniform array of thoracic segments is presumed to be a basal characteristic of the crustaceans (Schram 1986). This pattern, seen in *Artemia* and *Triops* among the branchiopod crustaceans, is also seen in some maxillopodans, and in the basal branches of the malacostracans (e.g. *Nebalia*). However, maxillipeds have arisen repeatedly in many groups, providing an excellent test case to correlate changes in gene expression with the evolution of new morphology. In the case of the malacostracans, this change in segment patterning has occurred against a background of conserved segment numbers in the thoracic tagma, making it particularly clear that the process involves a transformation in segment specialisation, and not the intercalation of extra segments.

Averof and Patel found that the diverse species lacking maxillipeds express the Ubx/abd-A class proteins throughout the thorax, with a boundary at the first thoracic segment. Thus for crustaceans, it is reasonable to assume that this is the ancestral pattern. Species with maxillipeds show a different pattern — with the expression boundary for these Hox proteins being shifted more posteriorly. The extent of the shift correlates with the number of maxillipeds that will form (though this correlation, established in the embryo, relates to the form of the animal at hatching, and not directly to the adult morphology). Thus in this case we can infer that repeated shifts in Hox expression boundaries have occurred concomitant with the evolution of increasingly complex body plans.

5 Conclusions

At present the Hox genes provide one of the best examples where patterns of gene expression can be related to the diversity of form. I have considered only studies on arthropods, but analogous studies on vertebrates are already providing an understanding of form and diversity among the vertebrates (e.g. Burke et al. 1995; Cohn et al. 1997). Relating these molecular studies to the fossil record promises a fascinating synthesis.

The Hox genes are not unique. They are but one among many groups of genes that can provide insight into the origins of diversity. Other families of genes are involved in generating the diversity of tissue types and, within any one tissue, the diversity of behaviour and function exhibited by specialised cells among, for example, the nerves and muscles. Developmental genetics, coupled with comparative genomics, will provide a new and richer understanding of this diversity, as it arises both in ontogeny and phylogeny.

Acknowledgements

Work in the author's laboratory was supported by the Wellcome Trust and the Biotechnology and Biological Science Research Council of the United Kingdom.

References

Abouheif E, Akam M, Dickinson WJ, Holland PWH, Meyer A, Patel NH, Raff RA, Roth VL, Wray GA (1997) Homology and developmental genes. Trends Genet 13:432-433

Affolter M, Schier A, Gehring WJ (1990) Homeodomain proteins and the regulation of gene expression. Current Opinion Cell Biol 2:485-495

Aguinaldo AMA, Turbeville JM, Linford LS, Rivera MC, Garey JR, Raff RA, Lake JA (1997) Evidence for a clade of nematodes arthropods and other moulting animals. Nature 387:489-493

Akam M (1987) The molecular basis for metameric pattern in the *Drosophila* embryo. Development 101:1-22

Akam M (1989) *Hox* and HOM: Homologous gene clusters in insects and vertebrates. Cell 57:347-349

Arenas-Mena C, Martinez C, Cameron RA, Davidson EH (1998) Expression of the Hox gene complex in the indirect development of a sea urchin. Proc Natl Acad Sci USA 95:13062-13067

Averof M, Akam M (1995a) Hox genes and the diversification of insect and crustacean body plans. Nature 376:420-423

Averof M, Akam M (1995b) Insect-crustacean relationships: insights from comparative developmental and molecular studies. Phil Trans Roy Soc 347:293-303

Averof M, Patel NH (1997) Crustacean appendage evolution associated with changes in Hox gene expression. Nature 388:682-686

Bateson W (1894) Materials for the study of variation. MacMillan & Co, London

Beeman RW, Stuart JJ, Brown SJ, Denell RE (1993) Structure and function of the homeotic gene complex (HOM-C) in the beetle *Trilobium castaneum.* BioEssays 15:439-444

Bridges CB, Morgan TH (1923) Bithorax In The third chromosome group of mutant characters of *Drosophila melanogaster.* Carnegie Inst Wash Publ, pp 137-146

Brooke NM, Garcia-Fernandez J, Holland PWH (1998) The ParaHox gene cluster is an evolutionary sister of the Hox gene cluster. Nature 392:920-922

Brusca RC, Brusca GJ (1990) Invertebrates. Sinauer, Massachusetts

Bürglin TR (1995) The evolution of homeobox genes In: Arai R, Kato M, Doi Y (Eds) Biodiversity and evolution. The National Science Museum Foundation, Tokyo, pp 289-334

Burke AC, Nelson CE, Morgan BA, Tabin C (1995) Hox genes and the evolution of vertebrate axial morphology. Development 121:333-346

Castelli-Gair J, Akam M (1995) How the Hox gene *Ultrabithorax* specifies two different segments: the significance of spatial and temporal regulation within metameres. Development 121:2973-2982

Cohn MJ, Patel K, Krumlauf R, Wilkinson DG, Clarke JDW, Tickle C (1997) *Hox9* genes and vertebrate limb specification. Nature 387:97-101

Damen WGM, Hausdorf M, Seyfarth EA, Tautz D (1998) A conserved mode of head segmentation in arthropods revealed by the expression pattern of Hox genes in a spider. Proc Natl Acad Sci USA 95:10665-10670

Denell RE, Brown SJ, Beeman RW (1996) Evolution of the organization and function of insect homeotic complexes. Seminars in Cell & Developmental Biology 7:527-538

De Rosa R, Grenier J, Andreeva T, Cook C, Adoutte A, Akam M, Carroll S, Balavoine G (1999) Hox genes in brachiopoods and priapulids and protostome evolution. Nature 399:772-775

Digregorio AD, Spagnuolo A, Ristoratore F, Pischetola M, Aniello F, Brauno M, Cariello L, Di Lauro R (1995) Cloning of ascidian homeobox genes provides evidence for a primordial chordate cluster. Gene 156:253-257

Freidrich M, Tautz D (1995) rDNA phylogeny of the major extant arthropod classes and the evolution of myriapods. Nature 376:165-167

Fromental-Ramain C, Warot X, Lakkaraju S, Favier B, Haack H, Birling C, Dierich A, Dolle P, Chambon P (1996) Specific and redundant functions of the paralogous *Hoxa-9* and *Hoxd-9* genes in forelimb and axial skeleton patterning. Development 122:461-472

Gehring WJ (1987) Homeo boxes in the study of development. Science 236:1245-52

Grenier JK, Garber TL, Warren R, Whitington PM, Carroll S (1997) Evolution of the entire arthropod *Hox* gene set predated the origin and radiation of the onychophoran/arthropod clade. Curr Biol 7:547-553

Halder G, Callaerts P, Gehring WJ (1995) New perspectives on eye evolution. Current Opinion Genes Dev 5:602-609

Kelsh R, Weinzierl ROJ, White RAH, Akam M (1994) Homeotic gene expression in the Locust *Schistocerca*: an antibody that detects conserved epitopes in Ultrabithorax and abdominal-A proteins. Develop Genet 15:19-31

Kmita-Cunisse M, Loosli F, Bierne J, Gehring WJ (1998) Homeobox genes of the ribbonworm *Lineus sanguineus*: Evolutionary implications. Proc Natl Acad Sci USA 95:3030-3035

Lawrence PA, Morata G (1994) Homeobox genes: their function in *Drosophila* segmentation and pattern formation. Cell 78:181-189

Lewis EB (1963) Genes and developmental pathways. Am Zool 3:33-56

Lewis EB (1978) A gene complex controlling segmentation in *Drosophila.* Nature 276:565-570

Lewis EB (1995) The Bithorax complex: the first fifty years. Les prix Nobel 233-260

Mahaffey JW, Kaufman TC (1987) The homeotic genes of the antennapedia complex and the bithorax complex of *Drosophila* In: Malacinski GM (Ed) Developmental genetics of higher organisms. MacMillan, New York, pp 329-359

McGinnis W, Garber RL, Wirz J, Kuroiwa A, Gehring WJ (1984) A homologous protein-coding sequence in *Drosophila* homoeotic genes and its conservation in other metazoans. Cell 37:403-408

McGinnis W, Krumlauf R (1992) Homeobox genes and axial patterning. Cell 68:283-302

Minelli A, Bortoletto S (1988) Myriapod metamerism and arthropod segmentation. Biol J Linn Soc 33:323-343

Popadic A, Rusch D, Peterson M, Rogers BT, Kaufman TC (1996) Origin of arthropod mandible. Nature 380:395

Rogers BT, Kaufman TC (1997) Structure of the insect head in ontogeny and phylogeny: a view from *Drosophila.* Int Rev Cytol 174:1-84

Schram F R (1986) Crustacea. University Press, Oxford

Slack JMW, Holland PWH, Graham CF (1993) The zootype and the phylotypic stage. Nature 361:490-492

Smith ML (1998) An analysis of Hox genes in myriapods. Ph D thesis. University of Cambridge

Snodgrass RE (1931) Evolution of the insect head and the organs of feeding. In Smithsonian Report, pp 443-489

Snow P, Buss LW (1994) HOM/Hox-Type Homeoboxes from *Stylaria lacustris* (Annelida: Oligochaeta). Mol Phyl Evol 3:360-364

Tazima Y (1964) The genetics of the silkworm. Logos Press, London

Telford MJ, Thomas RH (1998) Expression of homeobox genes shows chelicerate arthropods retain their deutocerebral segment. Proc Natl Acad Sci USA 95:10671-10675

Von Allmen G, Hogga I, Spierer A, Karch F, Bender W, Gyurkovics H, Lewis E (1996) Splits in fruitfly Hox gene complexes. Nature 380:116

Weygoldt P (1985) Ontogeny of the arachnid central nervous system. In: Barth FG (Ed) Neurobiology of arachnids. Springer-Verlag, Berlin, pp 20-37

14 Developmental Mechanisms Underlying the Origin and Evolution of Chordates

NORI SATOH, KOHJI HOTTA, GOUKI SATOH, SHUNSUKE TAGUCHI, HITOYOSHI YASUO, KUNI TAGAWA, HIROKI TAKAHASHI, AND YOSHITO HARADA

Department of Zoology, Graduate School of Science, Kyoto University, Sakyo-ku, Kyoto 606-8502, Japan

Abstract

Chordates evolved from a common ancestor shared by two other non-chordate deuterostome groups, hemichordates and echinoderms. The notochord is the most defining feature of chordates, and therefore the elucidation of molecular mechanisms underlying the formation of notochord will lead to a better understanding of developmental mechanisms that permitted and/or accelerated the origin and evolution of chordates. *Brachyury* encodes a transcription factor with the T DNA-binding domain and exerts a master control over the formation of notochord in ascidian embryos. However, *Brachyury* is conserved by non-chordate deuterostomes which do not have a notochord. We show here that at least the hemichordate *Brachyury* has the potential to induce the formation of notochord when the gene is ectopically expressed in chordate (ascidian) embryos. This strongly suggests that an acquisition of new function(s) of the pre-existing developmental gene is closely associated with the occurrence of a chordate-specific organ. Furthermore, we present data on the upstream control of ascidian *Brachyury* genes that is required for notochord-specific expression of the genes. In addition, we described an isolation of downstream genes, of which expression is regulated by the ascidian *Brachyury* to form the notochord.

Key words. chordates, deuterostomes, cephalochordates, urochordates, hemichordates, echinoderms, evolution, developmental mechanisms, notochord, *Brachyury*, expression patterns, transcriptional factor, T-domain, master control, functional conservation

1 Introduction

The evolutionary pathway from advanced invertebrates through primitive chordates to vertebrates has been a subject of extensive investigation and vigorous discussion for more than a century (Haeckel 1868; Garstang 1928; Berrill 1955; Jefferies 1986). The phylum Chordata consists of three subphyla, Urochordata, Cephalochordata and Vertebrata. Chordates exhibit a notochord, a dorsal hollow nerve cord, and pharyngeal gill slits, which are hallmarks of their body plan. In addition, echinoderms, hemichordates and chordates may share a common ancestor and form the monophyletic group of deuterostomes (Brusca and Brusca 1990; Willmer 1990; Wada and Satoh 1994; Nielsen 1995). Therefore, chordates originated from an ancestor more than 550 million years ago by organizing their characteristic features as mentioned above. Because the notochord is the most prominent feature of chordates, the elucidation of molecular developmental mechanisms underlying the formation of notochord will lead to a better understanding of mechanisms underlying the origin and evolution of chordates (Satoh and Jeffery 1995).

Among a dozen genes that are implicated in the formation of the chordamesoderm of vertebrate embryos, *Brachyury* is of particular interest, because this gene is involved in notochord differentiation (Herrmann and Kispert 1994). Since its identification in 1927, the mouse *Brachyury* (*T*) locus has been implicated in mesoderm formation and notochord differentiation (Chesley 1935). Homozygotic mutant mouse embryos die on day 11 of gestation with deficiencies in the posterior mesoderm formation in the primitive streak, an absent notochord and severe reduction of the allantois. In 1990, this gene was cloned (Herrmann et al. 1990). Transient expression of *Brachyury* is detected in nascent and migrating mesoderm generated from the primitive streak, then continuously in the notochord. At later stages, during axis elongation, *Brachyury* is expressed in the tail bud and in the notochord. The cloning of mouse *Brachyury* was followed by isolation and characterization of its homologues in *Xenopus* (*Xbra*; Smith et al. 1991), zebrafish (*Zf-T* or *no tail*; Schulte-Merker et al. 1992) and chick (*Ch-T* ; Kispert et al. 1995b). Vertebrate *Brachyury* genes are expressed in the embryonic area from which mesendoderm is generated; that is, in the primitive streak in the mouse embryo, the marginal zone of the frog embryo, or the germ ring in the fish embryo, and later, in the notochord and in the tail bud during axis elongation. These domains of *Brachyury* expression reflect the dual function of the gene during vertebrate development, namely in the notochord differentiation and in the mesoderm formation. In addition, zebrafish *no tail* has been shown to be allelic with *Brachyury* (Schulte-Merker et al. 1994). Therefore, in vertebrates, not only the pattern of *Brachyury* expression but also its function are likely to be conserved (Herrmann and Kispert 1994; Smith 1997; Papaioannou and Silver 1998).

Recent studies revealed that vertebrate *Brachyury* genes encode proteins that share a domain of about 180 amino acids in the N-terminal half. This domain or T-domain serves DNA binding activity, and thus the Brachyury protein acts as a transcriptional factor (Kispert and Herrmann 1993; Kispert et al. 1995a). The T-domain itself, however, is not unique to *Brachyury* (Pflugfelder et al. 1992). A great deal of information has recently been accumulated regarding T-box genes other than the *Brachyury* subfamily members, and it is now thought that T-box genes constitute a novel family of transcription factors that play a crucial role in the development of various animal groups (Papaioannou and Silver 1998).

2 Expression Patterns of Invertebrate Deuterostome *Brachyury* Genes

Because, as mentioned above, vertebrate *Brachyury* genes are responsible for the formation of notochord, our laboratory as well as other laboratories attempted the isolation and characterization of invertebrate deuterostome *Brachyury* genes.

2.1 Cephalochordates

Cephalochordates (lancelets or amphioxus) are small, fishlike creatures, consisting of about two dozen species. Recent molecular phylogenic analysis (Wada and Satoh 1994) and expression patterns of developmental genes (Holland and Garcia-Fernàndez 1996) strongly support the notion that the invertebrate group closest to vertebrates is the group of cephalochordates. The amphioxus *Brachyury* genes have been isolated from two species, *Branchiostoma floridae* (Holland et al. 1996) and *B. belcheri* (Terazawa and Satoh 1997). Both species showed the same pattern of *Brachyury* expression.

Each species contains two copies of *Brachyury*, *AmBra1* and *AmBra2* in *B. floridae* and *Am(Bb)Bra1* and *Am(Bb)Bra2* in *B. belcheri*, suggesting an independent duplication of *Brachyury* in the phylogenetic lineage of amphioxus. However, the two duplicated genes showed an almost identical expression pattern (Fig. 1A, B). The *Am(Bb)Bra1* and *Am(Bb)Bra2* are initially expressed in the involuting mesoderm of the gastrula, then in the differentiating somites of neurulae (Fig. 1A), followed by the differentiating notochord and finally in the tail bud of ten-somite stage embryos (Fig. 1B). This spatial and temporal distribution of amphioxus *Brachyury* transcripts resembles that of vertebrate *Brachyury* genes, except that *Am(Bb)Bra* expression in the somite continues for a longer duration.

2.2 Urochordates

Ascidians are one of the three urochordate groups. Because fertilized ascidian eggs develop rather quickly into tadpole-type larvae with several distinct types of tissues and organs, and because the lineage of embryonic cells is well-documented, the ascidian embryo provides an appropriate experimental system to explore the expression and function of developmental genes (Satoh 1994). The notochord of the ascidian tadpole consists of 40 cells, aligned in single file along the center of the larval tail. Lineage of notochord cells is completely described (Conklin 1905; Nishida 1987). Thirty-two of the 40 notochord cells occupy the anterior four-fifths of the notochord and derive from 8 of the A-line primordial notochord blastomeres of the 110-cell embryo, while the other 8 posterior-most cells come from 2 of the B-line primordial blastomeres of the 110-cell embryo (Fig. 1C'). These 10 blastomeres subsequently divide twice to form 40 notochord cells.

By taking advantage of these features, we determined that the expression of *Brachyury* (*As-T*) of *Halocynthia roretzi* is transient, that *As-T* is expressed exclusively in blastomeres of the notochord lineage, and that the timing of *As-T* expression coincides with that of the developmental fate restriction of the blastomeres (Fig. 1C; Yasuo and Satoh 1993). Therefore, in contrast to vertebrate and cephalochordate *Brachyury* genes, the expression of the ascidian *Brachyury* is restricted to the differentiating notochord cells. This was confirmed in the *Brachyury* gene (*Ci-Bra*) of another ascidian *Ciona intestinalis* (Corbo et al. 1997a).

2.3 Hemichordates

Hemichordata is a small phylum, containing the enteropneusts (acorn worms) and pterobranchs. The relationship of the hemichordates and the echinoderms was recognized from their similar larval forms (Brusca and Brusca 1990; Nielsen 1995). On the other hand, from the time of Bateson (1885), the hemichordates have been linked to the chordates because they share chordate-specific structures and organization that can be related to the chordate body plan. However, there had been no reports on hemichordate developmental genes before 1997.

In 1998, Tagawa et al. determined the expression pattern of the *Brachyury* (*PfBra*) gene of the acorn worm *Ptychodera flava*. A distinct expression was first detected at the early gastrula stage, and the expression was restricted to the blastopore or the base of the invaginating archenteron (Fig. 1D). During the next phase of hemichordate gastrulation, the blastopore closed, and then opened again as an anus by the early tornaria larval stage. The expression remained at the base of the archenteron during gastrulation, and this pattern of the *PfBra* expression continued during formation of the larva (Fig. 1D, E), until it disappeared in the 10-day-old larvae. In addition, another distinct expression became evident as early as the middle gastrula stage in the region that eventually forms the mouth or stomodeum (Fig. 1E). The *PfBra* expression in the stomodeum remained during formation of the mouth, and this expression became undetectable prior to the disappearance of the expression at the base of the archenteron. Therefore, *PfBra* is expressed in the blastopore and the stomodeum of gastrulae and early tornaria larvae.

The stomochord is an organ in the adult proboscis, and its homology to chordate notochord has been suggested for a long time (Bateson 1885). This organ is formed during metamorphosis, which usually occurs after 3 to 5 months of larval swimming. Peterson et al. (1999a) examined whether or not *PfBra* is expressed in the stomochord. They found that *PfBra* is not expressed in the stomochord. Instead,

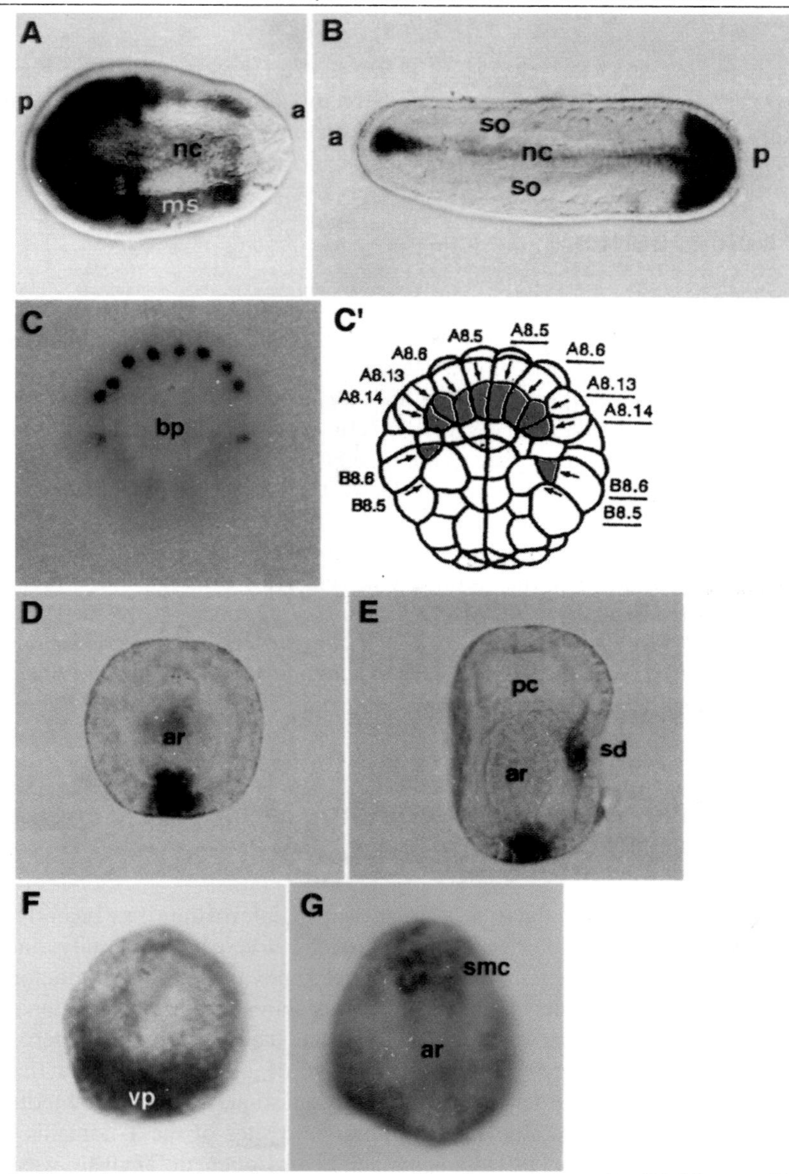

Fig. 1. Spatial expression of deuterostome *Brachyury* genes. (**A, B**) The amphioxus *Am(Bb)Bra1* is expressed in the paraxial mesoderm (ms) and notochord (nc) of the neurula (**A**), and in notochord (nc) and tail bud in the somitic-stage embryo (**B**). a, anterior; p, posterior of the embryo; so, somite. (**C**) The ascidian *As-T* is expressed in the primordial notochord cells of the 110-cell embryo. bp, blastopore. The primordial notochord cells of the 110-cell stage are diagrammatically shown in (**C'**). (**D, E**) The acorn worm *PfBra* expression in the archenteron invaginating region (ar) and stomodeum invaginating region (sd) of the gastrula (**D**) and tornaria larva (**E**). pc, protocoel. (**F, G**) The sea urchin *HpTa* expression in cells of the vegetal plate (vp) in the blastula (**F**) and secondary mesenchyme cells (smc) of the gastrula (**G**). ar, archenteron

PfBra is expressed in the mesoderm of the adult proboscis, collar and the very posterior region of the trunk. They suggested a role of *Brachyury* in the mesoderm formation of hemichordate adults, which, they thought, is relevant to the mesoderm (notochord) formation in chordate embryos.

2.4 Echinoderms

The phylum Echinodermata consists of five major groups, crinoids, asteroids, ophiuroids, echinoids and holothuroids. Among them, most of the developmental genes have been characterized in echinoid sea urchins (Davidson et al. 1998). The sea urchin *Brachyury* (*HpTa*) was isolated from *Hemicentrotus pulcherrimus* (Harada et al. 1995). In sea urchin embryos, the *HpTa* is expressed in the secondary mesenchyme founder cells, vegetal plate of the mesenchyme blastula (Fig. 1F), extending tip of the invaginating archenteron and, finally, the secondary mesenchyme cells at the late-gastrula stage (Fig. 1G). The secondary mesenchyme cells give rise to four types of cells, circum-esophageal muscles, pigment cells, basal cells and coelomic pouches. Differentiation of the secondary mesenchyme cells into these four types occurs after the prism stage. At this stage, the *HpTa* expression was undetectable. Therefore, we were unable to identify which components of the secondary mesenchyme are the *HpTa*-positive founder cells. A similar expression pattern of *Brachyury* is shown in another sea urchin *Strongylocentrotus purpuratus* (*SpTa*; Peterson et al. 1999b).

2.5 The Invertebrate Deuterostome *Brachyury* Genes are Orthologues of Vertebrate *Brachyury*

As mentioned in the Introduction, a great deal of information has recently been accumulated regarding T-box genes other than the *Brachyury* subfamily, and it is now thought that T-box genes constitute a novel family of transcription factors that play a crucial role in the development of various animal groups (Papaioannou and Silver 1998). Therefore, it remains possible that the invertebrate deuterostome *Brachyury* genes mentioned above are members of T-box genes other than the *Brachyury* subfamily. In order to determine this question, we performed molecular phylogenetic analysis. We aligned 146 amino acid sites of the T-domains based upon maximum similarity, by which molecular phylogenetic analysis was done using the neighbor-joining method (Saitou and Nei 1987). As shown in Fig. 2, the amphioxus *Am(Bb)Bra1* and *Am(Bb)Bra2*, ascidian *As-T* and *Ci-Bra*, acorn worm *PfBra*, and sea urchin *HpTa* formed a clade with the vertebrate *Brachyury* subfamily members, and this grouping was supported by the 100% bootstrap value. Therefore, it may be concluded that all of the invertebrate deuterostome *Brachyury* genes are orthologues of vertebrate *Brachyury* genes.

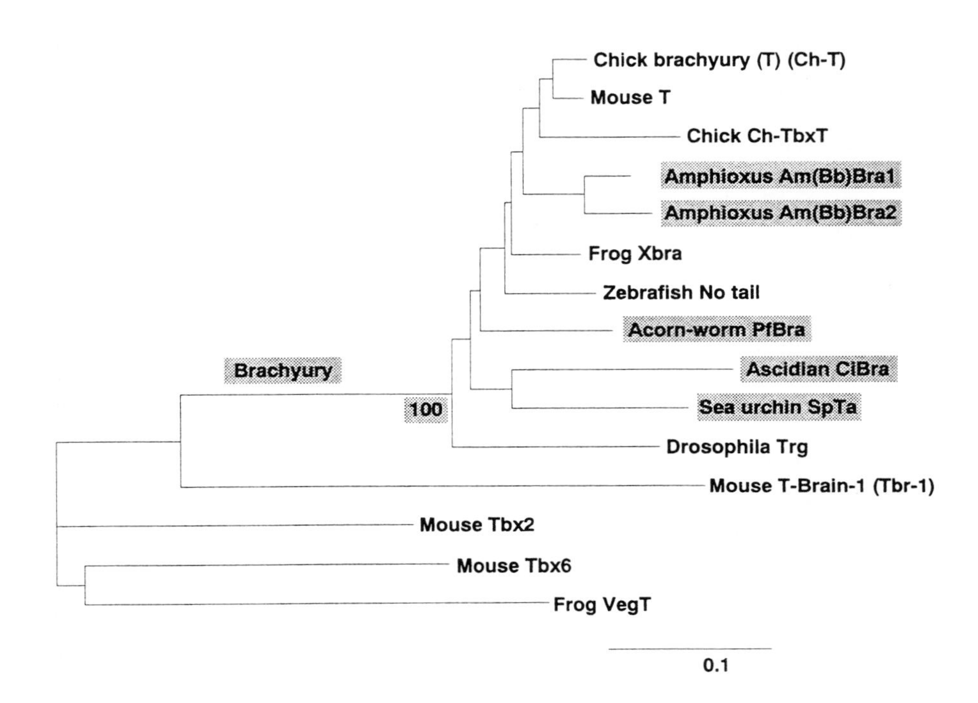

Fig. 2. A molecular phylogenetic analysis based on comparison of amino acid sequences of the T-domain showing that *Brachyury* of sea urchin (*SpTa*), acorn worm (*PfBra*), ascidian (*Ci-Bra*) and amphioxus (*Am(Bb)Bra1* and *Am(Bb)Bra2*) are members of the *Brachyury* subfamily of the T-box gene family. Branch length is proportional to the number of amino acid substitutions; the scale bar indicates 0.1 amino acid substitutions per position in the sequence. The grouping of deuterostome *Brachyury* genes is supported by 100% bootstrap pseudoreplications

3 The Ascidian *Brachyury* Exerts a Master Control over the Formation of Notochord

As in the case of vertebrate embryogenesis, the formation of notochord in ascidians is accomplished by cell-cell interaction during embryogenesis. Blastomere isolation and recombination studies indicated that the fate specification of the presumptive notochord blastomeres of the A-line is induced by a signal emanating either from adjacent primordial endoderm blastomeres or from neighboring presumptive notochord blastomeres (Nakatani and Nishida 1994). Induction begins during the late 32-cell stage, when the A-line presumptive notochord blastomeres still have potential to form notochord and spinal cord, and is completed by the 64-cell stage.

We first examined whether overexpression of *As-T* can induce notochord formation without cell-cell interaction (Yasuo and Satoh 1998). Synthetic *As-T* mRNA was injected into fertilized eggs, and blastomeres were dissociated as soon as the embryos reached the 32-cell stage. When the isolated blastomeres were cultured for about 12 hrs, they differentiated into notochord cells, suggesting that overexpression of *As-T* induces notochord cells without cell-cell interaction at the 32-cell stage. Then, we examined whether a mis-expression of *As-T* induces an ectopic differentiation of notochord cells in blastomeres of non-notochord lineage. The mis-expression of *As-T* induced an ectopic differentiation of notochord cells in at least endoderm and nerve cord precursor cells (Yasuo and Satoh 1998). These results strongly suggest that the ascidian *Brachyury* exerts a master control over the formation of notochord.

4 Functional Conservation of Deuterostome *Brachyury* Genes

As described above, chordate as well as non-chordate deuterostomes contain *Brachyury* genes, and the ascidian *Brachyury* exerts a master control over the formation of notochord. Therefore, an important question to be answered is on functional conservation among deuterostome *Brachyury* genes; namely, whether or not non-chordate deuterostome (hemichordate and echinoderm) *Brachyury* genes have a potential to induce notochord cells in chordate (ascidian) embryos. To determine this question, G. Satoh et al. (unpublished) took advantage of a *Ci-fkh* cassette described below (kindly provided from Mike Levine). *Ci-fkh* is *fork head/HNF-3*β gene of *C. intestinalis,* which is expressed in the endoderm, endodermal strand, notochord and nerve cord of the tailbud embryo (Corbo et al. 1997b). When a fusion gene construct Ci-fkh/lacZ was electroporated into *Ciona* fertilized eggs, the reporter gene is eventually expressed in the endoderm, endodermal strand, notochord and nerve cord, exactly reflecting the spatial expression pattern of *Ci-fkh*. We made a fusion gene construct Ci-fkh/Ci-Bra in which *Ci-Bra* cDNA was under the control of the *Ci-fkh* promoter (Fig. 3A). When this construct was electroporated into *Ciona* fertilized eggs, *Ci-Bra* was ectopically expressed in cells of the endoderm, endodermal strand, and nerve cord of the tailbud embryos. This caused an ectopic differentiation of notochord cells in the presumptive endoderm and endodermal strand cells, which was visualized by *in situ* hybridization with probes of notochord-specific genes (Fig. 3A). Because of vacuolation of notochord cells, *Ci-Bra*-overexpressing embryos swelled up in their ventral portion (Fig. 3A). Thus, an ectopic expression of *Ci-Bra* induces extra notochords in *Ciona* tailbud embryos, as in the case of ectopic expression of *As-T* in *Halocynthia* embryos (Yasuo and Satoh 1998).

To examine the potential of the amphioxus *Am(Bb)Bra2* to induce notochord cells in ascidian embryos, a fusion gene construct Ci-fkh/Am(Bb)Bra1 in which *Am(Bb)Bra2* cDNA was fused with the *Ci-fkh* promoter, was electroporated into

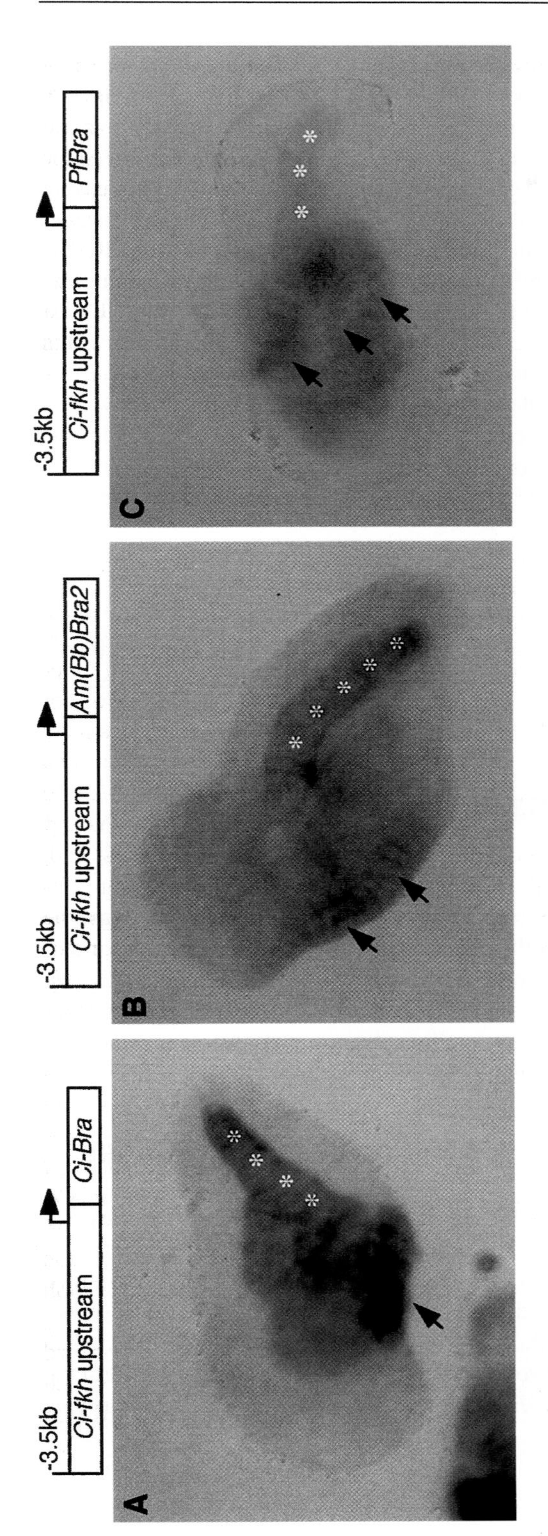

Fig. 3. Formation of extra notochords by an ectopic expression of chordate and non-chordate deuterostome *Brachyury* genes. **(A)** Induction of extra notochord (arrow) by an ectopic expression of the ascidian *Ci-Bra* with the aid of Ci-fkh/Ci-Bra vector electroporation. Formation of notochord was visualized by whole-mount ***in situ*** hybridization with a notochord-specific gene probe. White asterisks indicate the notochord formed normally by *Ci-Bra* expression (A-C). **(B)** Induction of extra notochords by an ectopic expression of the amphioxus *Am(Bb)Bra2* with the aid of Ci-fkh/Am(Bb)Bra2 vector electroporation. **(C)** Induction of extra notochords by an ectopic expression of the acorn worm *PfBra* with the aid of Ci-fkh/PfBra vector electroporation

Ciona eggs (Fig. 3B). In this case, *Am(Bb)Bra2* was expressed in the presumptive endoderm and endodermal strand cells. Whole-mount *in situ* hybridization with the probe for the notochord-specific gene revealed differentiation of extra notochord cells in the ventral region of the embryo (Fig. 3B). We obtained the same result for *Am(Bb)Bra1*. Therefore, *Brachyury* of lower chordate amphioxus is able to form notochord in tunicate embryos.

Then, we examined *PfBra* of non-chordate deuterostome hemichordates. We made a fusion gene construct Ci-fkh/PfBra in which *PfBra* cDNA was fused with the *Ci-fkh* promoter (Fig. 3C). When this construct was electroporated into *Ciona* eggs, *PfBra* was expressed in the presumptive endoderm and endodermal strand cells. As is evident in Fig. 3C, whole-mount *in situ* hybridization with the notochord-specific gene probe revealed the formation of extra notochords in the ventral region of the embryo. Therefore, *Brachyury* of non-chordate hemichordates is able to form notochord when the gene was expressed in tunicate embryos. As present, we are examining the potential of the sea urchin *SpTa* by a fusion gene construct Ci-fkh/SpTa.

The present results therefore provide convincing evidence for the functional conservation of *Brachyury* between non-chordate and chordate deuterostomes. The results also emphasize an essential role of *Brachyury* in the occurrence of notochord during the evolution of chordates. Namely, the occurrence of notochord during chordate evolution was not brought about by the appearance of new developmental genes but by an acquisition of a new function of a pre-existing gene.

5 Upstream and Downstream Regulation of Ascidian *Brachyury*

The above-mentioned results suggest that comparative studies, between chordates and non-chordate deuterostomes, of the genetic circuitry of *Brachyury*, in particular downstream or target genes, may lead to further profound understanding of molecular developmental mechanisms underlying the notochord formation and thus the evolution of chordates. At present, we are investigating the upstream and downstream regulation of ascidian *Brachyury* genes.

5.1 Upstream Control of the Ascidian *Brachyury* Genes

Both *As-T* of *H. roretzi* and *Ci-Bra* of *C. intestinalis* are exclusively expressed in notochord cells (Yasuo and Satoh 1993; Corbo et al. 1997a). How is this restricted expression of ascidian *Brachyury* genes controlled? Corbo et al. (1997a) examined a minimal promoter for the notochord-specific expression of *Ci-Bra*. As is shown in Fig. 4, they demonstrated that the 434 bp of a minimal enhancer of *Ci-Bra* contains three distinctive regions. From distal to proximal, a negative control region (from -434 to -299) excludes *Ci-Bra* expression from inappropriate embryonic

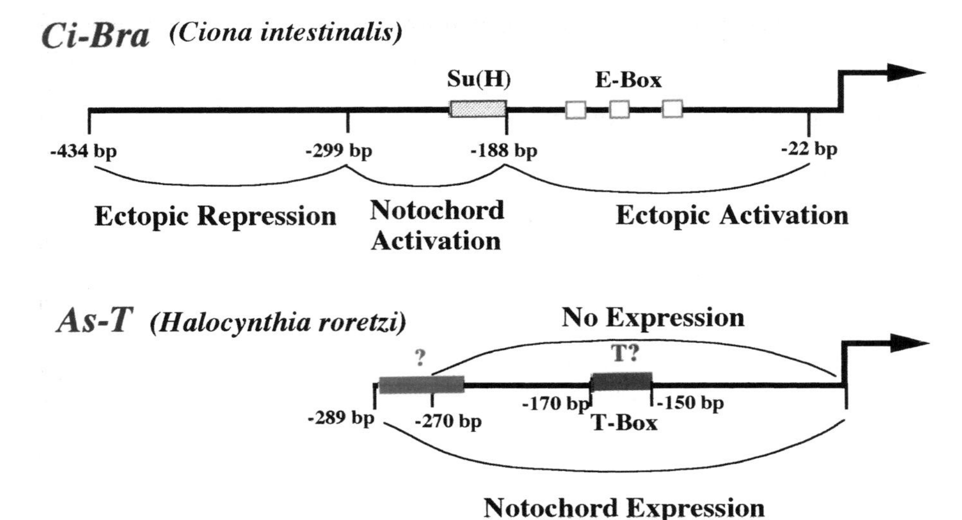

Fig. 4. Comparison of *Ci-Bra* minimal promoter (upper; Corbo et al. 1997a) with *As-T* minimal promoter (lower) that are associated with notochord-specific expression of the ascidian *Brachyury* genes

lineages, including the trunk mesenchyme and the tail muscle. This region contains *snail* binding sites (Corbo et al. 1997a, b). The Ci-snail can act as repressor of *Ci-Bra* to other mesoderm and define the boundary of the notochord (Fujiwara et al. 1998). The second region (from -299 to -188) is responsible for notochord enhancement and the enhancer is activated by a regulatory element which is closely related to the recognition sequence of *Suppressor of Hairless*-encoded transcription factor. The proximal region (from -188 to -1) is associated with enhancement of *Ci-Bra* expression in mesenchyme and muscle, and contained three E-boxes (Fig. 4).

On the other hand, Takahashi et al. (unpublished) determined the proximal *cis*-elements necessary for the notochord-specific expression of *As-T*, with various deletion constructs of As-T/lacZ. Results of deletion constructs were rather clear-cut. Namely, fusion constructs with 289 bp and more upstream sequences from the transcription start site of *As-T* derived notochord-specific reporter gene expression (Fig. 4). In contrast, fusion genes with 270 bp and less 5' flanking sequences of *As-T* derived no reporter gene expression (Fig. 4). Expression of *lacZ* was sometimes found in embryonic cells other than notochord. However, usually, the ectopic expression of *lacZ* was detected only in epidermal cells, different from the pattern of the ectopic expression of *Ci-Bra* deletion constructs in mesenchyme and muscle cells (Corbo et al. 1997a).

Kispert and Herrmann (1993) examined specific DNA binding of the mouse Brachyury (T) protein among DNA fragments which were selected from a mixture of random oligomers. They found a 20-bp palindrome as a possible consensus sequence that binds to the *Brachyury* DNA-binding domain. However, this proposed sequence has not been reported yet from *in vivo* studies. Interestingly, the promoter of *As-T* contained the palindrome between -170 and -150 from the transcription start site (Fig. 4). When this consensus sequence was deleted from As-T/lacZ, the *lacZ* expression completely diminished, suggesting that this consensus sequence is essential for *As-T* expression. These results suggest that the T-binding sequence of *As-T* is important for enhancement of the gene expression or autogenetic activation of *As-T* itself.

Therefore, the results of the promoter elements of *As-T* described above was quite in contrast to that of *Ci-Bra*. Therefore, we are examining the reporter gene expression when As-T/lacZ is injected into *Ciona* eggs or when Ci-Bra/lacZ is injected into *Halocynthia* eggs.

5.2 Downstream Control of the Ascidian *Brachyury* Gene

As described in section 4, *Ci-Bra* ectopic expression caused an ectopic differentiation of notochord cells in the endoderm and endodermal strand precursor cells. Northern blot analysis revealed that at least a few hundred times *Ci-Bra* mRNAs were present in Ci-fkh-controlled *Ci-Bra* overexpressing embryos. Takahashi et al. (unpublished) were able to make a cDNA library of mRNAs of *Ci-Bra* overexpressing embryos subtracted with mRNAs of normal embryos. The subtractive library contained 923 cDNA clones. Sequencing about 500 bp of both 5' and 3' ends of all of these clones, revealed that 599 of 923 clones were independent. Then we examined whether the corresponding genes were actually up-regulated by *Ci-Bra* overexpression. Each of the 599 clones was dot-hybridized with filters loaded with the same amount of total RNAs of control and *Ci-Bra* overexpressing embryos. As a result, 501 of the 599 cDNA clones represented genes that were up-regulated by *Ci-Bra* overexpression.

Finally, Hotta et al. (unpublished) examined the spatial expression of all of the 501 clones by whole-mount *in situ* hybridization. As a result, transcripts of 50 of the 501 clones were specific and/or predominant to the notochord cells. Sequence analysis of 5' and 3' regions suggested that polypeptides encoded by the notochord-related genes include aryl hydrocarbon receptor nuclear translocator, reticulocalbin precursor, ras GTPase-activating-like protein, GTP-binding protein, protein tyrosine phosphatase, myosin heavy chain (non-muscle type), extensin precursor, ezrin/moesin/radixin, tensin, P-selectin precursor, agrin/HSPG, collagen α1 (XI) chain, sulfate transporter, ATP-citrate (pro-S-)-lyase, sulfate adenylate transferase, N-acetyllactosamine synthase, prothrombinase precursor, calcium/calmodulin-dependent protein kinase type I, 3-hydroxyacyl-coa dehydrogenase type II, β-transduction repeat-containing protein, coatomer a subunit, SCO1 protein precursor, cytochrome P450 IIH1 (PB15). These genes may be associated with the construction of ascidian notochord.

Acknowledgments

The research of our laboratory was supported by a Grant-in-Aid from the Ministry of Education, Science, Sports and Culture, Japan for Specially Promoted Research (No. 07102012) to N.S.

References

Bateson W (1885) The later stages in the development of *B. kowalevskii*, with a suggestion as to the affinities of the Enteropneusta. Quart J Microsc Sci 25 (Suppl.):81-122

Berrill NJ (1955) The Origin of the vertebrates. Oxford Univ Press, Oxford

Brusca RC, Brusca GJ (1990) Invertebrates. Sinauer Associates, Inc., Sunderland, MA

Chesley P (1935) Development of the short-tailed mutant in the house mouse. J Exp Zool 70:429-459

Conklin EG (1905) The organization and cell lineage of the ascidian egg. J Acad Nat Sci B 1:1-119

Corbo JC, Levine M, Zeller RW (1997a) Characterization of a notochord-specific enhancer from the *Brachyury* promoter region of the ascidian, *Ciona intestinalis*. Development 124:589-602

Corbo JC, Erives A, Di Gregorio A, Chang A, Levine M (1997b) Dorsoventral patterning of the vertebrate neural tube is conserved in a protochordate. Development 124:2335-2344

Davidson EH, Cameron RA, Ransick A (1998) Specification of cell fate in the sea urchin embryo: summary and some proposed mechanisms. Development 125:3269-3290

Fujiwara S, Corbo JC, Levine M (1998) The *Snail* repressor establishes a muscle/notochord boundary in the *Ciona* embryo. Development 125:2511-2520

Garstang W (1928) The morphology of the tunicata, and its bearings on the phylogeny of the Chordata. Q J Microsc Sci 72:51-187

Haeckel E (1868) Naturliche Schopfungsgeschichte. Reimet, Berlin

Harada Y, Yasuo H, Satoh N (1995) A sea urchin homologue of the chordate *Brachyury* (*T*) gene is expressed in the secondary mesenchyme founder cells. Development 121:2747-2754

Herrmann BG, Kispert A (1994) The *T* genes in embryogenesis. Trends Genet 10:280-286

Herrmann BG, Labeit S, Poustka A, King TR, Lehrach H (1990) Cloning of the *T* gene required in mesoderm formation in the mouse. Nature 343:617-622

Holland PWH, Garcia-Fernândez J (1996) *Hox* genes and chordate evolution. Dev Biol 173:382-395

Holland PWH, Koschorz B, Holland LZ, Herrmann BG (1996) Conservation of *Brachyury* (*T*) genes in amphioxus and vertebrates: developmental and evolutionary implications. Development 121:4283-4291

Jefferies RPS (1986) The Ancestry of the vertebrates. Brit Mus Nat Hist, London

Kispert A, Herrmann BG (1993) The *Brachyury* gene encodes a novel DNA binding protein. EMBO J 12:3211-3220

Kispert A, Koschorz B, Herrmann BG (1995a) The T protein encoded by *Brachyury* is a tissue-specific transcription factor. EMBO J 14:4763-4772

Kispert A, Ortner H, Cooke J, Herrmann BG (1995b) The chick *Brachyury* gene: developmental expression pattern and response to axial induction by localized activin. Dev Biol 168:406-415

Nakatani Y, Nishida H (1994) Induction of notochord during ascidian embryogenesis. Dev Biol 166:289-299

Nielsen C (1995) Animal evolution. Interrelationships of the living phyla. Oxford Univ Press, Oxford

Nishida H (1987) Cell lineage analysis in ascidian embryos by intracellular injection of a tracer enzyme. III. Up to the tissue restricted stage. Dev Biol 121:526-541

Papaioannou VE, Silver LM (1998) The T-box gene family. BioEssays 20:9-19

Peterson KJ, Cameron RA, Tagawa K, Satoh N, Davidson EH (1999a) A comparative molecular approach to mesodermal patterning in basal deuterostomes: the expression pattern of *Brachyury* in the enteropneust hemichordate *Ptychodera flava*. Development 126:85-95

Peterson KJ, Harada Y, Cameron RA, Davidson EH (1999b) Expression pattern of *Brachyury* and *Not* in the sea urchin: comparative implications for the origins of mesoderm in the basal deuterostomes. Dev Biol 207:419-431

Pflugfelder GO, Roth H, Poeck B (1992) A homology domain shared between *Drosophila optomotor-blind* and mouse *Brachyury* is involved in DNA binding. Biochem Biophys Res Commun 186:918-925

Saitou N, Nei M (1987) The neighbor-joining method: a new method for reconstructing phylogenetic trees. Mol Biol Evol 4:406-425

Satoh N (1994) Developmental Biology of Ascidians. Cambridge Univ Press, New York

Satoh N, Jeffery WR (1995) Chasing tails in ascidians: developmental insights into the origin and evolution of chordates. Trends Genet 11:354-359

Schulte-Merker S, Ho RK, Herrmann BG, Nüsslein-Volhard C (1992) The protein product of the zebrafish homologue of the mouse *T* gene is expressed in nuclei of the germ ring and the notochord of the early embryo. Development 116:1021-1032

Schulte-Merker S, van Eeden FJM, Halpern ME, Kimmel CB, Nüsslein-Volhard C (1994) *no tail* (*ntl*) is the zebrafish homologue of the mouse *T* (*Brachyury*) gene. Development 120:1009-1015

Smith JC, Price BMJ, Green JBA, Weigel D, Herrmann BG (1991) Expression of a *Xenopus* homolog of *Brachyury* (*T*) is an immediate-early response to mesoderm induction. Cell 67:79-87

Smith J (1997) *Brachyury* and the T-box genes. Curr Opin Genet Dev 7:474-480

Tagawa K, Humphreys T, Satoh N (1998) Novel pattern of *Brachyury* gene expression in hemichordate embryos. Mech Dev 75:139-143

Terazawa K, Satoh N (1997) Formation of the chordamesoderm in the amphioxus embryo: Analysis with *Brachyury* and *fork head/HNF-3* genes. Develop Genes Evol 207:1-11

Wada H, Satoh N (1994) Details of the evolutionary history from invertebrates to vertebrates, as deduced from the sequences of 18S rDNA. Proc Natl Acad Sci USA 91:1801-1804

Willmer P (1990) Invertebrate Relationships. Cambridge Univ Press, Cambridge

Yasuo H, Satoh N (1993) Function of vertebrate *T* gene. Nature 364:582-583

Yasuo H, Satoh N (1998) Conservation of the developmental role of *Brachyury* in notochord formation in a urochordate, the ascidian *Halocynthia roretzi*. Dev Biol 200:158-170

15 The Regulation of Dorsiventral Symmetry in Plants

Stefan Gleissberg[1], Minsung Kim[2], Judy Jernstedt[3], and Neelima Sinha[2]

[1] Institute of Systematic Botany, University of Mainz, 55099 Mainz, Germany
[2] Section of Plant Biology, University of California, Davis, CA 95616, USA
[3] Department of Agronomy and Range Science, University of California, Davis, CA 95616, USA

Abstract

The higher plant shoots are generally radially symmetrical; leaves produced at the shoot apex are dorsiventral while axillary shoots again show radial symmetry. Recently analyzed mutants in different plants indicate that the proper definition of adaxial and abaxial identities is necessary to generate a leaf margin and dorsiventral symmetry. Two genes important in the regulation of transsectional leaf symmetry are PHANTASTICA (a MYB (Myeloblastosis oncogene)-domain transcription factor) and KNOTTED1-like genes (homeodomain transcription factors). We review these results in light of hypotheses about the evolutionary origin of leaves and discuss similarities of mutant phenotypes to unifacial leaves occurring in extant taxa. Related symmetry phenomena of flattened shoots and parallels to the acquisition of dorsal and ventral identities in *Drosophila* wing imaginal discs are pointed out.

1 Introduction

The higher plant shoot system is generally radially symmetrical; leaves produced on the shoot apex are bilaterally symmetrical while axillary shoots usually show radial symmetry. In a developing leaf primordium sub-organ identities are delimited along three axes. Acquisition of characteristic features along these three axes gives the leaf a shape peculiar to the species. This review will focus on the acquisi-

Key words. shoot and leaf development, radial symmetry, dorsiventrality, KNOX genes, leaf evolution, *PHANTASTICA*, leaf blades, abaxial, adaxial, *KNOTTED*, unifaciality, homeobox genes, *MYB* domain, wing development, *rough sheath* 2

tion of polarity along the ab-adaxial (dorsiventral) axis. Considerable confusion exists in the current literature about the terms dorsal and ventral. Differences between adaxial and abaxial leaf sides have often been described using dorsal and ventral that compare them to the upper and lower sides of animals. However, while a leaf's abaxial (lower) side was traditionally denominated its "dorsal" and the adaxial (upper) its "ventral" side (Goebel 1928; Napp-Zinn 1973), the use of these terms in just the opposite way in recent publications has created terminological confusion. We will, therefore, use instead the more unequivocal terms adaxial (oriented towards the shoot tip) and abaxial (pointing to the base of the shoot) in the description of transsectional asymmetry. Recently, several interesting studies have revealed possible genetic regulation of symmetry in leaves and other plant organs. Here, we review these results in light of data about the evolutionary and ontogenetic origin of abadaxiality and of related symmetry phenomena in other organs and organisms.

2 Evolutionary History of Leaves

Some of the earliest known vascular plants, dating from the mid-Silurian to lower Devonian, 395–420 million years ago, lacked leaves. These usually diminutive plants (*Cooksonia, Rhynia*) had cylindrical photosynthetic stems which branched dichotomously several to many times. However, in a relatively short time by evolutionary standards, a diversity of multicellular emergences and flattened appendages appeared on stems. Some of these are now recognized as early leaves.

The modern taxon *Psilotum*, of uncertain evolutionary affinity, bears small scale-like structures helically arranged on the upper part of the aerial stem. Internally, the appendage consists of photosynthetic parenchyma cells which are continuous with similar tissue in the stem. There is no vascular tissue in the appendages, although in *P. complanatum*, a vascular strand terminates at the base of the foliar structure. This latter arrangement, the termination of "leaf traces" at the bases of veinless enations, was also found in extinct lower Devonian lycopsids such as *Asteroxylon*. An increase in the size of the foliar appendage is correlated with extension of a vascular bundle into the lamina. Indeed, Bower (1935) proposed that some of the earliest leaves evolved from nonvascularlized epidermal outgrowths by extension of the trace into the lamina of the enation. Fossil and modern members of the Lycophyta are characterized by a particular type of leaf called a microphyll, in which a single leaf trace extends as a (typically) unbranched midvein into each of the blades. By the lower to middle Devonian, microphylls were distinctly dorsiventral and arranged in a definite phyllotactic pattern. Stomata were found on one or both epidermal surfaces. Anatomical differentiation into palisade and spongy mesophyll layers is present in some (e.g., *Selaginella*) but not all (e.g., *Lycopodium*) extant microphyllous taxa (Gifford and Foster 1989). Microphylls range in size from a length of a few millimeters in modern *Selaginella* spp. up to 1 meter in length in *Sigillaria*, an extinct Carboniferous lepidodendrid. Some microphylls (e.g.,

Selaginella spp.) bear an enigmatic structure known as a ligule on the adaxial surface, and others are associated with sporangia in the axils or on the adaxial surfaces and are referred to as sporophylls. Thus, microphylls exhibit ab-adaxial asymmetry, some with varying degrees of anatomical and functional specialization.

The most common leaf type in the vascular plants is termed a megaphyll. Megaphylls are flattened structures with a relatively complex system of veins; they may be simple or compound. Megaphylls of the type represented by fern fronds are hypothesized to have evolved from lower Devonian to lower Carboniferous "pre-fern" ancestors possessing a three-dimensional branching system with small "sterile appendages." The flattened highly compound fronds of ancient and modern ferns can be derived by invoking the hypothetical processes of overtopping, planation and webbing of Zimmermann's "telome theory" (Zimmermann 1965). Accordingly, the phylogenetic and morphological homology of fern fronds would be to the branch systems of the earliest vascular plants.

The typical fern leaf is dorsiventral, with stomata generally occurring on the abaxial side. The mesophyll may consist of homogeneous parenchyma with chloroplasts (chlorenchyma) or be organized into definite adaxial palisade and abaxial spongy parenchyma layers (Bower 1935; Ogura 1972; Gifford and Foster 1989). Sporangia may be borne singly, in clusters, in a row on or near the margin of the leaf, or more commonly, on the abaxial surface.

The megaphylls of modern seed plants (cycads, *Ginkgo*, conifers, gnetophytes, and angiosperms) are now thought likely to have had a different evolutionary origin than fern megaphylls, although some of the same elementary processes of the telome theory are invoked. Even within the gymnospermous seed plants, megaphylls of the cycadophyte line and of the coniferophyte line also may have had different evolutionary origins. According to Beck's interpretation, the compound cycadophyte leaf was derived from an entire lateral branch system of a progymnosperm ancestor, while only the ultimate segments of the lateral branch became a leaf in the coniferophyte line (Beck 1976, 1981; Stewart and Rothwell 1993). In both lines, planation and webbing of ultimate segments on a system of dichotomously branching axes can be envisioned to produce a compound or a simple leaf with branched venation.

The needlelike leaves of many modern conifer genera (e.g., *Pinus*, *Abies*) were considered by Florin (1950, 1951) to have resulted from simplification and reduction from a dichotomously branched appendage. In spite of their simple morphology, conifer leaves possess an intricate and complicated anatomy, including (in different genera), unifacial and bifacial mesophyll, specialized mesophyll parenchyma ("folded parenchyma"), and accessory conducting cells (transfusion tissue) and internal barriers (endodermis).

However, the theoretical derivations assumed by the telome theory are problematic, since they do not represent observable morphogenetic processes (Sattler 1998). Moreover, it has been questioned that radial organ symmetry was the primitive condition for vascular plants. According to Hagemann (1976), ancestral land plants could have had a two-dimensional organization, as seen in the thalli of liverworts and fern prothalli. From this perspective, radial shoot axes would have arisen later in phylogeny as specialized structures. Hagemann (1976) noted that some extant ferns (e.g., Hypolepidaceae) still exhibit transversely flattened shoots that bear leaves in the same plane of symmetry. According to Sattler (1998), some primitive land plants (*Cooksonia, Zosterophyllum*) already had both radial and flattened organization.

Despite the probable differences in evolutionary origin and mature form and structure, enations, microphylls and megaphylls have many features in common. Among these are their determinate growth, multicellular origin, and initiation at the shoot apical meristem by cell divisions in the surface and underlying layers. Wardlaw (1957) reported that leaf inception is generally comparable in all vascular plants, regardless of leaf type. According to Wardlaw, the critical difference lies in the inherent growth potentials of enation, microphyll and megaphyll primordia. In megaphylls, prolongation of meristematic activity of leaf apical initials, especially in the tangential plane, eventually results in the formation of leaf marginal meristems. Marginal meristems contribute to the formation of the lamina and its venation, both of which may be quite elaborate. In contrast, the apical initials of a microphyll or enation primordium (e.g., *Lycopodium* or *Psilotum*, respectively) persist for a time as a group of distal meristematic cells without formation of marginal meristems. The meristematic properties of the initial cells eventually disappear after, depending on the duration, having given rise to a small lamina with a single vein (e.g., *Lycopodium*) or a non-laminate, non-vascularized scale (e.g., *Psilotum*) (Wardlaw 1957). Although microphylls and megaphylls apparently arose independently, they both still exhibit ab-adaxiality. This suggests that "intrinsic" features — the placement of primordia within the apical-basal gradient at the shoot tip — may be necessary to achieve transversal symmetry of a leaf. Gene products might have acquired the function to interpret this gradient and establish ab-adaxial polarity. It will be interesting to know if the same or different genes play this role in microphyllous and megaphyllous plant groups. However, ultimately radial symmetric organs like axillary meristems also form near the shoot apex, so interpretation of the gradient must be different in the various appendages of the shoot apical meristem.

A still unresolved question is at what point is a derivative organ on a shoot apex determined as a bilateral leaf or a radial axillary bud? Cutter (1958) suggested that there are at least two sets of factors controlling the placement and fate of lateral organs. The phyllotactic factors determine organ placement while the organogenic factors determine organ fate. When these two factors are separated in time it is possible to convert the fate of an organ from bilateral to radial (as is seen in ferns). However, when the two factors operate in an almost coincident fashion the resultant organ is almost always a bilaterally symmetric leaf. This occurs at position I_2 or earlier — as soon as its location is defined, the organ is also determined to become a bilaterally symmetric leaf.

3 Leaf-like Shoots/Flattened Stems

Leaves are not the only plant organs to exhibit a flattened or dorsiventral organization. Entire shoots of some flowering plant genera (and a single gymnosperm genus) show morphological modifications which result in strikingly flattened and often leaflike structures (Troll 1937).

Troll (1937) recognized two main types of flattened shoots: platyclades and phylloclades. Platyclades, such as *Homalocladium platycladium* (Polygonaceae), are flattened shoots which develop a number of conspicuous nodes and internodes. Platyclades exhibit indeterminate growth and often show heteroblastic leaf development. Anatomy in the distal regions of the *Homalocladium* platyclade shows chlorenchyma adjacent to both flattened epidermal surfaces and sclerenchyma at the margins of the flattened stem (Kaussmann 1955).

Phylloclades, on the other hand, are determinate and distinctly leaflike in form. The aerial shoots of *Ruscus* spp., one of the best known examples, bear lateral branches which resemble leaves in shape, anatomy and orientation. Nonetheless, the lateral phylloclades of *Ruscus* are initiated in the axil of the subtending scale leaf and show early development typical of other axillary shoots (Hirsch 1977). Phylloclades of *Ruscus* may be either fertile or sterile. Sterile phylloclades have no nodes and bear no leaves. Fertile phylloclades, on the other hand, bear a small inflorescence on the upper surface, subtended by a small bract. The inflorescence and the bract are situated at or near the center of the phylloclade surface (Hirsch 1977). Phylloclade anatomy is also strikingly leaflike, with parallel veins converging at the phylloclade tip and chlorenchyma at both adaxial and abaxial surfaces (Cooney-Sovetts and Sattler 1986).

4 Transsectional Symmetry — Correspondence of Mutant Phenotypes and Unifacial Leaves in Angiosperms

In leaves, usually, two distinct planes of asymmetry can be found in most plants: 1) basal-apical (longitudinal) asymmetry and 2) adaxial-abaxial (transsectional or transverse) asymmetry. Left-right (lateral) asymmetry can also be found in some species, i.e., *Ulmus*. How these three asymmetric aspects of leaves are controlled genetically is largely unknown. Several interesting mutants and genes responsible for the asymmetry of leaves have been reported recently and may allow us in the future to understand the generation and function of asymmetry in leaves and other organs.

The *phantastica* mutant in *Antirrhinum majus* has radially symmetric leaves. In *phantastica* leaves, adaxial cell fate is replaced by abaxial cell fate, suggesting that PHANTASTICA plays a role in establishing adaxial cell fate in leaf primordia (Fig. 1A and B; Waites et al. 1998). The loss of PHANTASTICA function in the

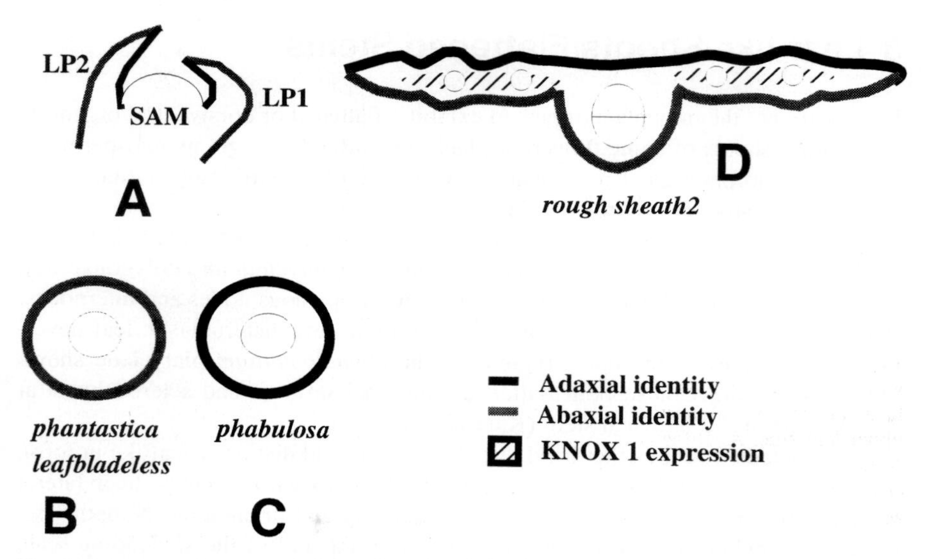

Fig. 1. Diagrams showing: A. The adaxial and abaxial regions of leaf primordia (LP1 and LP2) initiating at the shoot apex (SAM); B. The abaxialized nature of the *phan* and *lbl1* mutations; C. The adaxialized nature of the *phab* mutation; and D. Transverse section of a leaf of the *rs2* mutation showing the ectopic expression of *KNOX* genes

phantastica mutant also affects flower organs. Petals and sepals of *phantastica* flowers are also radially symmetrical, suggesting that the establishment of asymmetry in lateral organs shares a common mechanism. A similar radialized mutant, *phabulosa*, was recently reported in *Arabidopsis* (Fig. 1C; McConnell and Barton 1998). The *phabulosa* mutant lacks a leaf lamina like *phantastica* but has only adaxial cells. Another *phantastica*-like mutant, the *leafbladeless1* (*lbl1*) has radially symmetric leaves with only abaxial cell types, suggesting LBL1 is necessary for the establishment of adaxial cell identity in leaf primordia in maize (Fig. 1B; Timmermans et al. 1998).

The phenotypes of the recently described *phan*, *phb*, and *lbl1* mutants resemble the morphology of unifacial leaves that are known from many angiosperm species. Dating back to the beginning of this century, numerous investigators of unifacial leaves attempted to interpret their developmental origin (reviewed in Napp-Zinn 1973). This resemblance suggests that the corresponding genes may have been involved in the evolution of unifacial leaves, providing a source of morphological innovations in a variety of angiosperm groups. Here, we will discuss some of the results of the comparative morphological studies and relate these to the recent genetic and molecular analyses.

Unifacial leaves are characterized by the reduction of the adaxial leaf surface (Goebel 1928; Troll 1939). Their shape is often cylindrical, but may also be medially or transversely flattened. Epidermal and mesophyll tissues usually have only

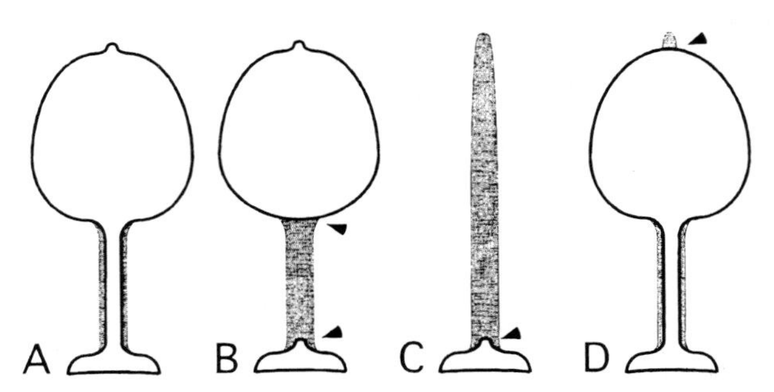

Fig. 2. Diagrams showing leaves with unifacial zones. Leaves are viewed from above (adaxial view). Leaf margins are shown as thick lines, adaxial leaf faces are in white, and abaxial faces are in grey. Arrowheads indicate cross zones where lateral margins fuse across the upper leaf side. A. Bifacial leaf differentiated into sheathing leaf base, petiole, and blade. Note that the adaxial face is narrowed in the petiole. B. A unifacial petiole is associated with cross zones at the sheath/petiole and the blade/petiole boundary. C. The entire upper leaf zone (not differentiated in blade and petiole) is unifacial. Note that the leaf base (lower leaf zone) retains bifaciality, resulting in a cross zone at the border with the upper leaf. D. Leaf with unifacial precursor tip. A cross zone is present at the border with the blade

abaxial characteristics (e.g., no palisade parenchyma). Vascular bundles in petioles of normal bifacial leaves form an adaxially open arc with the xylem poles oriented adaxially. In cylindrical unifacial leaves, vascular bundles are arranged in a ring closed by an adaxial median vein with the xylem of bundles pointing to the center. Medially flattened unifacial leaves may show bundles which alternate xylem and phloem oriented toward the center of the leaf. As a rule, only parts of a leaf are unifacial, most commonly the petiole or, in many monocotyledons, the distal-most portion of the blade, the forerunner or precursor tip. Unifacial precursor tips are found in *Hosta lancifolia*, while *Anthurium scherzerianum* provides an example in which only the petiole develops a unifacial, rounded shape. The cylindrical leaves in the genus *Juncus* and several species of *Allium* are unifacial along their length, with no differentiation of blade and petiole. However, even in these extreme cases, transverse asymmetry is maintained at the flattened base, where the leaf margins flare out and are attached to the stem at the node.

The externally visible course of the margin is the best indication of whether or not a given leaf segment of a plant species is unifacial (Napp-Zinn 1973). Whereas in bifacial leaves the margins extend continuously from the leaf base to the tip, separating both leaf sides (Fig. 2A), this continuity is interrupted at the border between bifacial and unifacial sectors. These borders are generally characterized by so-called "cross zones", where the margins merge on the adaxial side of the leaf. Leaves with unifacial petioles therefore have two sites of joining margins at the sheath/petiole and at the petiole/blade boundaries (Fig. 2B), whereas only one cross zone is observed in leaves where the distal part (leaf tip or whole upper leaf) is unifacial (Fig. 2C, D). In the leaf of *Alocasia macrorrhiza* both petiole and leaf tip are unifacial, giving rise to three cross zones at their respective borders to the bifacial blade and sheath. The phenotypes of *phan* and *lbl1* mutants, in which ectopic lamina ridges are formed at the borders of abaxial and adaxial tissue sectors, parallel the development of cross zones at the borders between bifacial and unifacial leaf segments.

The adaxially fused margins at the borders of bifacial and unifacial leaf sectors would seem to provide morphogenetic potential to enhance diversity during evolution. Cross zones between leaf sheath and petiole may develop median (instead of paired lateral) stipules, as in *Bergenia crassifolia*. From the cross zone located at the base of a blade adjacent to a unifacial petiole, median leaflets can arise (*Rodgersia aesculifolia*), or, if this region is included in surface growth of the blade, peltate leaves can form (*Tropaeolum majus*). Morphologically more derived forms such as ascidiate leaves found in the carnivorous plants *Sarracenia* or *Nepenthes* are also based on peltate leaves.

The reason why the course of the leaf margin is stressed by many authors in judging the unifaciality syndrome is that it is frequently only partly reflected in the configuration of tissues. In those cases, unifacial leaf sectors retain the open bundle arc and xylem orientation typical for bifacial leaves (petiole of *Acer mono*). On the other hand, many bifacial leaves develop identical epidermis and mesophyll on both sides (equifacial leaves, *Eucalyptus globosa*) or show an inverse arrangement with palisade parenchyma on the abaxial side (*Alstroemeria pelegrina*). This indicates that the determination of adaxial and abaxial identities may have an early morphogenetic role (in designating the course of the leaf margin which in turn allows for oriented surface development) and a later histogenetic role in defining tissue identities in the respective domains.

5 The Ontogenetic Origin of Unifaciality

The point in time after initiation that a leaf acquires ab-adaxial symmetry has been the subject of controversy. Although some authors viewed leaves as initially radial protuberances that acquire flattening only after emergence (Cutter 1958; Steeves and Sussex 1989), others believe that leaf primordia are initially subjected to adaxial-abaxial polarity resulting from their position in a polar field at the flanks of the shoot apex (Wardlaw 1968; Hagemann 1970). Hence, unifacial leaf parts that lack transverse asymmetry would arise, in principle, postgenitally. The latter point of view is also indicated by the bifacial leaf base present in all unifacial leaves.

It has been observed that increased thickening growth of the leaf primordium can ultimately produce a rounded, unifacial segment (e.g., petiole, forerunner tip). At the same time, development of the margin is suppressed so that lamina outgrowth does not take place. The adaxial bundles of the vascular circle in such unifacial sectors differentiate from tissue produced by the adaxial thickening meristem. However, both adaxially and abaxially located tissue of the primordium contribute to thickening (Roth 1949; Kaplan 1975).

From the evidence of developmental studies it remains unclear how thickening growth ultimately leads to the loss of adaxial and marginal identity. According to Kaplan (1975), adaxial meristem activity rounds out the initially flattened organ of a unifacial petiole, and may leave the margins cryptically present along the flanks of the mature petiole. However, in this case a cross zone should not occur at the border between a bifacial and a unifacial sector; instead, the margins (e.g., at the leaf base-petiole border) are expected to fade out laterally. According to Hagemann (1970), extensive thickening growth may completely interrupt the continuity of the margins in prospective unifacial leaf segments. This would allow the free ends of the marginal ridges (called marginal blastozones if they exhibit organogenetic competence) to fuse across the adaxial surface, forming cross zones.

A third possibility is indicated by petioles where the abaxial surface comprises most of the circumference, while the adaxial surface is narrowly wedged between

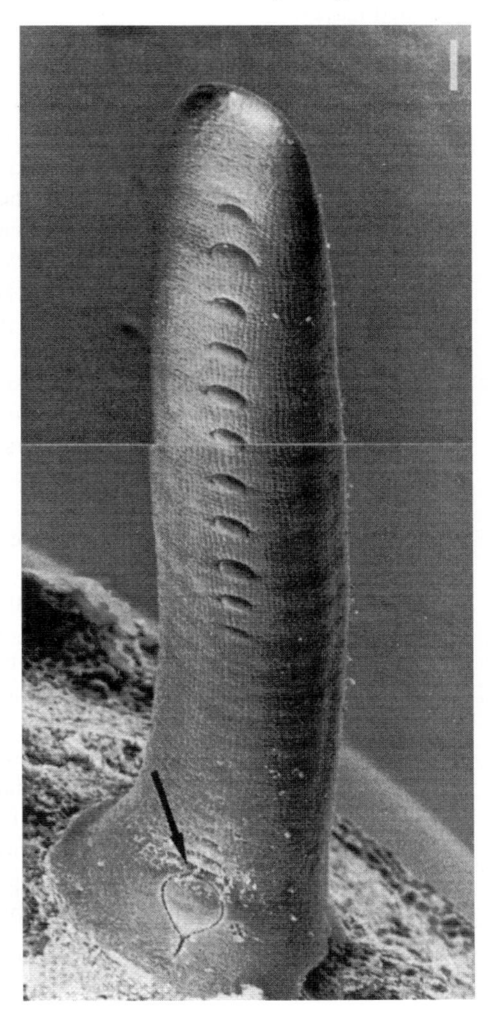

Fig. 3. Scanning electron micrograph (SEM) of a young leaf of *Oxypolis filiformis* (SEM kindly provided by Dr. R. Eberwein, Vienna). The leaf has a unifacial upper leaf with a single adaxial-median blastozone where transversely oriented leaflet primordia are initiated. Arrow indicates the cross zone at the border of lower leaf and upper leaf. The lower leaf is also fused transmedianly opposite of the apical meristem. Bar = 100 μm

the closely approached marginal ridges. These subunifacial structures with a narrow adaxial furrow could be converted into unifacial ones by a congenital longitudinal fusion of the contacting marginal ridges. Although this process has not been documented developmentally, the morphology of some mature leaves suggests this. Leaves in *Oxypolis filiformis* (Fig. 3) produce a single median row of pinnae indicating the presence of a single adaxial marginal blastozone (Fig. 3) (Kaplan 1970a; Eberwein 1995). Kaplan (1980) suggested that the adaxial ridge of the *Acacia* phyllodes corresponds to longitudinally fused margins as they occur in bifacial, leaflet-producing blade segments. *Acacia* phyllodes correspond developmentally to sword-like leaves known from many monocotyledons (*Iris*, *Acorus*), where adaxial meristem activity is more extensive than in cylindrical unifacial leaves, leading to medially flattened leaves (Troll 1939; Kaplan 1975). However, the adaxial ridge of sword leaves is generally not homologized with the leaf margins of transversely flattened leaves.

Yet another model for the developmental origin of abaxialized leaf segments was proposed by Goebel (1905) and later elaborated by Thielke (1948) and Roth (1949). According to these authors in, for example, *Iris*, a secondary leaf apex is formed on the abaxial side of the primordium early in ontogeny; this subsequently forms the new leaf axis, while the activity of the primary apex ceases. As a result of this "sympodial" mode of development, the leaf portion formed by the secondary apex is entirely surrounded by abaxial tissue. Adaxial tissue remains confined to a sector below the primary apex, which develops into the ligule. This model was criticized initially by Troll and Meyer (1955) because the model implied that unifacial segments arise postgenitally from bifacial primordia. Kaplan's (1970b) studies indicated that the supposed secondary apex was in fact the primary apex that becomes dislocated abaxially by early onset of adaxial meristem activity.

The phenotype of the *phantastica* mutant in *Antirrhinum* led Waites and Hudson (1995) to propose a new model for transverse leaf asymmetry. In these mutants, leaf margins do not differentiate in the absence of adaxial surface identity. Completely abaxialized leaves formed at higher nodes in *phantastica* plants are cylindrical in shape and lack a margin. Early formed leaves are bifacial, but show patches of abaxial tissue on the upper surface that are surrounded by ectopic marginal ridges. In leaves showing abaxialization only proximally, this region is separated from the bifacial distal part by an adaxially fused margin. This is similar to cross zones developed at the border of unifacial and bifacial leaves and is also seen in *lbl1* mutants of maize. These data suggest that the determination of margins in the primordium requires a border of adaxial and abaxial domains and that marginal identity of bifacial leaves cannot be established in the absence of adaxial identity. It suggests, furthermore, that unifacial, abaxialized leaf sectors can form by a localized and controlled downregulation of a factor conferring adaxial identity. This could be correlated with enhanced thickening growth, the dedifferentiation (loss) of the developing margin, and the subsequent fusion of the remaining marginal ridges at the new borders of adaxial and abaxial identity, forming cross zones.

6 Adaxial, Abaxial, and Marginal Identities

In bifacial leaves, ad-abaxial leaf polarity is cytohistologically recognizable soon after initiation, when earlier vacuolation of the larger abaxial portion leads to acrovergent curvature of the primordium (Hagemann 1970; Hagemann and Gleissberg 1996). Cross sections of early-stage primordia typically show the densely staining meristematic marginal domains that are connected by a band of procambium. Cells adjacent to this band in both the abaxial and adaxial domains are already more vacuolated at that time. It is possible that adaxial and abaxial domain identities specify not only the sites of the meristematic margins but also the densely staining band of procambial cells by which the developing margins are connected (Médard 1988). This procambial sheet develops into the arc-shaped row of vascular bundles typical for petioles of bifacial leaves.

A marginal blastozone and the sheet of procambial tissue can also be seen in leaf sectors that become unifacial at maturity. The circular arrangement is achieved by the additional differentiation of adaxially located bundles from the adaxial meristem. It remains to be shown how mutant leaves with abaxialized sectors at maturity behave developmentally in this respect. Early stages of *phan* mutant leaves as seen in the scanning electron micrograph (SEM) (Waites and Hudson 1995) show external flattening. More detailed examination is necessary to determine (1) if early leaf development of cylindrical, abaxialized mutant leaves exhibit densely cytoplasmic marginal and procambial domains; (2) if an adaxial meristem is active; (3) if the adaxial components of the vascular circle differentiate later than their abaxial counterparts; and (4) if mature mutant leaves retain bifaciality at their bases.

7 Marginal Identity and Proximo-Distal Growth

The specification of adaxial versus abaxial domains may occur as early as in the preprimordium (leaf anlage) (Timmermans et al. 1998; Waites and Hudson 1995). This may coincide with and could result in the formation of a crescent-shaped, small-celled meristematic zone at the shoot apex flank that precedes the outgrowth of the leaf primordium (Hagemann 1970). Hagemann and Gleissberg (1996) considered the leaf apex to be an integrated part of the marginal blastozone by which the ridge-shaped growth direction of the leaf is specified. If the determination of a marginal blastozone requires the presence of a contact zone of adaxial and abaxial identity, one would suggest at least an initial brief imposition of adaxial identity in the preprimordium. In contrast, Timmermans et al. (1998) concluded that the formation of a proximal/distal growth axis can still occur in the absence of adaxial and marginal domains in the *lbl1* mutant. The same is suggested by the *phan* mutant studied by Waites and Hudson (1995) and Waites et al. (1998), and the adaxialized leaves of the *phabulosa* mutant of *Arabidopsis* (McConnell and Barton 1998). Similarly, the "sympodial leaf model" of unifaciality suggested that the growth axis of the leaf can be uncoupled from the site of its margin. It is, however, possible that the divergence of the margin from the leaf apex occurs as a postgenital event as in the formation of cross zones below a precursor tip (sepals of *Passiflora caerulea*, Kunze 1985). Critical developmental studies in conjunction with studies on expression of the responsible genes in preprimordia and early post-initiation primordia will help to decide the issues of proximal/distal versus marginal growth directions.

8 Adaxialized Leaves — the Exception?

In almost all species exhibiting unifaciality it is the adaxial leaf surface that is suppressed (Napp-Zinn 1973). Loss-of-function mutations associated with radiality and loss of transverse polarity like *phan* in *Antirrhinum* (Waites and Hudson 1995) and *lbl1* in *Zea* (Timmermans et al. 1998) also show abaxialized leaves. This may indicate that adaxial identity relies on a more labile developmental pathway that is perturbed by knock-outs of a single gene or a few genes. Known mutants with adaxialized leaves like *phb1* in *Arabidopsis* (McConnell and Barton 1998) and *Rld1* and *Ce1* in maize (Timmermans et al. 1998) are dominant and could thus represent gain-of-function mutants of a factor conferring adaxial identity. The recessive *ago1* mutant of *Arabidopsis* (Bohmert et al. 1998) could, however, represent a loss-of-function mutant in which abaxial identity is lost. *ago1* leaves are surrounded by trichomes typically developed on the adaxial surface of rosette leaves.

Adaxialized (inverse unifacial) leaf organs are known from the bracts of *Peperomia* (Napp-Zinn 1973). A particularly interesting case is the leaves of two varieties of *Codiaeum variegatum*, that exhibit both normal and inverse unifaciality in a single leaf (Baum 1952). This curious phenotype is in accordance with the idea that symmetry is controlled by the variation of the expression domain of a single adaxializing factor.

9 Experimental Investigations of Ab-Adaxiality in Leaves

As a leaf initiates on the shoot apical meristem it quickly arches over the SAM (shoot apical meristem) due to more growth on the abaxial face. These cells are usually more vacuolated and larger than cells on the adaxial face of the organ. These features give the leaf a dorsiventral symmetry almost from the point of inception at the apex. Is the acquisition of dorsiventrality in a leaf primordium conditioned by proximity to the apical meristem? Wardlaw (1949) suggested that an inhibitory influence from the shoot apical meristem may prevent growth on the adaxial side of the leaf primordium. Early investigators tried to answer this question by using experimental approaches like microsurgery. Sussex (1955) used fine incisions to isolate the incipient leaf primordium (I_1 or P_0) from the shoot meristem.

The I_1 primordium subsequently developed into a determinate radially symmetrical structure indicating that influences from the SAM may be important for acquisition of ab-adaxiality in developing leaf primordia. Hanawa (1961) tangentially bisected a young leaf primordium on a *Sesamum* shoot meristem. The adaxial half developed into a bilaterally symmetrical leaf while the abaxial half formed a radially symmetrical structure lending further support to the hypothesis about SAM control of leaf symmetry. In a similar study on *Manihot esculenta*, Médard (1988) used microsurgery to either restrict the contact between adaxial and abaxial regions of leaf primordia or to remove the more delicate adaxial region of primordia. Various abnormal structures with radially symmetrical leaves and leaves with more rounded lobes resulted. It was suggested that contact between the adaxial and abaxial regions of the leaf primordium controls organogenesis of the leaf lamina. In a non-invasive experiment, Fleming and coworkers (1997) placed EXPANSIN-loaded beads onto the I_2 site of tomato shoot apices. The regions where the beads were placed grew out and formed radially symmetric lateral organs and also often resulted in a reversal in phyllotaxis. These results suggest that biophysical constraints prevent outgrowths at the shoot apex and that at the I_2 site, primordium determination into an organ with ab-adaxility has not yet taken place. However, as discussed above, radial leaf portions may form postgenitally from initially dorsiventral primordia.

10 Genetic and Molecular Analyses of Asymmetry in Lateral Organs

Recently, several studies have investigated the genetic controls of symmetry in certain plant organs. The molecular studies have focused on the establishment of transsectional symmetry in leaves, and radial symmetry in flowers. Also, studies on the establishment of dorsiventrality in lateral organs and the consequences of this on lateral organ development in insects and other animals have provided additional insights into the mechanisms by which gene expression cascades can set up gradients and establish two opposing cell fates.

10.1 Studies in Leaves

Radially symmetric leaves in *phan* mutants are abaxialized while *phab* mutant leaves are adaxialized (Waites et al. 1998; McConnell and Barton 1998). Based on the *phantastica*, *phabulosa*, and *lbl1* mutant phenotypes, we can hypothesize that both adaxial and abaxial cell fates are necessary for proper margin and lamina development. These results also suggest that adaxial and abaxial cell fates are mutually antagonistic.

PHANTASTICA mRNA accumulates in lateral organ primordia such as leaf primordia and flower primordia (Fig. 4). Interestingly, this *PHANTASTICA* mRNA

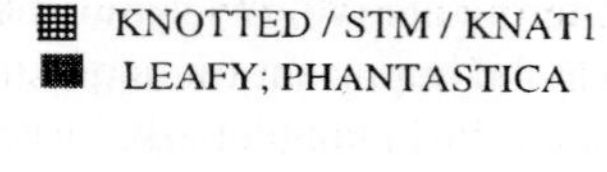

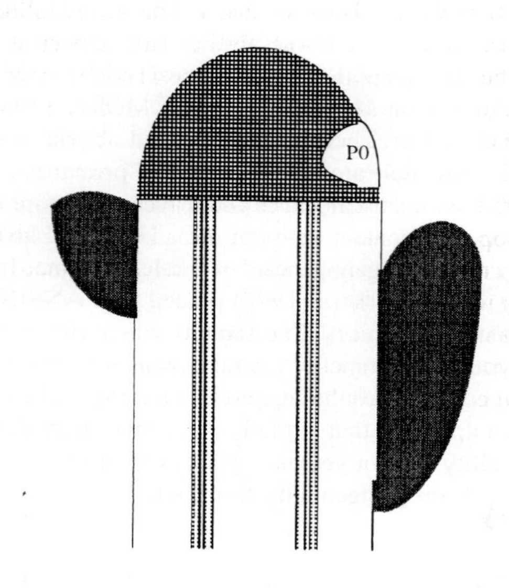

Fig. 4. Line diagram showing expression of the *KNOX* genes *KN1* and *STM1* in the shoot apical meristem, and *PHAN* expression in developing leaf primordia. These two expression patterns are complementary in nature. PO, incipient leaf primordium

expression pattern is complementary to *SHOOTMERISTEMLESS* (a *KNOTTED1* like gene in *Arabidopsis*) mRNA accumulation, which is downregulated in lateral organ primordia (Fig. 4; Waites et al. 1998). This expression pattern of *PHANTASTICA* suggests that PHAN could be a negative regulator of *SHOOTMERISTEMLESS*. Support for the regulatory relationship of *PHANTASTICA* and *KNOTTED1* was reported in maize recently (Schneeberger et al. 1998). These authors found that a mutation resembling *KNOTTED1*-like gain of function mutants (*roughsheath 2*) is caused by the failure to successfully downregulate *KNOTTED1* and other related genes like *ROUGHSHEATH1* in the leaf primordium. The *rs2* mutation has been reported to be at a *PHAN*-like locus in maize (Fig. 1D; Schneeberger et al. 1998). The *leafbladeless1* (*lbl1*) mutation has leaves with only abaxial identity (Timmermans et al. 1998). A direct or indirect role of *Lbl1* to downregulate *KNOTTED1* was also proposed. *KNOTTED1* and its homologs are well known for their function in leaf, meristem and compound leaf development (Jackson et al. 1994; Sinha et al. 1993; Smith et al. 1992; Hareven et al. 1996; Janssen et al 1998; Chen et al. 1997). Thus, intense analysis of *PHANTASTICA, KNOTTED1,* and their homologs, and interactions between these two classes of genes will provide insight into fundamental developmental regulation in plants.

The possibility that homeobox genes can directly play a role in the establishment of ab-adaxiality was suggested recently (Tamaoki et al. 1997). Tobacco plants transformed with an antisense construct of the homeobox gene *NTH15* showed loss of dorsiventrality in the midrib, suggesting that *NTH15* plays a role in establishing adaxial cell fate of the midrib. Bifaciality is maintained in mutant leaves.

The *argonaute1* mutant in *Arabidopsis* also has radially symmetric cauline leaves and filamentous flower organs, which lack ab-adaxiality (Bohmert et al. 1998). However, cotyledons and rosette leaves of *ago1* still have ab-adaxiality even though they are filamentous. This feature is very similar to *lam* in which ab-adaxiality of leaf primordia is established, but leaf lamina formation is defective (McHale 1992, 1993). *AGO1* was cloned and shown to be a novel protein of unknown function. *AGO1*-like genes have been found only in multicellular organisms suggesting the importance of *AGO1* in developmental organization relating to multicellularity. In tomato, several nonallelic *wiry* mutations have been described. Even though severity of the phenotype varies, all *wiry* mutants cause parts of the leaf to become radially symmetric. Flower organs in some *wiry* mutants are also radially symmetrical. The fact that the same phenotype can be caused by different *wiry* loci suggests that the specification of the adaxial cell fate may be caused by an interplay among several genes.

10.2 Flower Symmetry

In general, angiosperm flowers can be classified into four major types, 1) radially symmetrical (regular, actinomorphic), 2) bisymmetric (bilateral) [such as the flower of *Dicentra spectabilis*], 3) zygomorphic (dorsiventral, monosymmetrical), 4) asymmetrical flowers. A good example of genetic control of floral asymmetry was reported in *Antirrhinum majus*. Luo and coworkers (1996) reported the cloning of a gene, *CYCLOIDEA,* controlling floral asymmetry. In *cycloidea* mutants (in a certain genetic background, *dichotoma*) a ventralization occurs in the dorsal region of the flower transforming the zygomorphic flower into a radially symmetric structure. Two possible functions of *CYCLOIDEA* were suggested. *CYCLOIDEA* expression in the dorsal region of flower meristems at an early stage could increase growth rate and the initiation of primordia in the dorsal region. At a later stage, the continuous expression of *CYCLOIDEA* in the dorsal region of flower primordia could affect petal and stamen morphology. However, to make a flower perfectly radially symmetrical another locus, *DICHOTOMA*, is needed, suggesting several loci control floral asymmetry in *Antirrhinum.* Support for this idea comes from the fact that no mutants have been reported in symmetric flower species which cause asymmetric flower formation. Mutants which change zygomorphic into actinomorphic flowers can be found easily. To test the two traditional hypotheses of the acquisition of flower symmetry (multiple vs. single origin of zygomorphy) an evolutionary study on *CYCLOIDEA* in different species was suggested by Coen and Nugent (1994).

The number of planes of symmetry in a flower can vary depending on species. Typical *Arabidopsis* flowers are bisymmetric (bilateral) with 4 sepals, 4 petals, 6 stamens, and 2 carpels. Running and Meyerowitz (1996) reported a mutant, *perianthia*, in *Arabidopsis* which alters symmetry of the flower and also changes floral organ number. The *perianthia* mutant has flowers that are radially symmetrical, with five sepals, petals, and stamens, and two carpels. This pentamerous pattern resembles flowers of other dicot species, i.e, *Prunus* and *Vinca*. Genetic analysis shows that *PERIANTHIA* acts downstream of floral meristem identity genes, but independent from flower organ identity genes and meristem size genes such as *CLAVATA, FASCIATA* and *REVOLUTA*. It will be interesting to see how bilateral symmetry in the Brassicaceae evolved from pentamerous flowers and the role genes like *PERIANTHIA* may have played in this process.

11 Asymmetry in Other Organisms

The establishment of asymmetry has been shown to be important for later developmental processes in animals, too. Dorsoventrality has been well studied in *Drosophila*. Wing development in *Drosophila* and leaf development in plants share several similarities. The wing imaginal disc in *Drosophila* is symmetrical at the early stage of development. However, this imaginal disc establishes clear dorsoventrality in later stages, producing an expanded wing along the region where dorsal and ventral cell fates encounter each other. This situation is very similar to that seen during leaf development in plants because a symmetric leaf primordium acquires ab-adaxiality, leading to the formation of leaf laminae along the region where abaxial and adaxial cells meet. The expression of *APTEROUS* (a *LIM* homeodomain protein) first establishes dorsoventrality in the *Drosophila* wing imaginal disc (Cohen et al. 1992). The role of *APTEROUS* in establishing dorsal cell fate is similar to that of *PHANTASTICA* in establishing adaxial cell fate in a leaf primordium. *APTEROUS* activity in the dorsal region activates *FRINGE* and *SERRATE* in the dorsal region. In the ventral region of the imaginal disc, *DELTA* is activated by *SERRATE* and *APTEROUS*. The activity of *FRINGE* on one hand to activate *DELTA* which in turn activates *NOTCH* in the dorsal region, and on the other hand to activate *SERRATE* which in turn activates *NOTCH* in the ventral region, establishes *NOTCH* expression in the boundary between the dorsal and ventral region (Klein and Arias 1998). Also, a mutual feedback regulation of *NOTCH* and *SERRATE* (or *DELTA*) exaggerates the *NOTCH* expression along the region where dorsal and ventral cell fates meet (Pannin et al. 1997). This *NOTCH* expression is narrowed down into sharp small area, usually, 2–3 layers of cells by *NUBBIN* (Neumann and Cohen 1998). At a later stage, *NOTCH* activates *WINGLESS* and *VESTIGIAL*, leading to formation of the wing structure in the boundary region (Neumann and Cohen 1996).

Thus, in the wing imaginal disc of *Drosophila*, a cascade of genes sets up a dorsal and ventral boundary. Activation of specific genes in this boundary region leads to wing formation. Are there any parallels between the molecular controls involved in wing development in insects and leaf blade formation in plants? The fact that multiple non-allelic loci in maize and tomato can lead to similar phenotypes with radially symmetric leaves suggests that a gene expression cascade may be involved in the establishment of ab-adaxiality and blade formation. Two identified players appear to be PHAN (a MYB-domain transcription factor) and KNOTTED1-like proteins (homeodomain transcription factors). Other key genes have yet to be identified. Information gained from a survey of diversity in plant form, including such alternate morphologies as unifacial leaves and flattened stems, will be a great future resource in establishing the links between gene expression and phenotype.

Acknowledgments

The authors thank Prof. Donald R. Kaplan for comments on an earlier draft of this manuscript, Tom Goliber and Sharon Kessler for many helpful suggestions, and Dr. R. K. Eberwein, Vienna, for the use of Fig. 3.

References

Baum H (1952) Normale und inverse Unifazialität an den Laubblättern von *Codiaeum variegatum*. Oester Bot Zh 95:421-451

Beck CB (1976) Current status of the Progymnospermopsida. Rev Palaeobot Palynol 21:5-23

Beck CB (1981) *Archaeopteris* and its role in vascular plant evolution. In: Niklas KJ (Ed) Paleobotany, paleoecology, and evolution. vol. 1. Praeger, New York, pp 193-230

Bohmert K, Camus I, Bellini C, Bouchez D, Caboche M, Benning C (1998) AGO1 defines a novel locus of *Arabidopsis* controlling leaf development. EMBO J 17:170-180

Bower FO (1935) Primitive land plants. Macmillan, London

Chen J-J, Janssen B-J, Williams A, Sinha N (1997) A gene fusion at a homeobox locus: alteration in leaf shape and implications for morphological evolution. Plant Cell 9:1289-1304

Coen ES, Nugent JM (1994) Evolution of flowers and inflorescences. Development (Suppl. 1994):107-116

Cohen B, McGuffin M, Pfeifle C, Segal D, Cohen S (1992) Apterous: a gene required for imaginal disk development in *Drosophila* encodes a member of the LIM family of developmental regulatory proteins. Genes Dev 6:715-729

Cooney-Sovetts C, Sattler R (1986) Phylloclade development in the Asparagaceae: an example of homeosis. Bot J Linnean Soc 94:327-372

Cutter EG (1958) Studies on morphogenesis in Nymphaeaceae. Phytomorphology 8:74-95

Eberwein RK (1995) Bau und Ontogenese unkonventioneller Blätter des Typs 'unifaziale Phyllome' und deren Beitrag zur Theorie des Spermatophytenblattes. Diss. RWTH Aachen

Fleming A, McQueen-Mason S, Mandel T, Kuhlemeier C (1997) Induction of leaf primordia by the cell wall protein expansin. Science 276:1415-1418

Florin R (1950) Upper Carboniferous and Lower Permian conifers. Bot Rev 16:258-282

Florin R (1951) Evolution in cordaites and conifers. Acta Horti Bergiani 15:285-388

Gifford EM, Foster AS (1989) Morphology and evolution of vascular plants. WH Freeman, New York

Goebel K (1905) Organography of plants. Part 2. Clarendon, Oxford

Goebel K (1928) Organographie der Pflanzen. Erster Teil. 3rd ed. Fischer, Jena

Hagemann W (1970) Studien zur Entwicklungsgeschichte der Angiospermenblätter. Ein Beitrag zur Klärung ihres Gestaltungsprinzips. Bot Jahrb 90:297-413

Hagemann W (1976) Sind Farne Kormophyten? Eine Alternative zur Telomtheorie. Plant Syst Evol 124:251-277

Hagemann W, Gleissberg S (1996) Organogenetic capacity of leaves: the significance of marginal blastozones in angiosperms. Plant Syst Evol 199:121-152

Hanawa J (1961) Experimental studies of leaf dorsiventrality in *Sesamum indicum* L. Bot Mag Tokyo 74:303-309

Hareven D, Gutfinger T, Parnis A, Eshed Y, Lifschitz, EM (1996) The making of a compound leaf: genetic manipulation of leaf architecture in tomato. Cell 84:735-744

Hirsch AM (1977) A developmental study of the phylloclades of *Ruscus aculeatus* L. Bot J Linnean Soc 74:355-365

Jackson D, Veit B, Hake S (1994) Expression of maize *KNOTTED1* related homeobox genes in the shoot apical meristem predicts patterns of morphogenesis in the vegetative shoot. Development 120:405-413

Janssen B, Lund L, Sinha N (1998) Overexpression of a homeobox gene, *LeT6*, reveals indeterminate features in the tomato compound leaf. Plant Physiol 117:771-786

Kaplan DR (1970a) Comparative development and morphological interpretation of "rachis-leaves" in Umbelliferae. Bot J Linnean Soc 63 (Suppl 1):101-125

Kaplan DR (1970b) Comparative foliar histogenesis in *Acorus calamus* and its bearing on the phyllode theory of monocotyledonous leaves. Am J Bot 57:331-361

Kaplan DR (1975) Comparative developmental evaluation of the morphology of unifacial leaves in the monocotyledons. Bot Jahrb Syst 95:1-105

Kaplan DR (1980) Heteroblastic leaf development in *Acacia*. Morphological and morphogenetic implications. La Cellule 73:137-196

Kaussmann B (1955) Histogenetische Untersuchungen zum Flachsprossproblem. Bot Studien 3:1-136

Klein T, Arias M (1998) Interaction among *Delta, Serrate* and *Fringe* modulate *Notch* activity during *Drosophila* wing development. Development 125:2951-2962

Kunze H (1985) Studien zur Blattmetamorphose. Beitr Biol Pflanzen 61:49-77

Luo D, Carpenter R, Vincent C, Copsey L, Coen E (1996) Origin of floral asymmetry in *Antirrhinum*. Nature 384:794-799

McConnell JR, Barton MK (1998) Leaf polarity and meristem formation in *Arabidopsis*. Development 125:2935-2942

McHale N (1992) A nuclear mutation blocking initiation of the lamina in leaves of *Nicotiana sylvestris*. Planta 186:355-360

McHale N (1993) LAM-1 and FAT genes control development of the leaf blade in *Nicotiana sylvestris*. Plant Cell 5:1029-1038

Médard R (1988) La dorsoventralité initiale de l'ébauche foliaire du *Manihot esculenta*. Can J Bot 66:273-2284

Napp-Zinn K (1973) Anatomie des Blattes. 2.A. Entwicklungsgeschichtliche und topographische Anatomie des Angiospermenblattes. In: Zimmermann W (Ed) Handbuch der Pflanzenanatomie. 2. Auflage Bd 8. T 1-2. Gebrüder Bornträger, Berlin

Neumann CJ, Cohen SM (1996) A hierarchy of cross-regulation involving *Notch*, *wingless*, *vestigial* and *cut* organizes the dorsal-ventral axis of the *Drosophila* wing. Development 122:3477-3485

Neumann CJ, Cohen SM (1998) Boundary formation in *Drosophila* wing: *Notch* activity attenuated by the POU protein *Nubbin*. Science 281:409-413

Ogura Y (1972) Comparative anatomy of vegetative organs of the pteridophytes. Gebrüder Bornträger, Berlin

Pannin VM, Papayannopoulos V, Wilson R, Irvine KD (1997) *fringe* modulates Notch ligand interactions. Nature 387:908-912

Roth I (1949) Zur Entwicklungsgeschichte des Blattes, mit besonderer Berücksichtigung von Stipular- und Ligularbildungen. Planta 37:299-336

Running M, Meyerowitz E (1996) Mutation in the *PERIANTHIA* gene of *Arabidopsis* specifically alter floral organ number and initiation pattern. Development 122:1261-1269

Sattler R (1998) On the origin of symmetry, branching and phyllotaxis in land plants. In: Jean RV, Barabé D (Eds) Symmetry in plants. World Scientific, Singapore, pp 775-793

Schneeberger R, Tsianti M, Freeling M, Langdale J (1998) The *rough sheath2* gene negatively regulates homeobox gene expression during maize leaf development. Development 125:2857-2865

Sinha NR, Williams RE, Hake S (1993) Overexpression of the maize homeobox gene, *KNOTTED-1*, causes a switch from determinate to indeterminate cell fates. Genes Dev 7:787-795

Smith L, Green B, Veit B, Hake S (1992) A dominant mutation in the maize homeobox gene, *KNOTTED-1*, causes its ectopic expression in leaf cells with altered fates. Development 116:21-30

Steeves TA, Sussex IM (1989) Patterns in plant development. 2nd ed., Cambridge Univ Press, Cambridge

Stewart WN, Rothwell GW (1993) Paleobotany and the evolution of plants. Cambridge Univ Press, Cambridge

Sussex I (1955) Morphogenesis in *Solanum tuberosum* L. Apical structure and developmental pattern of the juvenile shoot. Phytomorphology 5:253-273

Tamaoki M, Sato Y, Matsuoka M (1997) Dorsoventral pattern formation of tobacco leaf involves spatial expression of a tobacco homeobox gene, *NTH15*. Genes Gen 72:1-8

Thielke C (1948) Beiträge zur Entwicklungsgeschichte unifazialer Blätter. Planta 36:154-177

Timmermans MCP, Schultes NP, Jankovsky, JP, Nelson T (1998) *Leafbladeless1* is required for dorsoventrality of lateral organs in maize. Development 125:2813-2823

Troll W (1937) Vergleichende Morphologie der höheren Pflanzen. Bd I, 1. Gebrüder Bornträger, Berlin

Troll, W. (1939) Vergleichende Morphologie der höheren Pflanzen. Bd I, 2. Gebrüder Bornträger, Berlin

Troll W, Meyer HJ (1955) Entwicklungsgeschichtliche Untersuchungen über das Zustandekommen unifazialer Blattstrukturen. Planta 46:286-360

Waites R, Hudson A (1995) *phantastica*: a gene required for dorsoventrality of leaves in *Antirrhinum majus*. Development 121:2143-2154

Waites R, Selvadurai, HRN, Oliver IR, Hudson A (1998) The *PHANTASTICA* gene encodes a MYB transcription factor involved in growth and dorsoventrality of lateral organs in *Antirrhinum*. Cell 93:779-789

Wardlaw CW (1949) Experiments on organogenesis in ferns. Growth 13 (Suppl):93-131

Wardlaw CW (1957) Experimental and analytical studies of Pteridophytes XXXVII. A note on the inception of microphylls and macrophylls. Ann Bot (n.s.) 21:427-437

Wardlaw CW (1968) Morphogenesis in plants. Methuen, London

Zimmermann W (1965) Die Telomtheorie. Gustav Fischer, Stuttgart

16
Evolution of Reproductive Organs in Vascular Plants

MITSUYASU HASEBE[1] AND MOTOMI ITO[2]

[1] National Institute for Basic Biology, 38 Nishigonaka, Myodaiji-cho, Okazaki 444-8585, Japan
[2] Department of Biology, Faculty of Science, Chiba University, 1-33 Yayoicho, Inage-ku, Chiba 263-0022, Japan

Abstract

A flower is the most complicated reproductive organ of a plant and is composed of four floral organs, namely carpels, stamens, petals and sepals. Development of the floral organs is mainly regulated by the members of the MADS transcription factors, and the evolution of reproductive organs is related to the evolution of the MADS genes. Based on recent studies of MADS genes in gymnosperms and ferns it has been speculated that the common ancestor of vascular plants had only a few MADS genes and the patterns of expression were ubiquitous. (1) The increase of MADS genes by gene duplication and (2) the recruitment of some MADS genes to be expressed in specific tissues were likely to have been important for the evolution of complicated reproductive organs, such as angiosperm flowers, from simple reproductive organs like fern sporangia.

1 Evolution of Reproductive Organs in Vascular Plants

A flower is the most complex reproductive organ in land plants and is usually composed of a receptacle and four kinds of floral organs arranged in four whorls; sepals, petals, stamens (microsporophylls), and carpels (megasporophylls) as shown in Fig. 1. Haploid cells corresponding to micro- and megaspores are formed by meiosis in a stamen and a gynoecium. Microspores are enclosed in a microspo-

Key words. MADS, LEAFY, homeotic gene, flower development, molecular evolution, flower evolution, angiosperm, gymnosperm, *Gnetum*, *Ginkgo*, conifers, fern, evolutionary development

rangium (a pollen sac) composed of multi-layered cells. Megaspores develop in a nucellus (a megasporangium) covered with two integuments. Lower vascular plants have simpler reproductive organs than angiosperms and lack petals and sepals (Fig. 1). The gymnosperm bears monoecious or dioecious strobili instead of flowers. Male and female strobili are composed of microsporophylls and megasporophylls, respectively. Microsporangia on a microsporophyll contain microspores, while a megaspore is formed in a nucellus (a megasporangium) covered with an integument (Fig. 1). Most ferns are homosporous and form uniform spores in sporangia born on the abaxial side of a sporophyll (Fig. 1).

The most recent common ancestor of ferns and seed plants is thought to be a member of the Trimerophytophyta, which has dichotomously branched stems and lacks leaves and roots (Stewart and Rothwell 1993). In this group a sporangium containing homosporous spores terminates at the tip of a stem and no other reproductive organs are associated with the sporangium. The spores are thought to be haploid, as are those of living vascular plants, because triradiate markings thought to be caused by meiosis have been observed on the surface of the spores (Stewart and Rothwell 1993). Therefore, the spores and sporangia of ferns and seed plants (the nucellus and pollen sacs of angiosperms) are likely to be homologous. Developmental similarity of spores and sporangia in ferns and seed plants (Gifford and Foster 1988) also supports the homology of these organs. However, other reproductive organs are not likely to be homologous between ferns and seed plants, because

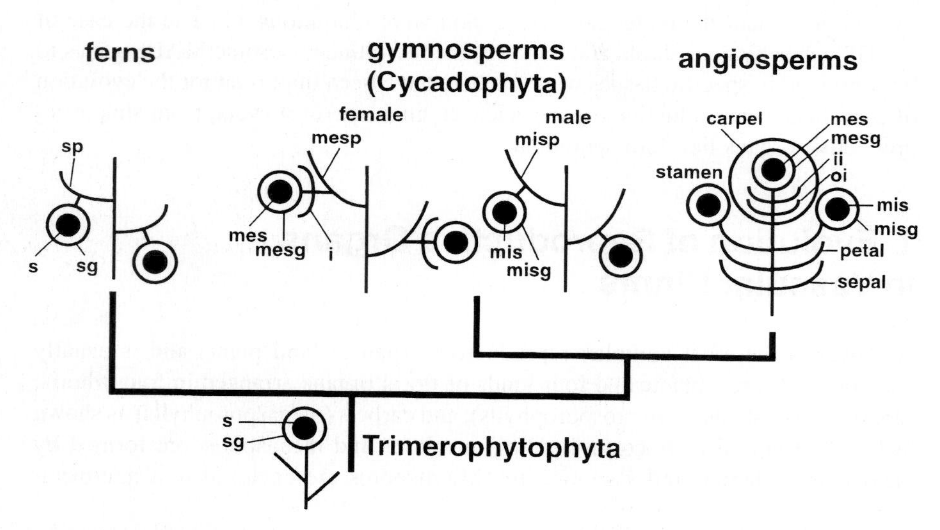

Fig. 1. Diagrams of reproductive organs in representative vascular plants. s, spores; sg, sporangium, sp, sporophyll; mes, megaspores; mesg, megasporangium; mesp, megasporophyll; mis, microspores; misg, microsporangium; misp, microsporophyll

the Trimerophytophyta did not have any adjunctive organs to the sporangium. Fern-sporophylls evolved in a fern lineage, while integuments, seed plant-megasporophylls (carpels in angiosperms), and seed plant-microsporophylls (stamens in angiosperms) were established in a seed plant lineage. The inner integument of the angiosperm and the gymnosperm is thought to be homologous (Doyle 1996), but the relationships of other reproductive organs among different lineages (e.g., the outer integument of the angiosperm, the ovuliferous scales of conifers, the male and female sporophylls of living cycads, the colors of *Ginkgo,* and the envelopes of the Gnetales) are unclear. Although the evolutionary relationships among these reproductive organs are not clear, the increase of organs covering the sporangium during the course of seed plant evolution, especially in the angiosperm lineage, is remarkable.

Evolution of morphological characters is likely to have been caused by mutations of genes, although details of the genes that cause morphological changes have not yet been revealed because of insufficient knowledge of the molecular basis of plant development. Recent progress in molecular developmental biology has begun to make possible an approach to the evolution of morphological characters. In this chapter, a hypothesis on the genetic changes that caused the evolution of reproductive organs in vascular plants is discussed.

2 Genes Related to Floral Organ Development

Molecular genetic studies using *Arabidopsis* and *Antirrhinum* have revealed that three classes of genes (A-, B- and C-function genes) play important roles in the development of the four kinds of floral organs: sepals, petals, stamens, and gynoeciums (reviewed in Weigel and Meyerowitz 1994). The A- and C-function genes manage the development of sepals and gynoecium, respectively. The combination of the A- and B-function genes and that of the B- and C-function genes results in the development of petals and stamens, respectively. The cloning of the A-, B-, and C-function genes revealed that most of them belong to the MADS gene family whose members encode transcription factors (Shore and Sharrocks 1995). The MADS genes have two well-conserved domains: the MADS and K domains. While MADS genes have been reported from metazoans and fungi, the MADS genes encoding the K domain are specific to the plant kingdom (Theissen et al. 1996; Hasebe and Banks 1997).

Angiosperm MADS genes have been classified into more than 10 groups (reviewed in Theissen et al. 1996; Hasebe and Banks 1997) and four of them correspond to A- (AP1 group), B- (AP3 and PI groups) and C- (AG group) function genes. As expected from the functions of the A-, B-, and C-function genes, they are expressed in the first-second, second-third, and third-fourth whorls, respectively (reviewed in Weigel and Meyerowitz 1994). Some of the other MADS genes are also expressed in floral organs, suggesting the involvement of the genes in floral organ development, although the function of the genes is unclear. MADS genes

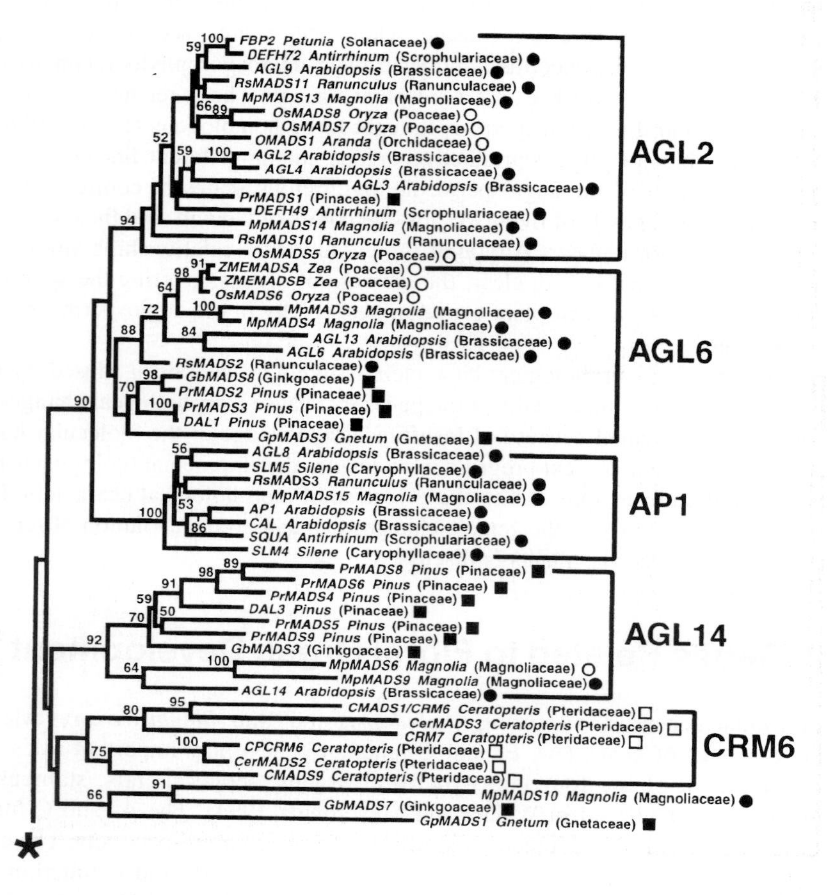

Fig. 2. The neighbor joining tree (Saito and Nei 1987) of land plant MADS genes. Amino acid sequences of representative MADS genes were aligned using the CLUSTAL W program (Thompson et al. 1994) and then revised manually. The 146 amino acid residues corresponding to positions 8-61, 63-77, 85-88, 92, 94-110, 116-143, and 148-174 from the initial methionine codon of *AP1* (Mandel et al. 1992) were used to calculate evolutionary distances with the PROTDIST program (Felsenstein 1993). The tree was obtained with the neighbor joining method (Saito and Nei 1987) using the NEIGHBOR program (Felsenstein 1993). Branch length is proportional to distances. Scale bar = 0.1 amino acids per residue. Bootstrap values calculated with the SEQBOOT program (Felsenstein 1993) are indicated for nodes supported in more than 50% of 100 bootstrap replicates. This tree is rooted with the *PpMADS1* gene of the moss *Physcomitrella patens*. Genera and families (in brackets) are indicated after gene names. Symbols following gene names indicate the origin of the genes: dicots (*black circles*), monocots (*white circles*), gymnosperms (*black squares*), or ferns (*white squares*). Brackets on the right indicate the different groups and subgroups of the plant MADS gene family. The upper and basal portions of the tree are connected by an *asterisk*

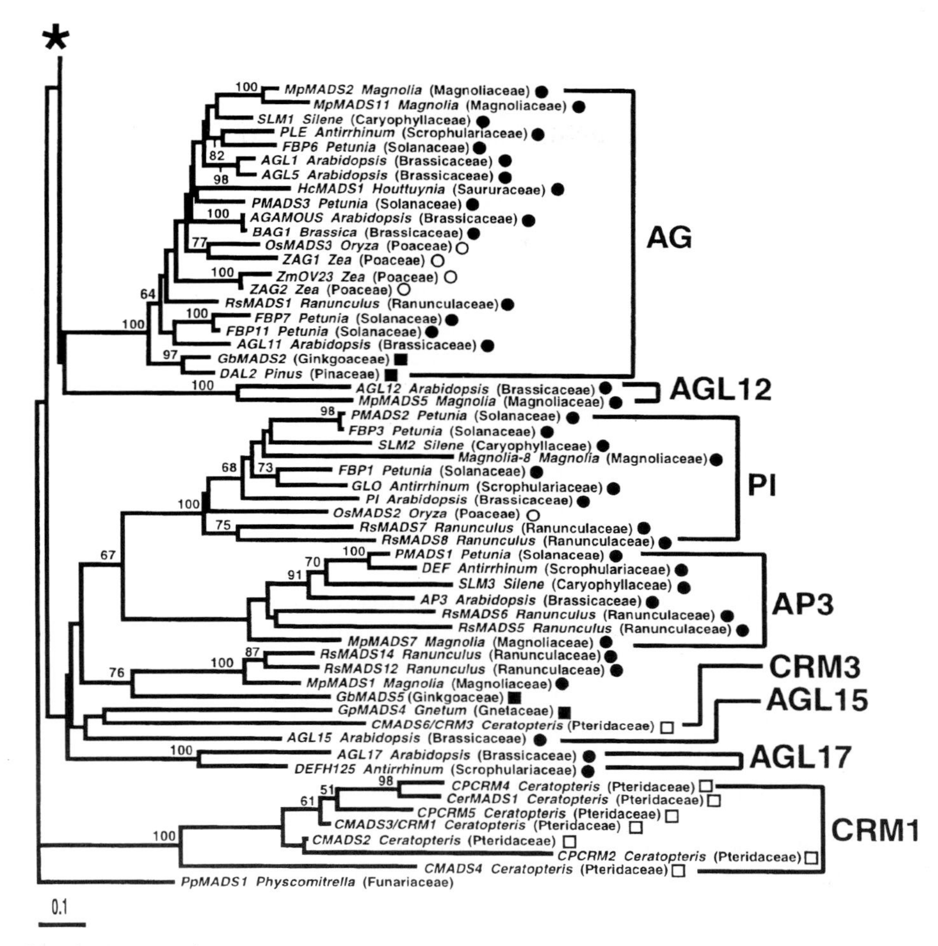

Fig. 2. *Continued*

expressed in vegetative organs including stems, leaves, and/or roots have also been reported but their function is not known. The A-, B-, and C-function genes in the MADS gene family are regarded as homeotic selector genes of flower organ development, and the changes of the spatial and temporal patterns of expression of these MADS genes are likely to be related to the evolution of floral organs.

MADS genes are related to the initiation of ovules but not directly involved in the development of integuments (reviewed in Gasser et al. 1998), although some MADS genes have been observed to be expressed in ovules (Rousley et al. 1995; Savidge et al. 1995). Therefore, evolution of the four floral organs (sepals, petals, stamens and carpels) and the integuments in the seed plant lineage are likely to have independently occurred based on mutations in different genetic cascades.

3 Divergence Time of MADS Genes

The MADS genes with the K box have been cloned from angiosperms, gymnosperms, ferns, and a moss (Purugganan et al. 1995; Theissen et al. 1996; Hasebe and Banks 1997; Purugganan 1997; Münster et al. 1997; Hasebe et al. 1998; Mikami and Ito, unpublished; Kofuji and Hasebe, unpublished). The gene tree of representative plant MADS genes is shown in Fig. 2. Six clades (AGL2, AGL6, AGL14, and AG groups) include both angiosperm and gymnosperm MADS genes, indicating that these clades were already diverged from other clades when angiosperms and living gymnosperms diverged. The relationships between fern, moss, and seed plant clades were not well resolved with any statistical confidence, and the divergence time among these clades could not be estimated based on the tree topology. Purugganan et al. (1995) postulated the monocot/dicot divergence date as 135 million years ago and calculated the evolutionary rate of MADS genes. The evolutionary clock was used to infer the "average" divergence time between the seed plant MADS gene clades as about 340 million years ago (Purugganan et al. 1995). As ferns and seed plants are estimated to have diverged approximately 400 million years ago based on paleobotanical studies (Stewart and Rothwell 1993), the number of seed plant MADS genes is likely to have increased after the divergence of ferns and seed plants due to gene duplication in the common ancestor of seed plants. Recently, Purugganan (1997) used more data than in her previous study (Purugganan et al. 1995) and estimated the time of divergence between the AG clade and the clade including AGL2, AGL6, and AP1 (the AGL2-AGL6-AP1 clade) to be approximately 480 million years ago, which is before the time of divergence between ferns and seed plants. Most genes in the AG and CRM6 clades encode proteins with additional amino-terminal amino acids to the MADS domain, while the initiation codons of other MADS genes are located adjacent to the 5' end of the sequences encoding the MADS domain (Hasebe and Banks 1997). Therefore, the AG and CRM6 groups are likely to form a sister relationship (Hasebe et al. 1998), supporting the estimate by Purugganan (1997). This suggests that the Trimerophytophyta, the common ancestor of ferns and seed plants, had at least two MADS genes (ancestral genes to the AGL2-AGL6-AP1 and AG clades), and that other seed plant MADS genes later diverged from these genes around 340 million years ago (Fig. 3).

4 Recruitment of Ancestral MADS Genes for Reproductive Organ Development

Most of the angiosperm MADS gene groups have their own patterns of expression and functions. For example, the AP1, AP3, PI, and AG groups in Fig. 2 correspond to the A-, B- and C-function genes of floral organ development, and they are expressed in specific floral organ primordia. On the other hand, most of the fern

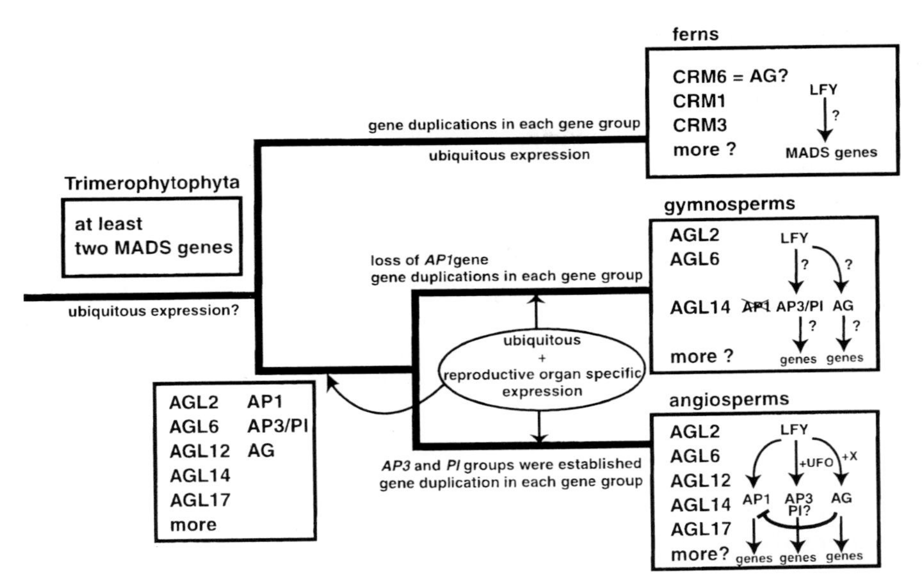

Fig. 3. Hypothesis on the evolution of MADS genes in vascular plants

Ceratopteris richardii MADS (CMADS) genes have been shown to have similar patterns of mRNA expression (Hasebe et al. 1998). The *CRM3* gene is an exception and mainly expressed in gametophyte tissue, while mRNA of other CMADS genes included in the CRM1 and CRM6 groups (Fig. 2) is similarly detected in both sporophytic and gametophytic tissues. This suggests that functions of MADS genes are not as diversified in ferns as they are in angiosperms. Based on the results of in situ hybridization, four CMADS genes (CMADS1, 2, 3, and 4) of the two groups are expressed similarly in the meristematic regions of sporophyte shoots and roots, vascular tissue, as well as in reproductive structures, including sporophylls and sporangial initials, although the amount of expression in each tissue is slightly diversified in each gene. In the angiosperm lineage, temporal and spatial diversification of expression occurs more extensively than in the fern lineage. In angiosperm MADS genes, some genes are expressed in both vegetative and reproductive tissues based on the results of the northern hybridization, and the ubiquitous pattern of expression is similar to that of CMADS genes. Based on these observations, it is hypothesized that MADS genes were ubiquitously expressed in the common ancestor of ferns and seed plants as observed in both CMADS genes and some angiosperm MADS genes, and that restriction of the expression of some MADS genes into specific tissues of reproductive organs occurred in the seed plant lineage (Fig. 3; Hasebe et al. 1998). The recruitment of MADS genes expressed in each floral organ primordium is likely to have been an important event in floral organ evolution in angiosperms. Conversely, the simple reproductive organs of ferns are likely to reflect the undifferentiated status of fern MADS genes.

When did the specification of mRNA expression evolve? A recent report on gymnosperm MADS genes in *Pinus radiata* (Mouradov et al. 1998a) gives a partial answer to the question. A female strobilus, a pine cone, is composed of numerous modified shoots. The shoot comprises two ovules, an ovuliferous scale, and a bract. The ovule is covered with an integument. A male strobilus is made of microsporophylls bearing microsporangia. The PrMADS1 gene of the AGL2 group and the PrMADS2 gene of the AGL6 group were not expressed in any vegetative organs examined, but similarly expressed in the primordia of ovules and ovuliferous scales in the seed cone and sporogenous cells in the pollen cone (Mouradov et al. 1998a). This means that specification of MADS gene expression into reproductive organs occurred in gymnosperms. Therefore, the recruitment of MADS genes as reproductive-organ specific genes took place in the common ancestor of seed plants or in parallel in gymnosperm and angiosperm lineages.

5 A-, B- and C-Function Genes Already Diversified in the Common Ancestor of Gymnosperms and Angiosperms

Gymnosperms have more complex reproductive organs than ferns, but simpler organs than in angiosperms (Fig. 1). The common ancestor of gymnosperms and angiosperms diverged from ferns 400 million years ago, and gymnosperms and angiosperms diverged from each other 200 million years ago. Recently, more than ten MADS genes were cloned from conifers (Mouradov et al. 1998a), *Ginkgo* (Mikami and Ito, unpublished), and *Gnetum* (Shindo and Hasebe, unpublished). In the MADS gene tree (Fig. 2), the AGL2, AGL6, AGL14, and AG groups include both angiosperm and gymnosperm MADS genes. In other words, these groups had already diverged in the most recent common ancestor of gymnosperms and angiosperms (Fig. 3). Monophyly of the AGL2, AGL6, and AP1 groups is supported with a high bootstrap value (91%) and all fern MADS genes cluster outside of the AGL2-AGL6-AP1 clade, suggesting that the common ancestor of gymnosperms and angiosperms is also likely to have had the ancestral gene of the AP1 group. This means that the orthologous gene of the AP1 clade is likely to have been lost in the gymnosperm lineage, although it is possible that gymnosperm AP1 orthologs may be found by conducting further studies. The sister relationship between PI and AP3 groups is supported with moderately high bootstrap values (66%). The divergence time of these two groups is likely to be subsequent to the divergence of gymnosperm and angiosperm lineages, because one gene cloned from *Ginkgo* clustered basal to AP3 and PI groups with moderately high bootstrap values (M. Ito, data not shown). Other angiosperm MADS groups (AGL12, AGL15, AGL17) clustered outside of the AGL2-AGL6-AP1 and AP3-PI clades but the branching order among these groups and other MADS groups is not well resolved with any statistical confidence. A study using the molecular clock (Purugganan et al. 1995) suggests that the AGL12, AGL15, and AGL17 groups are likely to have diversified approximately 340 million years ago, indicating that these genes originated in a seed plant lineage after the branching of ferns from the Trimerophytophyta, approximately 400 million years ago. Therefore, at least nine (AGL2, AGL6, AP1, AGL14, AG, AP3-PI, AGL12, AGL15, and AGL17) groups had already diversified in the common ancestor of gymnosperms and angiosperms (Fig. 3).

The A-, B-, and C-function genes include members of AP1, AP3-PI, and AG groups, respectively, and manage development of sepals, petals, stamens, and gynoeciums in angiosperms. It is intriguing that orthologous genes of AP1 have not been reported from gymnosperms. Members of the AP1 group have indispensable roles in petal and sepal development. The lack of these organs in gymnosperms may be related to the lack of AP1 orthologs in the gymnosperm lineage (Fig. 3). In *Arabidopsis* and *Antirrhinum*, the loss of function mutants of genes in the *AP1* group form the flowers composed of only stamens and carpels, which are partly similar to gymnosperm reproductive organs. It should be noted that the mutant flowers are not completely the same as gymnosperm flowers (Mandel et al. 1992). For example, the *ap1* mutant flower is hermaphroditic and different from the monoecious strobilli of gymnosperms. The *ap1/cal* double loss-of-function mutant in *Arabidopsis* loses any floral organs (Kempin et al. 1995). These results suggest that other genes are involved in the evolution of flowers in addition to the A-function gene.

6 Possible Changes Needed for Flower Evolution

Based on the gene tree (Fig. 2) and the discussion above, all of A-, B- and C-function genes are likely to have had already diverged in the most recent common ancestor of gymnosperms and angiosperms. It is odd that the common ancestor did not have floral organs notwithstanding that it had all the orthologs of the angiosperm A-, B- and C-function genes. The two most probable explanations are (1) that the ancestral A-, B- and C- function genes of the common ancestor regulated down-stream genes in different ways from the present A-, B- and C- function genes of angiosperms and/or (2) that the A-, B- and C- function genes themselves were regulated differently in the common ancestor from that in angiosperms. The former possibility should be investigated by experiments in which angiosperm B- or C-function genes are swapped for their gymnosperm orthologs. If the gymnosperm genes can complement the angiosperm B- or C-function genes in the angiosperm loss-of-function mutants (e.g., *apetala3*, *pistilata*, and *agamous* mutants of *Arabidopsis*), the down-stream genes regulated by the B- or C-function genes should be similar between angiosperms and living gymnosperms, and this will demonstrate that their most recent common ancestor possibly had similar genetic cascades. Recently Rutledge et al. (1998) and Tandre et al. (1998) revealed that ectopic expression of the gymnosperm AG ortholog converted sepals to carpels and petals to stamens, suggesting that the gymnosperm AG ortholog can control the same down-stream genes in *Arabidopsis* as the angiosperm C-function gene can. If the gymnosperm orthologs of down-regulated genes of angiosperm C-function genes are coded in the gymnosperm genome, and if the gymnosperm AG orthologs can regulate the down-stream orthologs in gymnosperms, it is highly probable that the gene cascade regulated by the C-function gene was established in the common ancestor of angiosperms and living gymnosperms.

As for the latter possibility, recent studies give us some insights. The A-, B- and C-function MADS genes are positively regulated by the *LEAFY* (*LFY*) gene in the floral meristem of *Arabidopsis* (Weigel and Meyerowitz 1993). The interactions among MADS genes, *LFY*, and some additional genes characterize the patterns of expression of the A-, B- and C-function genes in each whorl (Parcy et al. 1998). The A-function gene *AP1* is directly induced by *LFY* in all four of the whorls and repressed in the inner two whorls by the C-function gene *AG*. Consequently, *AP1* expression is confined to the outer two whorls. The B-function gene is likely to be activated by the combination of *LFY* and other factors, such as the *UNUSUAL FLOWER* (*UFO*) gene. While the *LFY* gene is ubiquitously expressed in all four whorls, the *UFO* expression is restricted to the second and third whorls, suggesting that *UFO* regulates the region where B-function genes are expressed. The activation of the C function-gene *AG* in the inner two whorls seems to rely on a combination of *LFY* and other unknown factors.

LFY orthologs have been reported from some gymnosperms (Frohlich and Meyerowitz 1997; Mouradov et al. 1998b; Mellerowicz et al. 1998). The gymnosperm *LFY* genes show global sequence similarity to *LFY* and its angiosperm orthologs, although the gymnosperm *LFY* genes lack the proline-rich and acidic motifs well conserved in angiosperm orthologs. The transformation of *Pinus radiata LFY* gene, *NEEDLY* (*NLY*), connected to the *Arabidopsis LFY* promoter could complement the *Arabidopsis lfy* mutant which has lost the *LFY* function (Mouradov et al. 1998b), indicating that the function of *LFY* orthologs is conserved between *Arabidopsis* and *Pinus*. The over-drive phenotypes of *NLY* in *Arabidopsis* also mimicked those of *LFY* in *Arabidopsis*, supporting the conservation of their functions between the two taxa (Mouradov et al. 1998b). Therefore, *LFY* orthologs are not likely to be related to the evolution of region-specific expression of MADS genes in the angiosperm lineage. If so, interactions between A- and C-function genes, or regulation of B-function gene by *UFO*, or regulation of C-function genes by unknown genes have possibly evolved in the angiosperm lineage. The gymnosperm C-function gene is likely to have similar functions to the angiosperm C-function genes, because transformed *Arabidopsis* with ectopic expression of the conifer *AG* ortholog mimics the phenotype of the transformed *Arabidopsis* ectopically expressing angiosperm *AG* (Rutledge et al. 1998; Tandre et al. 1998). This suggests that the present function of angiosperm C-function genes was already established in the most recent common ancestor of angiosperms and living gymnosperms. The *UFO* orthologs have not been reported from gymnosperms, and the future characterization of these genes should be informative in revealing the genetic modifications related to the evolution of flowers.

7 Future Prospects

Based on the above discussion, the increases of the number of MADS genes and the recruitment of these genes in specific tissues are likely to have had an important role in the evolution of reproductive organs in vascular plants (Fig. 3), although the details are still unknown. Our present scenario for the evolution of MADS genes and reproductive organs is as follows. Trimerophytophyta had at least two MADS genes, which were likely to have been orthologous to the genes in the AG and AGL2-AGL6-AP1 clades and ubiquitously expressed in both vegetative and reproductive organs. MADS genes were not recruited to be specifically expressed in reproductive organs in the fern lineage, although the number of MADS genes was increased. After the seed plant lineage diverged from the fern lineage, the number of MADS genes dramatically increased in the seed plant lineage, and most members of the gene groups presently observed in angiosperms were established. Before the divergence of angiosperms and living gymnosperms, some MADS genes were recruited to be specifically expressed in reproductive organs. In the living gymnosperm lineage, the *AP1* ortholog was lost and the number of other MADS genes in each gene group was increased. The loss of the *AP1* ortholog is likely to be the reason for the simpler reproductive organs of living gymnosperms compared to angiosperms. In the angiosperm lineage, the ancestral B-function gene was duplicated and *AP3* and *PI* genes were established. Extensive gene duplications in each gene group occurred. The time at which organ-specific expression of A-, B- and C-function genes was established is unknown. Comparisons between gene cascades of angiosperm B- and C-function genes and those of their gymnosperm orthologs should reveal when and how B- and C-function genes were recruited to be expressed in a tissue-specific manner, and became homeotic selector genes of floral organ development. For example, it is important to know (1) whether gymnosperm *LFY* genes can positively regulate gymnosperm B- and C-function gene orthologs, (2) whether the gymnosperm *UFO*, if it exists, can regulate the expression of angiosperm B-function genes as the angiosperm *UFO* gene can do, (3) whether the gymnosperm B-function gene orthologs can regulate the down-stream genes of the angiosperm B-function gene, and (4) whether *CURLY LEAF* (Goodrich et al. 1997) and *APETALA2* (Drews et al. 1991; Jufuku et al. 1994) genes known to be involved in the regulation of the angiosperm C-function genes also regulate gymnosperm MADS genes.

Furthermore, analyses of MADS genes in lower plants including bryophytes and green algae will provide valuable information on the original function of MADS genes in the most recent common ancestors of land plants and green plants, respectively.

Acknowledgments

We express our appreciation to T. Nishiyama, S. Shindo and the editors for their valuable suggestions on the manuscript. This research was supported by grants from the Ministry of Education, Science, Sports and Culture, Japan.

References

Doyle JA (1996) Seed plant phylogeny and the relationships of Gnetales. Int J Plant Sci 157 (supplement):S3-S39

Drews GN, Bowman JL, Meyerowitz EM (1991) Negative regulation of the *Arabidopsis* homeotic gene *AGAMOUS* by the *APETALA2* product. Cell 65:991-1002

Felsenstein J (1993) *PHYLIP* (*Phylogeny Inference Package*) version 3.5c. Available from the author. Department of Genetics, University of Washington, Seattle

Frohlich MW, Meyerowitz EM (1997) The search for flower homeotic gene homologs in basal angiosperms and Gnetales: a potential new source of data on the evolutionary origin of flowers. Int J Plant Sci 158 (Supplement):S131-S142

Gasser CS, Broadhvest J, Hauser BA (1998) Genetic analysis of ovule development. Annu Rev Plant Physiol Plant Mol Biol 49:1-24

Gifford EM, Foster AS (1988) Morphology and evolution of vascular plants. Freeman, New York

Goodrich J, Puangsomlee P, Martin M, Long D, Meyerowitz EM, Coupland G (1997) A polycomb-group gene regulates homeotic gene expression in *Arabidopsis*. Nature 386:44-51

Hasebe M, Banks JA (1997) Evolution of MADS gene family in plants. In: Iwatsuki K, Raven PH (Eds) Evolution and diversification of land plants. Springer, Tokyo, pp 179-197

Hasebe M, Wen C-K, Kato M, Banks JA (1998) Characterization of MADS homeotic genes in the fern *Ceratopteris richardii*. Proc Natl Acad Sci USA 95:6222-6227

Jufuku KD, den Boer BGW, Montagu MV, Okamuro JK (1994) Control of *Arabidopsis* flower and seed development by the homeotic gene *APETALA2*. Plant Cell 6:1211-1225

Kempin SA, Savidge B, Yanofsky MF (1995) Molecular basis of the *cauliflower* phenotype in *Arabidopsis*. Science 267:522-525

Mandel MA, Gustafson-Brown C, Savidge B, Yanofsky MF (1992) Molecular characterization of the *Arabidopsis* floral homeotic gene *APETALA1*. Nature 360:273-277

Mellerowicz EJ, Horgan K, Walden A, Coker A, Walter C (1998) *PRFLL* – a *Pinus radiata* homologue of *FLORICAULA* and *LEAFY* is expressed in buds containing vegetative shoot and undifferentiated male cone primordia. Planta 206:619-629

Mouradov A, Glassick TV, Hamdorf BA, Murphy LC, Marla SS, Yang Y, Teasdale RD (1998a) Family of MADS-box genes expressed early in male and female reproductive structures of monterey pine. Plant Physiol 117:55-61

Mouradov A, Glassick TV, Hamdorf BA, Murphy LC, Fowler B, Marla SS, Teasdale RD (1998b) *NEEDLY*, a *Pinus radiata* ortholog of *FLORICAULA/LEAFY* genes, expressed in both reproductive and vegetative meristems. Proc Natl Acad Sci USA 95:6537-6542

Münster T, Pahnke J, Rosa AD, Kim JT, Martin W, Saedler H, Theissen G (1997) Floral homeotic genes were recruited from homologous MADS-box genes preexisting in the common ancestor of ferns and seed plants. Proc Natl Acad Sci USA 94:2415-2420

Parcy F, Nilsson O, Busch MA, Lee I, Weigel D (1998) A genetic framework for floral patterning. Nature 395:561-566

Purugganan MD, Rounsley SD, Schmidt RJ, Yanofsky MF (1995) Molecular evolution of flower development: diversification of the plant MADS-box regulatory gene family. Genetics 140:345-356

Purugganan MD (1997) The MADS-box floral homeotic gene lineages predate the origin of seed plants: phylogenetic and molecular clock estimates. J Mol Evol 45:392-396

Rousley SD, Ditta GS, Yanofsky MF (1995) Diverse roles for MADS box genes in *Arabidopsis* development. Plant Cell 7:1259-1269

Rutledge R, Regan S, Nicolas O, Fobert P, Côte C, Bosnich W, Kauffeldt C, Sunohara G, Séguin A, Stewart D (1998) Characterization of an *AGAMOUS* homologue from the conifer black spruce (*Picea mariana*) that produces floral homeotic conversions when expressed in *Arabidopsis*. Plant J 15:625-634

Saito N, Nei M (1987) The neighbor-joining method: a new method for reconstructing phylogenetic trees. Mol Biol Evol 4:406-425

Savidge B, Rousley SD, Yanofsky MF (1995) Temporal relationship between the transcription of two *Aravidopsis* MADS box genes and the floral organ identity genes. Plant Cell 7:721-733

Shore P, Sharrocks AD (1995) The MADS-box family of transcription factors. Eur J Biochem 229:1-13

Stewart WN, Rothwell GW (1993) Paleobotany and the evolution of plants. Cambridge Univ Press, Cambridge

Tandre K, Svenson M, Svensson ME, Engström P (1998) Conservation of gene structure and activity in the regulation of reproductive organ development of conifers and angiosperms. Plant J 15:615-623

Theissen G, Kim JT, Saedler H (1996) Classification and phylogeny of the MADS-box multigene family suggest defined roles of MADS-box gene subfamilies in the morphological evolution of eukaryotes. J Mol Evol 43:484-516

Thompson JD, Higgins DG, Gibson TJ (1994) CLUSTAL W: improving the sensitivity of progressive multiple sequence alignment through sequence weighting, positionspecific gap penalties and weight matrix choice. Nucleic Acids Res 22:4673-4680

Weigel D, Meyerowitz EM (1993) Activation of floral homeotic genes in *Arabidopsis*. Science 261:1723-1726

Weigel D, Meyerowitz EM (1994) The ABCs of floral homeotic genes. Cell 78:203-209

Part 4
Genetic Biodiversity

17 Genetic Diversity of Color Vision in Primates

WEN-HSIUNG LI[1,4], YING TAN[1,4], STEPHANE BOISSINOT[2], SONG-KUN SHYUE[3], AND DAVID HEWETT-EMMETT[1]

[1] Human Genetics Center, University of Texas, 6901 Bertner Ave., P.O. Box 20334, Houston, TX 77225, USA
[2] Laboratory of Molecular and Cellular Biology, NIDDK, NIH Bethesda, MD 20892-0830, USA
[3] Institute of Biomedical Sciences, Academia Sinica, Taipei 11529, Taiwan
[4] Current address: Department of Ecology and Evolution, University of Chicago, 1101 East 57th Street, Chicago, IL 60637, USA

Abstract

In this chapter we review the various genetic systems of color vision in primates, the origins and evolution of these systems, and the critical amino acid residues responsible for spectral sensitivity differences among primate photopigments (opsins). Much progress has been made on these topics in recent years, thanks to the application of molecular and biochemical techniques as well as the use of traditional approaches to color vision research. The progress has led to the revelation of some misconceptions and to several changes of view. We also discuss the interactions between lifestyle (nocturnal or diurnal) and color vision system, and the role of natural selection in the evolution of color vision in primates. Further, we review evidence of frequent gene conversion between X-linked opsin genes or alleles and describe how gene conversion has made it extremely difficult to infer the evolutionary history of the color vision systems in higher primates.

Key words. opsin genes, trichromacy, dichromacy, monochromacy, polymorphism, ancient alleles, gene conversion, overdominant selection, gene silencing, parallel changes, nocturnal, diurnal, spectral tuning, simian, prosimian

1 Introduction

Color vision is a truly multidisciplinary subject involving physics, psychology, neuroscience, ophthalmology, genetics, ecology, evolution, molecular biology, biochemistry, etc. It has intrigued philosophers and scientists since Plato. While past progress was made mainly from studies in physics, psychology, and vision science, recent progress has been mainly due to the advent of molecular and biochemical techniques such as gene cloning, DNA sequencing, site-directed mutagenesis, and biochemical assays of pigment proteins. Thanks to new techniques and traditional means, we now have a much better knowledge of the various genetic systems of color vision in primates, the origins and evolution of these systems, and the critical amino acid residues responsible for spectral sensitivity differences among primate photopigments. We shall review progress in these topics. In addition, we shall discuss how gene conversion has greatly complicated the inference of the evolutionary history of the systems. Finally, we shall discuss the role of lifestyle (nocturnality or diurnality) and the role of natural selection in the evolution of primate color vision systems.

2 Primate Taxonomy

We start with a sketch of the primate phylogeny (Fig. 1; see Goodman et al. 1998). The primates are traditionally classified into the simians (higher primates) and the prosimians. The prosimians are divided into two groups: (1) the lemurs, bushbabies, lorises, etc., and (2) the tarsiers. The tarsiers are actually more closely related to the simians than to the other prosimians. The simians consist of the New World monkeys (NWMs) and the Old World primates. The latter includes *Homo sapiens* (humans), the apes, and the Old World monkeys (OWMs). The NWMs are often classified into two families, Cebidae and Atelidae; alternatively, Atelidae is further divided into two families, Atelidae and Pitheciidae (Schneider et al. 1993, 1996).

3 Color Vision Systems in Primates

3.1 Old World Primates

Among the mammals, the Old World primates possess the most sophisticated color vision system. They are trichromatic because they possess three types of pigments: the blue (short wavelength), green (middle wavelength), and red (long wavelength) pigments (opsins). The blue opsin is encoded by an autosomal gene, while the red and green opsins are encoded by two tightly linked duplicate genes on the X chromosome (Nathans et al. 1986a,b). [A normal human may possess up to 5 green opsin genes, but since they are almost identical and since only the 5' most green

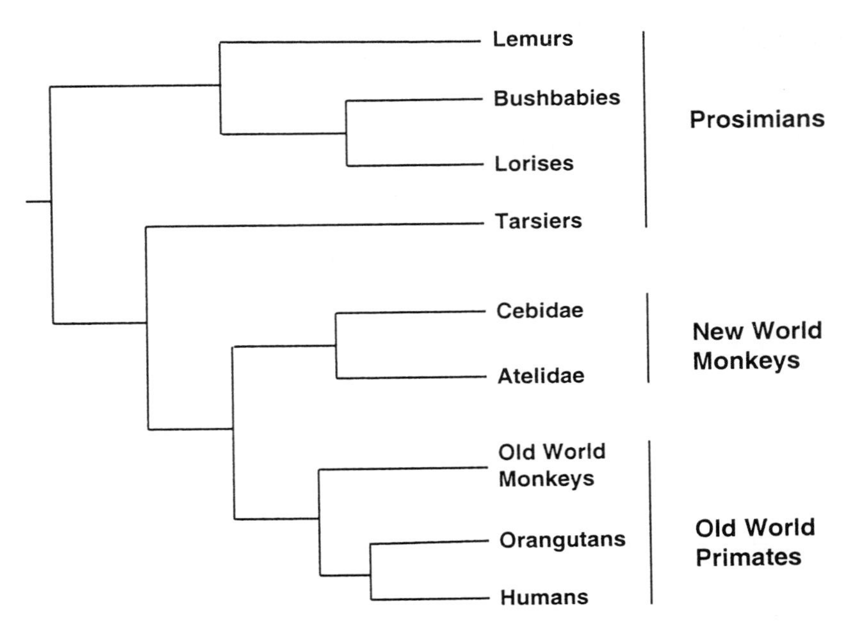

Fig. 1. Schematic representation of the primate phylogeny

gene appears to be expressed (Winderickx et al. 1992), we shall not be concerned with the multiplicity of green genes.] The red and green opsin genes contain 6 exons while the blue opsin gene contains only 5 exons. Molecular genetic studies revealed that the high frequencies of red-green color blindness in human populations are at least in part due to recombination or gene conversion between the red and green opsin genes (e.g., see Nathans et al. 1986b; Deeb et al. 1992). In addition to the three color photopigments, which are expressed in cone cells (cones), primates, like other animals, also have a rhodopsin, which is responsible for dim light vision and is expressed in rod cells (rods).

3.2 New World Monkeys

As it had long been firmly believed that all NWMs have only one X-linked opsin gene, it came as a great surprise when it was found that howler monkeys (*Alouatta* sp.) possess two X-linked opsin genes with spectral sensitivity peaks similar to those of human red and green pigments, and are trichromatic (Jacobs et al. 1996a; Boissinot et al. 1997; Hunt et al. 1998). However, all other NWMs studied possess only one X-linked opsin locus (see Jacobs 1996; Boissinot et al. 1997).

It has been known since the early 1980s that some NWMs such as the squirrel monkey and the marmoset have three high-frequency alleles at the single X-linked opsin locus (e.g., Jacobs 1984; Mollon et al. 1984; Travis et al. 1988). Because of the existence of this triallelic system, heterozygous NWM females are trichromatic, though males and homozygous females are dichromatic. Recently Shyue et al. (1998) and Boissinot et al. (1998) found that the triallelic system also exists in each of the three additional species studied: the capuchin, the tamarin, and the saki monkey. The capuchin and tamarin belong to the same family (Cebidae) as do the squirrel monkey and marmoset, but the saki monkey belongs to the other NWM family, Atelidae (or Pitheciidae, see above) and is one of the NWM species most divergent from the squirrel monkey and marmoset. Thus, the triallelic system appears to exist in the majority of NWMs.

The owl monkeys (a NWM genus) are the only nocturnal higher primate. Their blue opsin gene has been found to be defective (Jacobs et al. 1996b). So, they possess only one functional color vision gene (the X-linked opsin gene) and are monochromatic.

3.3 Prosimians

Recent studies of color vision in prosimians have clarified several issues. It was previously claimed that the ringtail lemur (*Lemur catta*), which is a diurnal prosimian, could make color discriminations, though the capacity was far from acute (Mervis 1974; Blakeslee and Jacobs 1985). However, Jacobs and Deegan (1993) found that ringtail lemurs and brown lemurs (also diurnal) have only a single class of photopigment in the middle to long wavelengths and a short wavelength pigment. They therefore concluded that both of these diurnal prosimians are dichromatic and speculated that the limited ability of ringtail lemurs to discriminate colors might have resulted from the ability of lemurs to jointly utilize signals from cones and rods. (Note that a mammal with a short wavelength pigment and a middle/long wavelength pigment cannot perceive the whole visible color spectrum and especially cannot distinguish between red and green, but can see only blue and yellow hues.)

Bushbabies are nocturnal prosimians. In apparent accord with this lifestyle, the early literature contained repeated claims that the bushbaby retina contains only rods, but no cones (see the review by Deegan and Jacobs 1996). Indeed, no cones were detected in a study using microspectrophotometric measurements of the absorbance properties of individual photoreceptors (Petry and Harosi 1990). However, labeling the photoreceptors in *Galago garnetti* with cone-specific antibodies, Wikler and Rakic (1990) estimated the proportion of cones in the bushbaby retina to be from ~1% to ~3%, depending on the retina location. Recently, Deegan and Jacobs (1996) found that the cones of the thick-tailed bushbaby (*Otolemur crassicaudatus*) contains a single type of photopigment with spectral peak near 545 nanometer (nm).

As color vision is presumably of no use to a strictly nocturnal animal, it has been commonly thought that color pigment genes in an animal with a long history of nocturnal life should have evolved rapidly and would become degenerate or nonfunctional because of relaxation in their functional constraints. In fact, the blue opsin gene in the bushbaby (*O. crassicaudatus*) has accumulated deleterious mutations and become nonfunctional (Jacobs et al. 1996b; Jacobs 1996). However, unexpectedly, sequencing work revealed that the X-linked opsin gene in both *Galago senegalensis* and *O. garnettii*, which is closely related to *O. crassicaudatus*, has been well conserved (Zhou et al. 1997); in fact, it has been even better conserved than the X-linked pigment genes in higher primates. It is possible that this opsin in combination with rhodopsin can provide the bushbaby with a wider light spectrum at dusk, during which the animal is active, than can rhodopsin alone (Deegan and Jacobs 1996). Moreover, since rhodopsin is saturated by daylight, the X-linked opsin might be the only functional opsin for the animal during daylight. Although bushbabies are usually not active in the daytime, they will occasionally need to move (e.g., in order to escape predation) and thus need to use the X-linked opsin. Furthermore, it has been suggested that this opsin gene might play a role in the circadian rhythm of mammals (Nei et al. 1997).

4 Critical Amino Acid Residues for Spectral Tuning

The phenotype of a pigment is commonly characterized by its spectral sensitivity peak (λ_{max}), which is usually estimated by electroretinogram or microspectrophotometry. Since the exact λ_{max} is very difficult to determine, only approximate values are obtained. For example, the λ_{max} values for the three alleles in squirrel monkeys were previously given as 538, 551, and 561 nm, respectively (Jacobs and Neitz 1987), but were estimated to be 535, 550, and 562 nm in a recent reanalysis of previous data (Jacobs 1996). These alleles are commonly denoted as P535, P550, and P562. The most common green and red pigments in humans have the λ_{max} values of ~530 and ~562 nm and are denoted as P530 and P562, respectively. Note that two pigments with the same λ_{max} value, e.g., human P562 and squirrel monkey P562, may have different origins and different amino acid sequences.

There has been much interest in knowing the amino acid residue sites that are involved in spectral tuning. One way to study this problem is to compare the amino acid sequences of closely related pigments with known λ_{max} values (Neitz et al. 1991; Shyue et al. 1998). Another way is to introduce mutations, singly or in combination, in an opsin cDNA by site-directed mutagenesis, express the mutant cDNA in animal cells, and measure the spectral sensitivity of each mutant sequence by spectrophotometry (Merbs and Nathans 1992, 1993; Asenjo et al. 1994). These studies have led to the identification of the following critical amino acid residue sites in the X-linked opsin sequences of higher primates:

Position	116	180	229	230	233	277	285	309
Change	Ser→Tyr	Ala→Ser	Ile→Phe	Ile→Thr	Gly→Ser	Phe→Tyr	Ala→Thr	Tyr→Phe
Shift (nm)	?	5	-2	?	-1	8	15	?

The above estimates of spectral shifts are only approximate. Positions 180, 277, and 285 are the major critical sites, causing shifts of 4–7, 6–10, and 10–16 nm, respectively. Positions 230 and 309 are considerably less important because according to Merbs and Nathans (1993), they cause shifts of 1 nm or less, though Asenjo et al. (1994) obtained estimates of 0–4 and 0–3 nm, respectively. The same comment applies to position 233 because Shyue et al. (1998) estimated a shift of -1 nm for Gly233Ser and Merbs and Nathans (1993) estimated a shift of -0.4 nm for Ala233Ser, though Asenjo et al. (1994) gave an estimate of -0-4 for Ala233Ser. The significance of position 116 is not certain. Shyue et al. (1998) found no effect of changes at this position, whereas Asenjo et al. (1994) found some effect when the change Tyr116Ser was introduced into a red (long wavelength) pigment. All other differences among the available primate X-linked opsin sequences do not appear to have any discernible effects on spectral tuning.

The above residues are those that are involved in the spectral tuning of X-linked pigments in higher primates. Sites 185, 277, and 285 have also been found to be major critical sites in the spectral tuning of red-green vision in other vertebrates (Yokoyama and Yokoyama 1990; Yokoyama and Radlwimmer 1998). Additional critical residues have been found in other mammals. For example, the changes His197Tyr and Ala308Ser were estimated to cause 28 and 18 nm shifts in the mouse, respectively (Sun et al. 1997). For more details, see Yokoyama and Radlwimmer (1998).

5 Frequent Gene Conversion Between X-linked Opsin Alleles or Genes

The tight linkage and high similarity (~98%) between the red and green opsin genes in Old World primates provide a favorable condition for gene conversion to occur between them. This possibility was first noted for exons 4 and 5 between the two genes in OWMs (Balding et al. 1992; Ibbotson et al. 1992) and for exon 3 between the two genes in humans (Winderickx et al. 1993). Deeb et al. (1994) and Reyniers et al. (1995) provided further evidence of frequent gene conversion between the two genes in humans and OWMs.

A remarkable example of gene conversion was found when introns 4 of human red and green opsin genes were sequenced (Shyue et al. 1994). The two introns were identical, though a > 8% divergence was expected between them because human red and green opsin genes arose from a duplication before the divergence of the OWM and human lineages, i.e., more than 25 million years ago.

The complete human red opsin gene and a large region of the human green opsin gene have been sequenced at the Sanger Center, Cambridge, UK, and a comparison reveals that all noncoding parts that are now available for comparison (the 3' part of intron 3, intron 4, intron 5 and the 3' flanking region) are identical or almost identical between the two genes (Table 1), suggesting frequent gene conversion between these two genes (Zhao et al. 1998). In contrast, the divergences in exons 4 and 5 are 3.8% and 4.7% (Table 1). It is likely that gene conversion has also occurred in exons. In fact, exons 1 and 6 of the red and green opsin genes have been completely homogenized (Table 1). This observation is not surprising because both exons 1 and 6 contain no critical amino acid residue. On the other hand, exons 2, 3, 4, and 5 each contain residues that are critical to the spectral differences between the red and green opsin peptides. A gene conversion event in any of these exons may reduce the spectral sensitivity differences between the two opsins, so it may be disadvantageous and eliminated from the population. This is probably the reason why exons 2, 3, 4, and 5 of the two genes have been maintained distinct.

Zhou and Li (1996) also found that the degree of divergence between the intron 4 sequences of the red and green opsin genes was only 0.3% in a chimpanzee and 0.9% in a baboon, indicating gene conversion in each of these two species. In comparison, exons 4 and 5 have diverged more than 6% at synonymous sites between the red and green opsin genes in each of these two species and in humans. When the synonymous divergences are used to infer the relationships among the human and baboon red and green opsin genes, human and baboon red opsin genes are clustered together, and so are human and baboon green opsin genes (Li 1997). This tree is in agreement with the view that the red and green opsin genes diverged before the divergence of the OWM and human lineages. However, when a tree is constructed from the intron 4 sequences, the two human genes are clustered together, and so are the two baboon genes (Li 1997). This example shows that gene conversion can drastically distort inferences on the evolutionary history of genes.

Table 1. Mean and standard error of the number of nucleotide substitutions per 100 sites between human red and green pigment genes in exons, introns, and 3' flanking sequences[a]

	No. of differences/ sequence length (bp)	Noncoding region (K)	Coding region	
			K_S	K_A
Introns 3	0/506	0.0 ± 0.0		
Exon 4	5/166		3.8 ± 3.0	3.3 ± 1.9
Intron 4	1/1554	0.1 ± 0.1		
Exon5	10/240		4.7 ± 2.9	3.9 ± 1.5
Intron 5	2/2282	0.1 ± 0.1		
Exon 6	0/108		0.0 ± 0.0	0.0 ± 0.0
3' Flanking	2/1256	0.2 ± 0.1		
Exons 4, 5, and 6	15/514		3.4 ± 1.6	3.0 ± 1.0

From Zhao et al. (1998).

[a] The K value was computed by Kimura's two-parameter method (Kimura 1980), and the K_S (synonymous) and K_A (nonsynonymous) values were computed by Li's method (Li 1993). Only about one-third of intron 3 was available for comparison.

```
        EXON 3  EXON 4                            INTRON 4                                   EXON 5
        --      ------  -----------------------------------------------------------------   ------
                                                         11111111111111111111
        55      666677                      1111122222222222233333345679900000022444467788889  888888
        35      589904   2233333333334491114590223444555580225596490782236992333482082666  022235
        85      256784  68901234567890193894630122149012796784929062731570481749273768237  136803

P562    TG      TTTGGG  TC-----------ACATCAGCTGGGGGT---GGACTGTTG-ACGC-TG-GGGCGAACAGCGGCGC  GAGAAA
                ||||||  | ||||||||||| | |||||| | ||||||||||||||||     ||||| ||||| |||||| |
P535c   GA      TTTGGG  TA-----------CC-TCAGCTAGAGGT---GGACTGTTGAGA-C-TG-CGGCGATCAGCGGAAC  AGAGTG
        ||               |                     | |                  |||     |     |        ||||||
P535    GA      CACAAT  CAGGGAAAGGGGGTGA-TGATCAAAAACCCATAGTCACCAAGAGTACTCC-AT-GTTCTAAACGG  AGAGTG
```

Fig. 2. Variable sites in exons 3, 4 and 5 and intron 4 of the P562, P535, and P535c alleles of the capuchin. The numbers at the top refer to the positions of the sites on the complete coding sequence (for exons 3, 4 and 5) and to the positions on an alignment of intron 4 sequences. The 5 positions in bold face are the critical amino acid residues, respectively, at positions 180, 229, 233, 277, and 285. From Boissinot et al. (1998)

Gene conversion has also been inferred to have occurred between alleles of the same locus. A very dramatic case was found between the P535 and P535c alleles in capuchins (Boissinot et al. 1998); P535c is a minor allele, and P535, P550, and P562 are the three major alleles. The P535 and P535c alleles are characterized by the same amino acids at positions 180, 277, and 285 (the three major critical positions, see above), and their exons 3 and 5 are identical (Fig. 2). Yet, they differ in exon 4 by 3.6% and in intron 4 by 2.0%. In contrast, P535c and P562 are identical in exon 4 and differ by only 0.4% in intron 4 (Fig. 2). A test by P.M. Sharp's modified method of Maynard Smith (1992) provides strong evidence ($p = 0.002$) that exon 4 and intron 4 of P535c have been converted by P562. Boissinot et al. (1998) have provided further evidence of frequent gene conversion between X-linked opsin alleles.

6 Origins of Color Vision Systems in Higher Primates

How did the various X-linked color vision systems in higher primates arise? In particular, were the red (P530) and green (P562) opsin genes in the Old World primates derived from two identical (or similar) alleles or from two alleles similar to the P535 and P562 alleles in New World monkeys? In the latter case, the two resultant duplicate genes, together with the blue opsin, would immediately confer trichromacy. Such an incorporation of two overdominant alleles into one chromosome has been proposed to be a possible advantage of gene duplication (Spofford 1972). Another intriguing question is whether the triallelic systems in different NWM species have a single origin or multiple origins?

To examine these issues, Shyue et al. (1995) sequenced exons 3, 4, and 5, and intron 4 of the three high-frequency alleles in the squirrel monkey and marmoset. Their data indicated that the human red and green opsin genes have an origin independent of that of the triallelic system in NWMs because an *Alu* repeat was present in the intron 4 sequences of all of the NWM alleles studied but absent from both human genes. This conclusion is further supported by the finding that the *Alu* repeat in intron 4 is also present in the three alleles from the capuchin, tamarin, and saki monkey (Fig. 3). In addition, the two howler monkey genes and the two human genes evidently have separated origins because the former contain the *Alu* repeat while the latter do not (Fig. 3)

Shyue et al.'s (1995) intron 4 sequence data favored the multi-origin hypothesis over the single-origin hypothesis because the three alleles in the squirrel monkey formed one monophyletic group, and so did the three alleles in the marmoset. The monophyly of the alleles in each species is also found in the capuchin, tamarin, and saki monkey (Fig. 3). This tree was inferred by the neighbor-joining method (Saitou and Nei 1987). For the parsimony analysis of insertions and deletions (indels, Fig. 3), the monophyly of the alleles in a species is supported by 6 indels in the marmoset, 1 indel in the tamarin, 5 indels in the squirrel monkey, and 7 indels in the capuchin. Therefore, both the neighbor-joining tree and the indel analysis strongly support the multi-origin hypothesis; in fact, they imply five independent origins for the triallelic system, one in each species (Fig. 3). Further, Fig. 3 implies that the two duplicate genes in the howler monkey had an origin independent of the triallelic system in the other NW monkeys.

However, the clustering of the alleles in each species is probably due to gene conversion. As described above, gene conversion occurs often between alleles in a population. Previously, it was thought that transferring of indels between alleles is unlikely to occur, so the clustering of alleles within each species was taken as evidence for an independent origin. However, the indel from positions 29 to 39 in intron 4 of capuchin P562 has obviously been transferred to P535c (Fig. 2). In Fig. 3, the indels within each species also suggest indel transferring. For example, in tamarin, 2 indels are shared by P543 and P562, 4 by P562 and P556, and 1 by P543 and P556. It is highly unlikely that so many indels arose independently in different alleles. So, this pattern suggests the transfer of indels between alleles by gene conversion, partially homogenizing the allelic sequences.

As gene conversion can drastically mislead phylogenetic analysis at noncritical sites for spectral tuning, the best data for inferring the evolutionary history of these X-linked color vision alleles and duplicate genes are probably the amino acid changes at the critical sites. Sites 180, 277, and 285 are the major critical sites, while sites 116, 229, 230, 233, and 309 have minor spectral tuning effects (see above). Sites 229 and 233 show a substitution pattern highly consistent with the three major sites (Fig. 3). However, sites 230 and 309 show very minor variation among sequences, so they are not informative for our purpose and will not be considered further. Site 116 is in exon 2, which is not under study. In terms of parsimony, Fig. 3 is highly implausible because it requires many parallel amino acid changes at the five critical sites considered. For example, at position 180, at least seven parallel changes between serine (S) and alanine (A) are required to explain the differences among alleles and genes at this site. When all five critical sites are considered together, the minimum number of substitutions required is 37 or 38, depending on whether one assumes that the amino acids at sites 180, 229, 233, 277, and 285 in the common

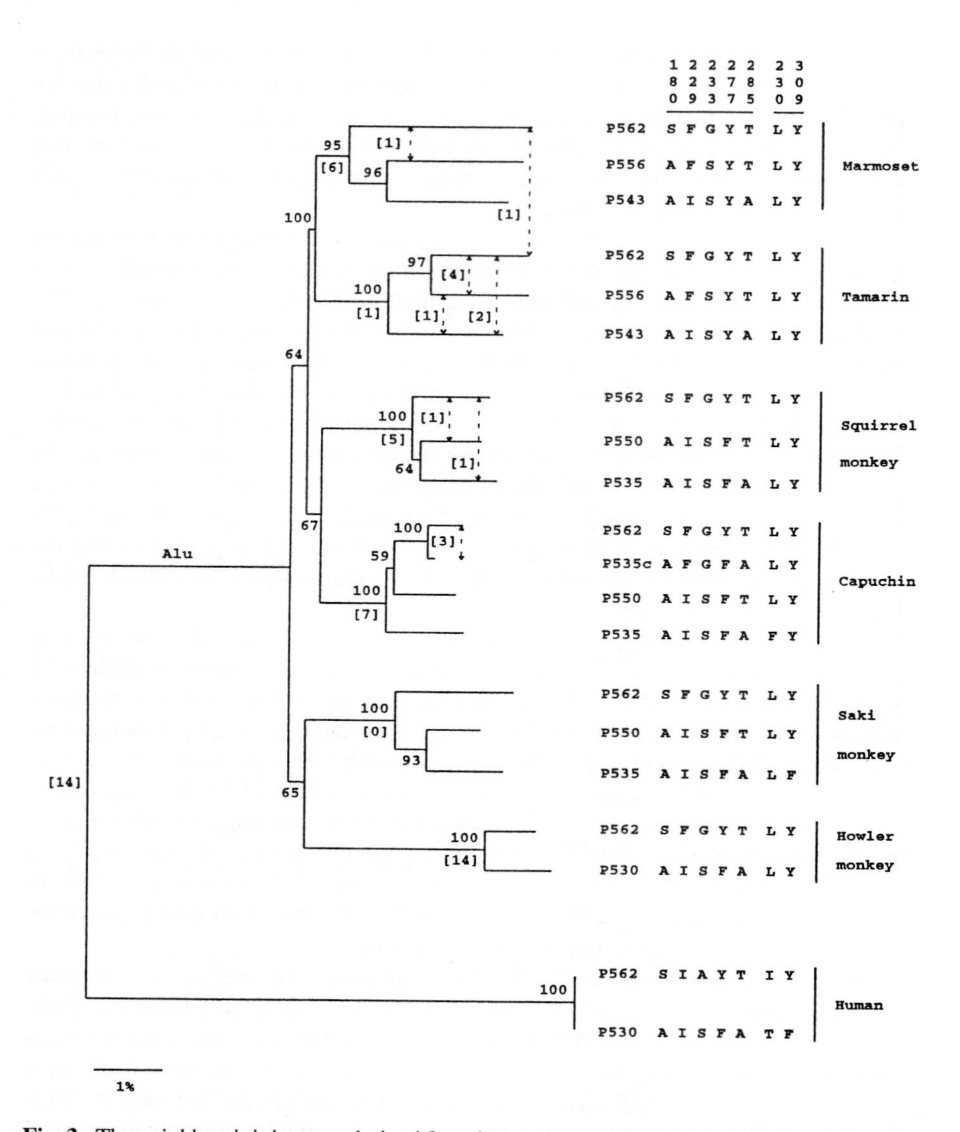

Fig. 3. The neighbor-joining tree derived from intron 4 sequences. The number at each node denotes the proportion of 500 bootstrap replicates that supported the subset of sequences. The number in brackets below a branch denotes the number of indels supporting the monophyly of the three alleles in each species. A dashed arrow with a number in brackets denotes the number of indels shared by two alleles. The 7 vertical numbers at the upper right of the tree refer to the positions of the 7 critical amino acid sites. From Boissinot et al. (1998)

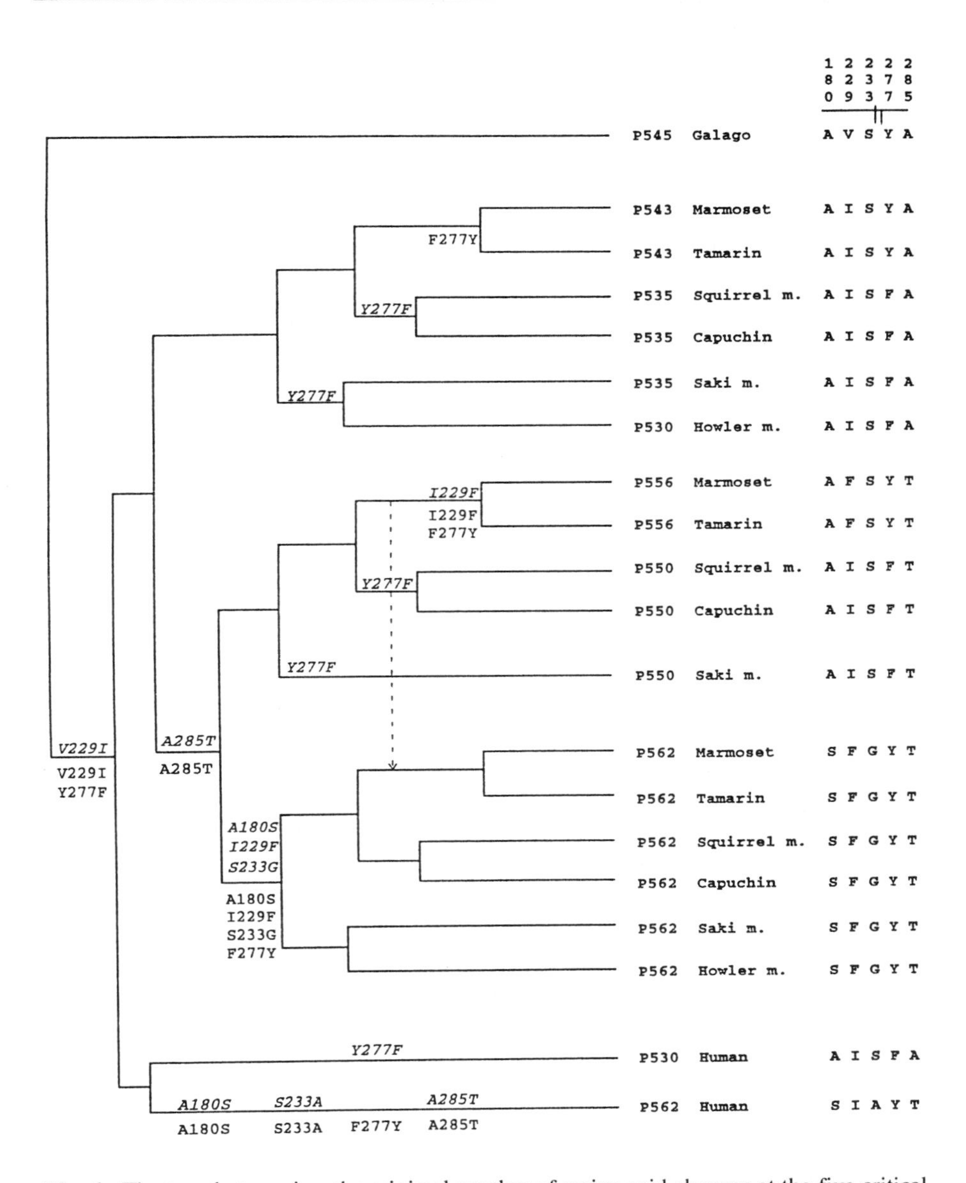

Fig. 4. The tree that requires the minimal number of amino acid changes at the five critical sites under the constraint of the species phylogeny in Fig. 3. Substitutions along branches are indicated as the ancestral amino-acid followed by its position on the complete coding sequence and then followed by the new amino-acid. Two equally parsimonious pathways are shown, one above branches (italic) and one below branches. A shorter tree (12 instead of 13 steps) is obtained if the marmoset-tamarin P556 allele has a recent origin and was derived from the P562 allele, as indicated by the dashed line and arrow. From Boissinot et al. (1998)

ancestor of all higher primates were S (Ser) F (Phe) G (Gly) Y (Tyr) T(Thr) (i.e., as in marmoset P562), or A (Ala) I (Ile) SFA (as in saki monkey P535, or AISYA as in marmoset P543). (Capuchin P535c is a minor allele and, for simplicity, is not considered.) In comparison, Fig. 4 requires only 11 or 12 amino acid substitutions in the higher primate sequences, assuming that the ancestral amino acids were AISFA (as in saki monkey, squirrel monkey and capuchin P535) or AISYA (as in marmoset and tamarin P543). Both scenarios require 13 substitutions for the entire tree (i.e., including also the two galago sequences) and are the most parsimonious scenarios subject to the constraint of the current knowledge of the phylogeny of these primates (i.e., the species phylogeny of Fig. 3). A slightly more parsimonious scenario (1 fewer change) is to assume that the marmoset-tamarin P556 was derived from the common ancestor of marmoset and tamarin P562. However, this scenario requires that the P550 allele has become lost in the common ancestor of the marmoset and tamarin (or in both species), despite the fact that it has persisted in the saki monkey, squirrel monkey, and capuchin.

In summary, since Fig. 4 requires less than one-third of the number of critical amino acid changes required by Fig. 3 (11 or 12 vs. 37 or 38), the single-origin hypothesis is far more plausible than the multi-origin hypothesis suggested by the intron 4 sequences. Moreover, a parsimony analysis with the galago sequences as outgroups suggests that the X-linked opsin alleles and duplicate opsin genes in the higher primates (including humans) were derived from a middle wavelength (green) opsin gene similar to either P543 (Jacobs 1993) or P535 (Winderickx et al. 1993), but not from a long wavelength (red) opsin gene as inferred without the galago sequences (Nei et al. 1997). Finally, the two howler monkey duplicate genes (P530 and P562) have a separate origin from the human red and green pigment genes, as suggested by Jacobs et al. (1996a), Boissinot et al. (1997), and Hunt et al. (1998). They were evidently derived from a combination of the P535 and P562 alleles. If this suggestion is substantiated by further data, it will provide the first example where the recombination of two overdominant alleles into one chromosome provides the selective advantage of gene duplication (Spofford 1972).

7 Evolutionary Mechanisms

It is now clear that trichromacy has arisen in higher primates in at least three different ways: (1) gene duplication and subsequent divergence in the common ancestor of the Old World primates, (2) a triallelic system in the NWMs except the howler monkeys, and (3) incorporation of two different alleles into one chromosome in the common ancestor of howler monkeys. These repeated occurrences of trichromacy point to the selective advantage of trichromacy over dichromacy. The advantage is commonly believed to be for detecting yellow or red fruits against a green foliage background, because dichromatic vision cannot distinguish between red and green colors.

Color vision is apparently of no use to a nocturnal animal. Indeed, all the nocturnal primates (the owl monkeys and several prosimians) that have been studied were found to have lost color vision — although their green\red opsin gene is functional and has been well conserved, their blue opsin gene has become nonfunctional. (Note that color vision requires at least two color photopigments.) In contrast, all diurnal prosimians studied were found to have two functional opsin genes (the blue and green\red opsin genes). Thus, for diurnal primates dichromacy is evidently more advantageous than monochromacy.

As noted above, the single-origin hypothesis for the triallelic system appears to be far more plausible than the multi-origin hypothesis. Note that the single-origin hypothesis implies that the triallelic system has persisted in these NW monkeys for more than 20 million years, because the divergence of the howler monkey lineage and the squirrel monkey-marmoset lineage has been estimated to be about 20 million years (Schneider et al. 1993, 1996). The antiquity of the system strongly suggests balancing selection for the maintenance of the system, because without balancing selection one or two of the three alleles would have become lost in a relatively short time (Kimura and Ohta 1969; Takahata and Nei 1990). Note that the existence of three rather than two polymorphic alleles at the X-linked opsin locus increases the chance of being heterozygous and thus of being trichromatic for a female NW monkey. On the other hand, if the triallelic system has multiple origins, then the repeated occurrences of the system and the numerous parallel amino acid substitutions required to explain the sequence differences within and between species (see Fig. 3) also suggest positive Darwinian selection. Actually, even the most parsimonious tree (Fig. 4) requires several parallel substitutions at some critical sites; e.g., at least 4 F→Y changes at site 277. Thus, regardless of which of the two hypotheses is true, the triallelic system probably has been maintained by natural selection.

8 Concluding Remarks

It is clear that much progress has been made in the evolutionary genetics of primate color vision. However, several issues remain unresolved. For example, although the single-origin hypothesis of the triallelic system in NWMs is more plausible than the multiorigin hypothesis, it remains to be substantiated by further data. The major difficulty is that gene conversion has drastically distorted the evolutionary history of genes and alleles. One way to overcome this problem is to have sequence data from many NWM species, so that the effect of gene conversion can be better understood and incorporated into phylogenetic analysis. As another example, the evolutionary genetics of prosimian color vision has not been well studied. Our preliminary data suggest that there is polymorphism at the X-linked locus in white sifakas (diurnal), but the extent of polymorphism is not clear and it is not known whether there is also polymorphism in other prosimian species, especially nocturnal ones. In addition, it is not clear how many times the blue opsin gene has become nonfunctional in prosimians. These studies will not only be useful for understanding the evolution of prosimian color vision genes, but also shed light on the lifestyle (nocturnal or diurnal) of prosimian ancestors.

Acknowlegements

This study was supported by NIH grants and the Betty Wheless Trotter Professorship.

References

Asenjo AB, Rim J, Oprian DD (1994) Molecular determinants of human red/green color discrimination. Neuron 12:1131-1138

Balding, DJ, Nichols RA, Hunt DM (1992) Detecting gene conversion: primate visual pigment genes. Proc Roy Soc London B 249:275-280

Blakeslee B, Jacobs GH (1985) Color vision in the ring-tailed lemur (*Lemur catta*). Brain Behav Evol 26:154-166

Boissinot S, Zhou Y-H, Qiu L, Dulai KS, Neiswanger K, Schneider H, Sampaio I, Hunt DM, Hewett-Emmett D, Li W-L (1997) Origin and molecular evolution of the X-linked duplicate color vision genes in howler monkeys. Zool Stud 36:360-369

Boissinot S, Tan Y, Shyue S-K, Schneider H, Sampaio I, Neiswanger K, Hewett-Emmett D, Li W-H (1998) Origin and antiquity of the X-linked triallelic color vision systems in New World monkeys. Proc Natl Acad Sci USA 95:13749-13754

Deeb SS, Jorgensen AL, Battist L, Iwasaki L, Motulsky AG (1994) Sequence divergence of the red and green visual pigments in great apes and humans. Proc Natl Acad Sci USA 91:7262-7266

Deeb SS, Lindsey DT, Hibiya Y, Sanocki E, Winderickx J, Teller DY, Motulsky AG (1992) Genotype-phenotype relationships in human red/green color vision defects: molecular and psychophysical studies. Am J Hum Genet 51:678-700

Deegan II JF, Jacobs GH (1996) Spectral sensitivity and photopigment of a nocturnal prosimian the bushbaby (*Otolemur crassicaudatus*). Am J Primatol 40:55-66

Goodman M, Porter CA, Czelusniak J, Page SL, Schneider H, Shoshani J, Gunnell G, Groves CP (1998) Toward a phylogenetic classification of primates based on DNA evidence complemented by fossil evidence. Mol Phyl Evol 9:585-598

Hunt D, Dulai KS, Cowing JA, Mollon JD, Bowmaker JK, Lee BB, Hewett-Emmett D, Li W-H (1998) Molecular evolution of trichromacy in primates. Vision Res 38:3299-3306

Ibbotson R, Hunt DM, Bowmaker JK, Mollon JD (1992) Sequence divergence and copy number of the middle- and long-wave photopigment genes in Old World monkeys. Proc Roy Soc London B 247:145-154

Jacobs GH (1984) Within-species variations in visual capacity among squirrel monkeys (*Saimiri sciureus*): Color vision. Vision Res 24:1267-1277

Jacobs GH (1993) The distribution and nature of color vision among the mammals. Biol Rev 68:413-471

Jacobs GH (1996) Primate photopigments and primate color vision. Proc Natl Acad Sci USA 93:577-581

Jacobs GH, Deegan JF II (1993) Photopigments underlying color vision in ringtail lemurs (*Lemur catta*) and brown lemurs (*Eulemur fulvus*). Am J Prim 30:243-256

Jacobs GH, Neitz J (1987) Inheritance of color vision in a New World monkey (*Saimiri sciureus*). Proc Natl Acad Sci USA 84:2545-2549

Jacobs GH, Neitz M, Deegan JF, Neitz J (1996a) Trichromatic colour vision in New World monkeys. Nature 382:156-158

Jacobs GH, Neitz M, Neitz J (1996b) Mutations in S-cone pigment genes and the absence of colour vision in two species of nocturnal primate. Proc Roy Soc London B 263:705-710

Kimura M (1980) A simple method for estimating evolutionary rates of base substitutions through comparative studies of nucleotide sequences. J Mol Evol 16:111-120

Kimura M, Ohta T (1969) The average number of generations until extinction of an individual mutant gene in a finite population. Genetics 63:701-709

Li W-H (1993) Unbiased estimation of the rates of synonymous and nonsynonymous substitution. J Mol Evol 36:96-99

Li W-H (1997) Molecular evolution. Sinauer Associates, Sunderland, MA

Maynard Smith J (1992) Analyzing the mosaic structure of genes. J Mol Evol 34:126-129

Merbs SL, Nathans J (1992) Absorption spectra of human cone pigments. Nature 356:433-435

Merbs SL, Nathans J (1993) Role of hydroxyl-bearing amino acids in differentially tuning the absorption spectra of the human red and green cone pigments. Photochem Photobiol 58:706-710

Mervis RF (1974) Evidence of color vision in a diurnal prosimian *Lemur catta.* Animal Learning and Behavior 2:238-240

Mollon JD, Bowmaker JK, Jacobs GH (1984) Variations of colour vision in a New World primate can be explained by a polymorphism of retinal photopigments. Proc Roy Soc London B 222:373-399

Nathans JD, Thomas D, Hogness DS (1986a) Molecular genetics of human color vision: The genes encoding blue, green, and red pigments. Science 232:193-202

Nathans J, Piantanida TP, Eddy RL, Shows TB, Hogness DS (1986b) Molecular genetics of inherited variation in human color vision. Science 232:203-210

Nei M, Zhang J, Yokoyama S (1997) Color vision of ancestral organisms of higher primates. Mol Biol Evol 14:611-618

Neitz M, Neitz J, Jacobs GH (1991) Spectral tuning of pigments underlying red-green color vision. Science 252:971-974

Petry HM, Harosi FI (1990) Visual pigments of the tree shrew (*Tupaia belangeri*) and greater galago (*Galago crassicaudatus*): a microspectrophotometric investigation. Vision Res 30:839-851

Reyniers E, Van Thienen M-N, Meire F, De Boulle K, Devries K, Kestelijn P, Willems PJ (1995) Gene conversion between red and defective green opsin gene in blue cone monochromacy. Genomics 29:323-328

Saitou N, Nei M (1987) The neighbor-joining method: a new method for reconstructing phylogenetic trees. Mol Biol Evol 4:406-425

Schneider H, Schneider MPC, Sampaio I, Harada ML, Stanhope M, Czelusniak J, Goodman M (1993) Molecular phylogeny of the New World monkeys (*Platyrrhini*, primate). Mol Phyl Evol 2:225-242

Schneider H, Sampaio I, Harada ML, Carroso CML, Schneider MPC, Czelusniak J, Goodman M (1996) Molecular phylogeny of the New World monkeys (*Platyrrhini* primates) based on two unlinked nuclear genes: IRBP intron 1 and ε-globin sequences. Am J Phyl Anthrop 100:153-179

Shyue S-K, Boissinot S, Schneider H, Sampaio I, Schneider MP, Abee CR, William L, Hewett-Emmett DL, Sperling HG, Cowing JA, Dulai KS, Hunt DM, Li W-H (1998) Molecular genetics of spectral tuning in New World monkey color vision. J Mol Evol 46:697-702

Shyue S-K, Hewett-Emmett D, Sperling HG, Hunt DM, Bowmaker JK, Mollon JD, Li W-H (1995) Adaptive evolution of pigment genes in higher primates. Science 269:1265-1267

Shyue S-K, Li L, Chang BH-J, Li W-H (1994) Intronic gene conversion in the evolution of human X-linked color vision genes. Mol Biol Evol 11:548-551

Spofford JB (1972) A heterotic model for the evolution of duplications. Brookhaven Symp Biol 23:121-143

Sun H, Macke KP, Nathans J (1997) Mechanisms of spectral tuning in the mouse green cone pigment. Proc Natl Acad Sci USA 94:8860-8865

Takahata N, Nei M (1990) Allelic geneology under overdominant and frequency-dependent selection and polymorphism of major histocompatibility complex loci. Genetics 124:967-978

Travis DS, Bowmaker JK, Mollon JD (1988) Polymorphism of visual pigment in a callitrichid monkey. Vision Res 28:481-490

Wikler KC, Rakie P (1990) Distribution of photoreceptor subtypes in the retina of diurnal and nocturnal primates. J Neurosci 10:3390-3401

Winderickx J, Battisti L, Hibiya Y, Motulsky AG, Deeb SS (1993) Haplotype diversity in the human red and green opsin genes: evidence for frequent sequence exchange in exon 3. Hum Mol Genet 2:1413-1421

Winderickx J, Battisti L, Motulsky AG, Deeb SS (1992) Selective expression of human X chromosome-linked green opsin genes. Proc Natl Acad Sci USA 89:9710-9714

Yokoyama S, Radlwimmer FB (1998) The "five-sites" rule and the evolution of red and green color vision in mammals. Mol Biol Evol 15:560-567

Yokoyama R, Yokoyama S (1990) Convergent evolution of the red- and green-like visual pigment genes in fish, *Astyanax fasciatus*, and human. Proc Natl Acad Sci USA 87:9315-9318

Zhao Z, Hewett-Emmett D, Li W-H (1998) Frequent gene conversion between human red and green opsin genes. J Mol Evol 46:494-496

Zhou Y-H, Li W-H (1996) Gene conversion and natural selection in the evolution of X-linked color vision in higher primates. Mol Biol Evol 13:780-783

Zhou Y-H, Hewett-Emmett D, Ward JP, Li W-H (1997) Unexpected conservation of the X-linked color vision gene in nocturnal prosimians: evidence from two bush babies. J Mol Evol 45:610-618

18
Avian Evolution During the Pleistocene in North America

ROBERT M. ZINK AND JOHN KLICKA

J. F. Bell Museum of Natural History, University of Minnesota, St. Paul, Minnesota 55108, USA

Abstract

Late Pleistocene glacial cycles (Wisconsinian, Illinoian) in North America are thought to have caused speciation in many songbirds. If true, such species pairs should have diverged little in mitochondrial DNA (mtDNA) sequence owing to the short time span involved. We compared mitochondrial DNA sequences of the cytochrome *b* gene for 21 such pairs. Gamma-corrected Kimura two-parameter distances averaged 7.1%, which, across a range of potential calibrations, is much larger than expected on average for speciation events that occurred within the last 250,000 years. Hence, most species presumed by previous authors to be of very recent origins are in fact much older. To discover effects of Late Pleistocene glaciations, we examined the genetic architectures of present day species. Predictions from coalescence theory were examined in light of our expectation that most bird populations have undergone recent population expansions. We examined the few available continent-wide phylogeographic surveys, finding unstructured haplotype trees, geographic patterns of nucleotide diversity, mismatch distributions and plots of the number of lineages vs time, all consistent with recent population expansions as species recolonized deglaciated areas.

1 Introduction

Historical patterns of speciation traditionally were inferred by associating events in earth history with particular phylogenetic hypotheses derived from morphology.

Key words. mitochondrial DNA, glacial cycles, speciation, phylogeography, genetic distances, molecular clock, Late Pleistocene Origins model, songbirds, sequence saturation, mismatch distribution, nucleotide diversity, likelifood ration tests, Song Sparrow, Timberline Sparrow, Red-winged Blackbird

These events in earth history share the common feature of providing geographic isolating barriers that allow for allopatric speciation. Typically barriers such as the uplift of a mountain range are invoked as the force driving speciation for specific taxonomic groups. Other barriers likely had more pervasive influences on the biota. The periodic glaciations of the Pleistocene are thought to have affected many organisms simultaneously. For example, in North America, taxonomists such as Mengel (1964, 1970) concluded that numerous bird species evolved as a result of geographic isolation induced by the last one or two cycles of glaciation. The models are often very specific as illustrated, for instance, by Hubbard's (1973) prediction that the speciation event leading to modern *Calcarius lapponicus* and *C. ornatus* was precipitated by the advance ca. 100,000 years before present (ybp) of the Laurentide Ice Sheet, the most recent major southward glacial advance. During this glacial period, Hubbard postulated that the common ancestral species was split into two refugia, allowing speciation to occur. Upon retreat of the glacier beginning 18,000 ybp, the new daughter species enlarged their ranges, eventually to their current sizes. Many other species pairs are thought to have arisen during this period or the previous glacial cycle, the Illinoian (beginning ca. 250,000 ybp). We (Klicka and Zink 1997) termed this hypothesis the Late Pleistocene Origins (LPO) model.

Morphological data do not allow testing such a hypothesis beyond the general phylogenetic pattern. That is, phenotypic patterns might be independent responses to environmental conditions over long periods rather than historical reflections of speciation at one or two points in the recent past. Molecular data offer tests of such historical theories by corroborating that species pairs are in fact sister species, and by documenting the degree of genetic differentiation between them. The degree to which DNA sequence evolution is clock-like is controversial (Avise 1994; Hillis et al. 1996); however, rough approximations can be made.

Klicka and Zink (1997) measured the mitochondrial DNA distance between 35 pairs of sister species of North American songbirds. If, as previously hypothesized, these species had originated as a result of the last one or two glacial cycles, then the mtDNA differences between sister species should be low. How "low" depends on the calibration of the molecular clock. Klicka and Zink (1997) used a widely accepted rate of 2% sequence divergence per million years, based on estimates for several avian orders (reviewed by Klicka and Zink 1997). More recently Fleischer et al. (1998) estimated a rate of 1.6% for cytochrome *b* sequences in Hawaiian songbirds. Given either of these rates, the 35 songbird species pairs should have had a mean mtDNA distance of less than 0.5%, if indeed speciation occurred during the Illinoian or Wisconsinian. However, the average divergence was 5.1%, a tenfold greater value than expected. Therefore we concluded that most of the 35 speciation events occurred much earlier than the Illinoian, some even in the Late Pliocene. Two other lines of evidence supported our rejection of the LPO model. The distribution of mtDNA distances between 13 additional songbird sister-species pairs, not previously hypothesized to fit the LPO model, did not differ significantly from that of the 35 pairs, indicating that what had been considered extremely recent species' origins did not differ from a "random" sample. We (Klicka et al. 1999) also discovered a single species pair, the Timberline Sparrow (*Spizella taverneri*) and Brewer's Sparrow (*S. breweri*), whose mtDNA characteristics were consistent with speciation occurring within the last 100,000 yr. In particular, this species pair exhibited a very low molecular distance, as well as a haplotype tree that did not exhibit reciprocal monophyly (that is the structure of the tree does not reflect species boundaries). None of the 35 species pairs exhibited these characteristics, which are genetic signatures of very recent speciation events, such as those predicted by the LPO.

Our conclusion stemming from the distribution of mtDNA distances was recently challenged by Arbogast and Slowinski (1998). In particular, these authors concluded that genetic distances derived from mtDNA sequence comparisons are biased (underestimated) because of "saturation" resulting from multiple base substitutions at single nucleotide positions. According to Arbogast and Slowinski, the songbird sister species originated much more recently than Klicka and Zink (1997) suggested, once corrections for saturation are made. Arbogast and Slowinski's preferred method of estimating the true evolutionary distance between species employed the gamma-HKY85 maximum likelihood model, which is becoming widely used (Swofford et al. 1996). Klicka and Zink (1998) suggested that this model likely overestimates the degree of saturation (most likely via inaccurate estimation of the alpha or shape parameter, not the model *per se*). For example, Arbogast and Slowinski (1998) derived a rate of 5%/million years (MY) for galliform birds from the cytochrome *b* data of Randi (1996). We analyzed the same data, finding, for example, that the uncorrected cytochrome *b* distance between *Gallus gallus* and *Alectoris magna*, 0.14, has a gamma-HKY85 distance of 1.02, suggesting a significant underestimate of the "true" evolutionary distance. Although we believe that these galliform sequences are saturated, this corrected distance is likely biologically meaningless. Thus, Arbogast and Slowinski's (1998) calibration based on these data (as well as one they derived from higher primates) is dubious. Nonetheless, even accepting these recomputed distances and the new calibration (5%/MY), 90% of their gamma-HKY85 divergence dates exceed one million years, a decisive refutation of the LPO model (Klicka and Zink 1998).

In this chapter we use a different maximum likelihood model to estimate the degree to which saturation (multiple hits) affects mtDNA distance estimates and we employ two different calibrations on the resultant distances. The results of this new analysis do not change the conclusion of Klicka and Zink (1997, 1998) that most species pairs of North American songbirds originated earlier than most previous authors believed.

Nonetheless, major effects of glaciers on the genetics of birds and other organisms seem unarguable. It is clear that major biomes were displaced and often greatly reduced in size (Pielou 1991). Such massive environmental perturbations must have left genetic "imprints" at some taxonomic level. If these imprints were not the speciation events themselves, the search for such imprints should proceed to lower taxonomic levels, namely that within species (termed phylogeography; Avise 1994).

Because many bird species must have recently experienced population increases associated with recolonization of deglaciated areas, we are interested in whether avian species show the predicted imprints (Fig. 1), such as geographically unstructured haplotype trees (Nee et al. 1995), a northward decrease in nucleotide diversity (Hewitt 1996), and particular shapes of mismatch distributions (Rogers and Harpending 1992; Rogers 1996) and plots of the ln number of lineages versus time (Nee et al. 1995). Unfortunately, few North American species have been studied thoroughly enough to test the theoretical predictions in a general way; we focus on data from a recent phylogeographic study (Fry and Zink 1998) of the Song Sparrow (*Melospiza melodia*).

2 Methods

2.1 Interspecific Analyses

We use the same mtDNA sequence data as Klicka and Zink (1997); restriction site data were ignored because of the difficulty in making corrections for saturation (nonetheless, correlation coefficients for sequence and restriction site distances are often high). To account for potential saturation effects, we computed gamma-corrected Kimura (1980) two-parameter (K2P) distances in Paup*, following the protocol of Fleischer et al. (1998). We derived a gamma value of 0.4 using the Sullivan et al. (1995) parsimony method in Paup* (Swofford 1996) using a minimum length tree for the 21 species for which over 1000 bp of cytochrome *b* existed. To be conservative, we subtracted 0.01 from these pairwise distances to account for ancestral haplotype diversity; Moore (1995) estimated a value of 0.007. To estimate divergence dates (i.e., "lineage sundering" as defined by Avise and Walker 1998), we divided the distances by 0.016, the gamma-K2P rate estimated for Hawaiian songbird cytochrome *b* evolution (Fleischer et al. 1998). We also considered data (Randi 1996) from galliform birds, which were used by Arbogast and Slowinski (1998) to derive a rate of 5%/MY. Using a calibration date of 20 MY (Randi 1996), and an alpha value of 0.4 (from the Sullivan et al. parsimony method in Paup*) we obtained a rate of 0.9%/MY. It has been suggested that galliforms and non-passerines evolve at a slower rate than passerines, although the validity of this claim is uncertain. If the rate were 2%/MY, then the average distance should be closer to 40%, whereas the gamma-K2P distance is ca. 20%. Using a calibration date of 20 MY (Randi 1996), and an alpha value of 0.08 (calculated under the gamma-HKY85 model) we obtained a gamma-K2P rate of 3.5%/MY. It appears likely that there is not a single rate for all birds. Hence, to test the LPO model as broadly as possible, we used both the songbird rate of Fleischer et al. (1998) and the gamma-K2P galliform rate of 3.5%.

The use of molecular dating methods involves several assumptions. First, the sequences should be evolving in a selectively neutral manner. We have no evidence that our sequences are under strong selection (Fry and Zink 1998). Most changes in

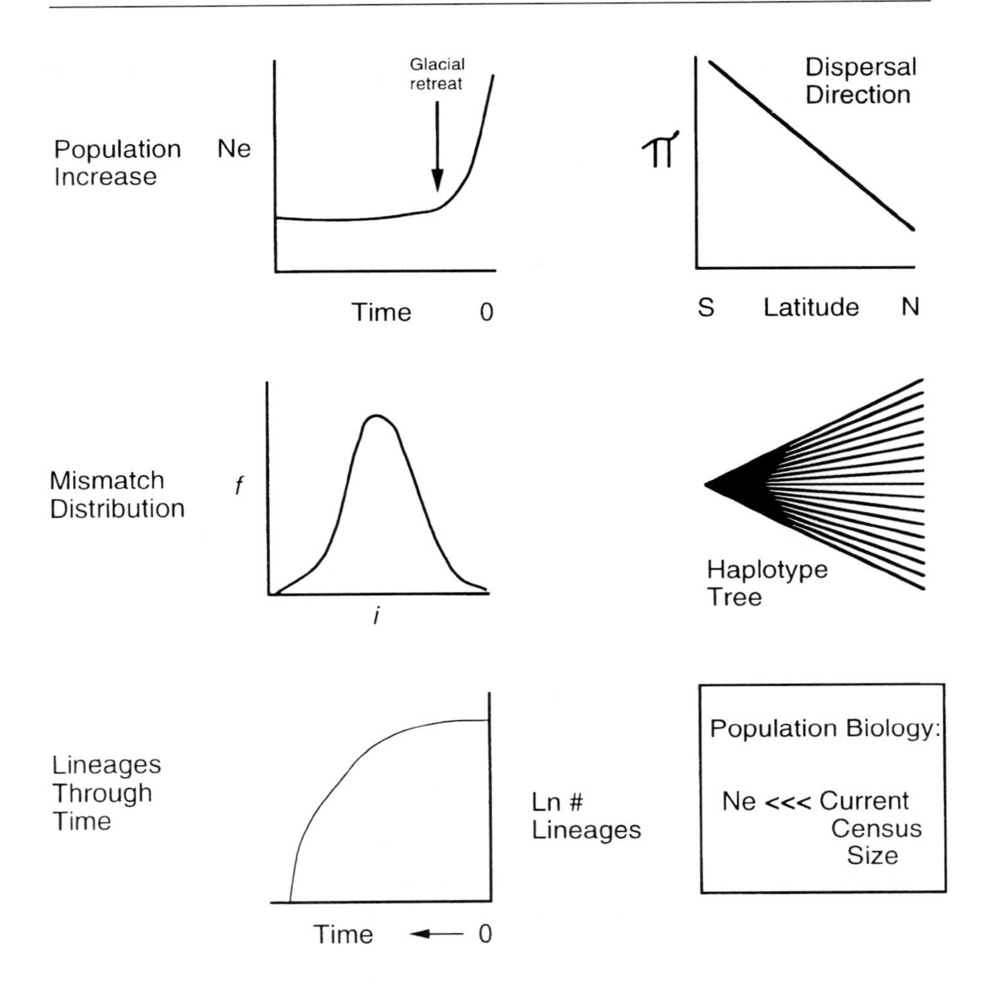

Fig. 1. Population genetic expectations derived from paleontological data and coalescence theory. Upper left: prediction that populations of birds increased over the past 18,000 yr. Middle left: expected shape of mismatch distribution (see text) for expanding population. Bottom left: shape of curve expected for expanding population. Upper right: predicted northward decrease in π as populations expand from Pleistocene refugia and colonize deglaciated areas. Center right: diagrammatic unstructured haplotype tree typical of expanded populations. Bottom right: *Ne* estimated from genetic data often far less than current census sizes (see Avise 1994)

our data set are silent, third-position transitions, which should be approximately selectively neutral. Secondly, sequences should be evolving at a constant rate. We tested for rate heterogeneity by constructing a gamma-HKY85 (alpha = 0.22) maximum likelihood tree, both with and without a molecular clock assumption, for the 21 species that are represented by ca. 1000 bp of cytochrome *b*. For the analysis in which a molecular clock was enforced, we rooted the tree at the midpoint; rooting on an ingroup taxon results in a spurious result (Klicka and Zink 1998). To assess whether the sequences are evolving in a clock-like fashion, a log-likelihood ratio test (LRT) can be performed by comparing twice the difference between the log-likelihoods for each tree to a chi-squared distribution with n-2 df, where n is the number of taxa (Huelsenbeck and Rannala 1997). A nonsignificant chi-squared value indicates no significant departure from a clock-like pattern of sequence evolution. We recognize that this test has been questioned (Goldman 1993), and we also question the computation of the df; however, it is the most commonly used test available.

Table 1. Gamma-corrected Kimura two-parameter mtDNA distances between 21 pairs of avian sister species. These values are not corrected for within species variation

Piranga olivacea-ludoviciana	0.07
Passerina cyanea-amoena	0.08
P. cyanea-versicolor	0.08
Sialia mexicana-sialis	0.06
Cardinalis cardinalis-sinuatus	0.11
Calcarius lapponicus-mccownii	0.11
C. lapponicus-ornatus	0.12
Oporonis philadelphia-tolmiei	0.02
O. philadelphia-agilis	0.09
Spizella breweri-pallida	0.07
Pheucticus melanocephalus-ludovicianus	0.05
Cyanocitta cristata-stelleri	0.17
Polioptila melanura-nigriceps	0.06
P. melanura-californica	0.05
Pipilo aberti-crissalis	0.03
P. aberti-fuscus	0.03
Toxostoma rufum-longirostre	0.08
T. lecontei-redivivum	0.07
T. lecontei-crissalis	0.07
T. redivivum-crissalis	0.06
T. bendirei-cinereum	0.02

2.2 Intraspecific Analyses

Data for the Song Sparrow comprise 700 bp of the "control region" for 95 individuals sampled from throughout the breeding range (Fry and Zink 1998). Computation of nucleotide diversity (π) and the neighbor joining (NJ) tree is described in Fry and Zink (1998). We computed a mismatch distribution, defined as the distribution of pairwise raw sequence differences between all pairs of 95 individuals. We computed the mismatch distribution expected for an exponentially growing population as a poisson distribution with the same mean as the observed distribution (Slatkin and Hudson 1991). We compared distributions with a Kolmogorov-Smirnov one-sample test. On the NJ tree we computed the ln number of lineages vs. time, with time taken as the overall mtDNA distance (Nee et al. 1995).

3 Results

3.1 Interspecific Analyses

The LRT for the molecular clock revealed no evidence of deviations from clock-like sequence evolution (ln-likelihood with clock = 6788.0, without clock = 6773.2, $\chi = 29.6$, df = 19, $P > 0.05$). The average gamma-K2P distance was 7.1% (Table 1). Use of either calibration (Fig. 2) provides no support for the LPO model; over 90% of values exceed 500,000 years.

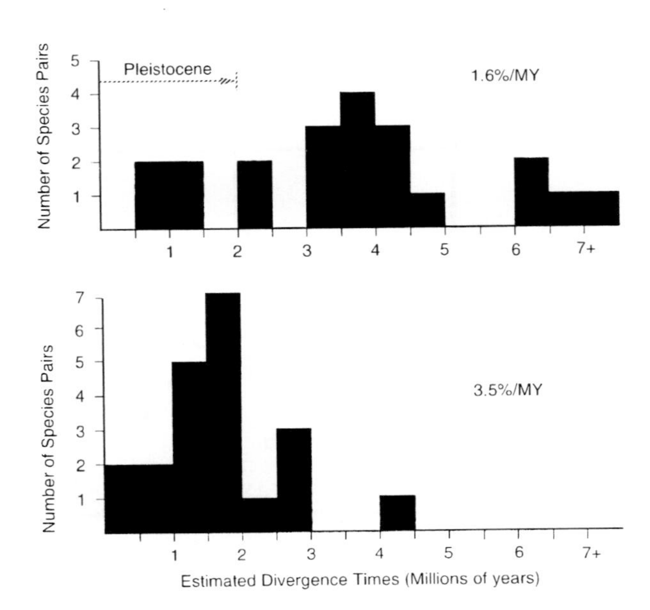

Fig. 2. Divergence dates (data from Table 1) plotted according to calibrations for songbirds (top) and galliform birds (bottom)

3.2 Intraspecific Studies

Geographic patterns of π are shown for two species (Fig. 3), the Song Sparrow and the Red-winged Blackbird (*Agelaius phoeniceus*). The latter is consistent with the predicted pattern whereas that for the Song Sparrow is not (Fig. 1). The mismatch distribution (Fig. 4) has the shape of a growing population although the observed distribution differed significantly from expected ($P < 0.01$). The plot (Fig. 5) of number of lineages vs time is consistent with that expected for a growing population. The NJ tree (see Fry and Zink 1998, their Fig. 2) shows relatively little structure.

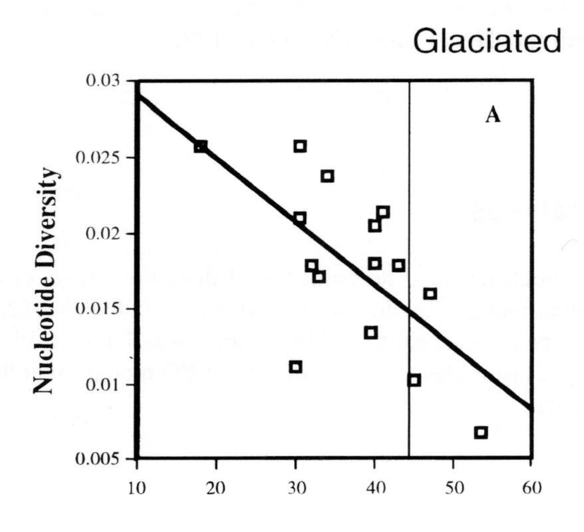

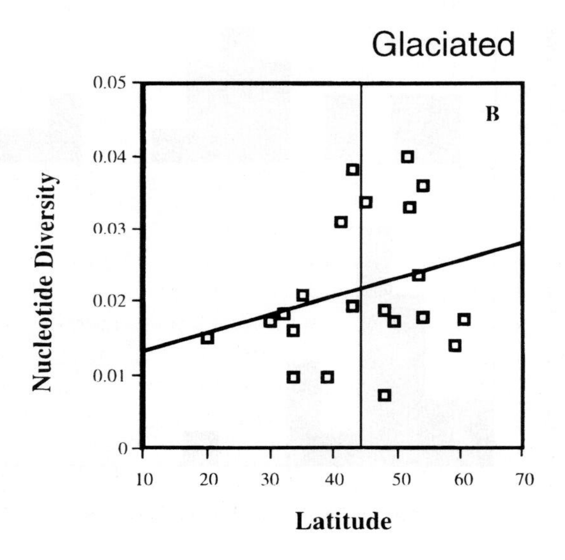

Fig. 3. Plot of latitude vs nucleotide diversity for Red-winged Blackbird (top; data from Ball et al. 1988, plotted by Fry and Zink 1998) and Song Sparrow (restriction site data from Fry and Zink [1998]). The vertical line denotes the approximate southern limit of the Laurentide Ice Sheet 18,000 ybp (Pielou 1991)

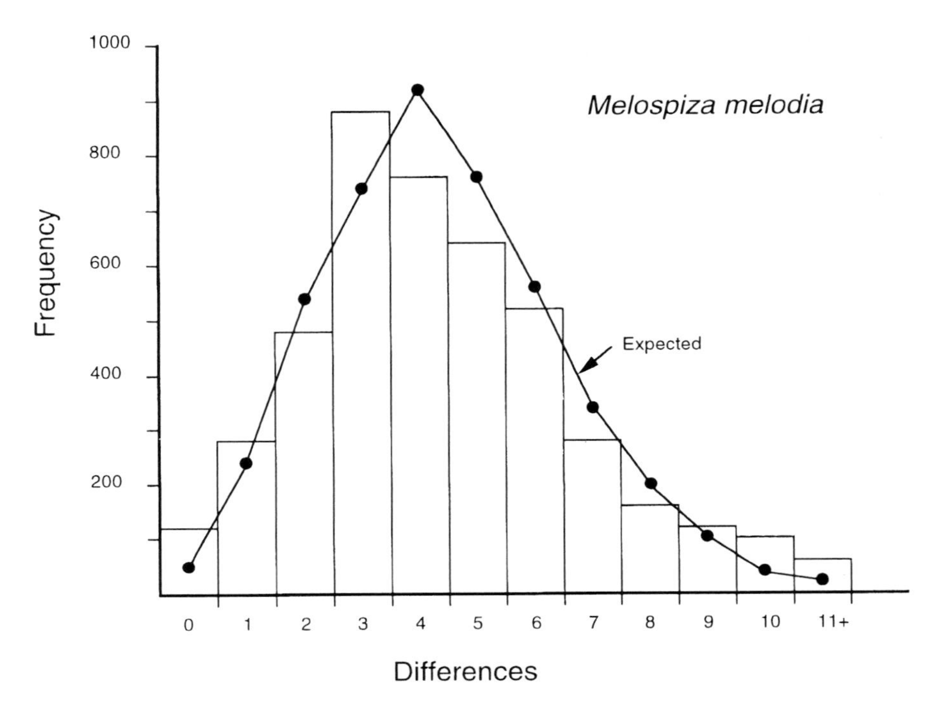

Fig. 4. Mismatch distribution for control region (700 bp) sequence data for 95 individual song sparrows (Fry and Zink 1998). Superimposed is expected distribution for an exponentially expanding population

4 Discussion

4.1 Tests of the LPO Model

Morphological and distributional data provided tempting hypotheses about the evolutionary consequences of Pleistocene glaciations. For example, the geographic distributions of many species pairs with eastern and western representatives abut in the Great Plains (Rising 1983). One would expect just such a result if common ancestors had been split by the Laurentide Ice Sheet into disjunct refugia, and the newly evolved daughter species had just attained a parapatric distribution. Lack of sympatry could be a result of insufficient time, or insufficient divergence in ecological requirements since deglaciation (≤18,000 years). However, the (uncorrected) mtDNA distances for the Great Plains species pairs averaged over 5%, and the values range from very small to over 10%, suggesting that these species evolved neither recently nor contemporaneously (Fig. 1 of Klicka and Zink 1997).

If the LPO model were correct, divergence values should cluster within the last 250,000 yr. Furthermore, we might expect two peaks of values, corresponding to

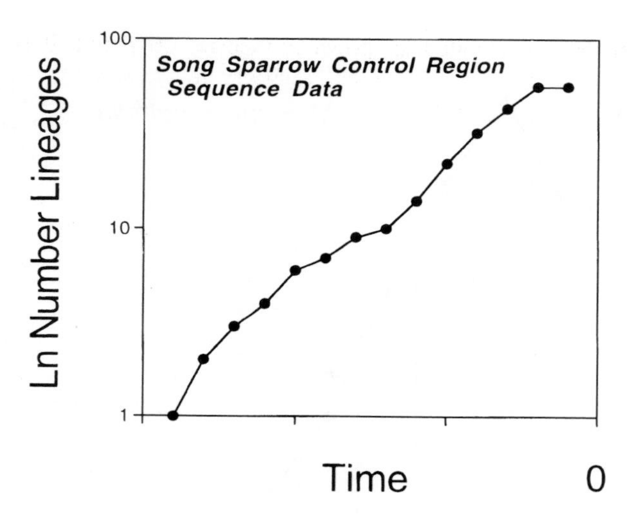

Fig. 5. Plot of ln number lineages vs time for Song Sparrow control region sequences derived from NJ tree in Fry and Zink (1998)

each of the last glacial advances (we recognize that variances around mtDNA genetic distances and relatively small sample sizes might prevent recovery of a bimodal distribution with closely spaced peaks). Although confidence limits around divergence dates are large (Hillis et al. 1996) we find it implausible that either of these distributions (Fig. 2) could be explained by speciation events occurring within the past 250,000 yr. For example, if it were true that our original distribution of 35 sister-species mtDNA distances with a mean of 5.1% "fit" into the preceding 250,000 yr time interval, it would require rethinking the field of molecular systematics. Such a result would indicate levels of saturation that possibly would prevent phylogenetic inference owing to the homoplasy resulting from the multiple hits that would predominate at variable positions. Rates would have to be of the order of 20%/MY. Alternatively, if all 35 species pairs did originate less than 250,000 ybp, many less than 100,000 ybp, it would reveal an astonishing amount (more than tenfold) of molecular rate heterogeneity in different sister species. However the LRT for a molecular clock revealed no rate heterogeneity. Thus, it seems clear that these species did not evolve contemporaneously, and that their evolution was more ancient than previously considered (Klicka and Zink 1997, 1998).

Modern paleoecological information suggests that glaciations have influenced the North American continent for 2.4 million years (Webb and Bartlein 1992). Clearly glaciers can provide potent isolating barriers, and we have no doubt that many speciation events occurred because of glaciation. We simply extend previous ideas about the role of glacial cycles in speciation over an extended time period. Although prevailing views of the Pleistocene, at least for birds, is one of much diversification, it is possible that Late Pleistocene glaciations played a major role in the extinction of birds, by displacing, fragmenting and reducing habitats. Zink and Slowinski (1995) concluded that passerine bird diversification in fact slowed during the Pleistocene, rather than accelerated, as previous theories predicted. It is clear that we have much to learn about avian evolution during the recent past (see Avise and Walker 1998).

4.2 Glaciers and Phylogeography

Simple comparisons of current breeding ranges of birds, and the extent of the Laurentide Ice Sheet (Pielou 1991), reveal that bird distributions have changed dramatically over the past 15,000 years. Although the precise distribution of habitats during the height of the last glacial period is unknown, it seems likely that most habitats (perhaps excluding marsh) expanded from refugia as the glacier receded. Consequently we predict that many bird species experienced population increases as they recolonized deglaciated areas. Thus our null prediction is that the genetic imprints of Late Pleistocene glaciations will be those associated with population increases.

Many studies have found unstructured haplotype trees within North American species (Zink 1996, 1997), including the Song Sparrow (Fry and Zink 1998). Such trees are consistent with recent population increases. Some species, such as the Fox Sparrow (*Passerella iliaca*), show strong phylogeographic structure, but it is not clear whether this is in fact one species owing to the relative depth of the mtDNA divisions (Zink 1994). If widespread panmictic species were indeed fractured by the Laurentide Ice Sheet, it is possible that we could not as yet detect genetic divisions (Avise and Walker 1998). It takes approximately 4Ne generations for a lineage, once split to yield daughter species that have mtDNA haplotypes that exhibit reciprocal monophyly (Ne = effective population size). This process could require greater than 18,000 yr. For example, Ball and Avise (1992) suggested an Ne value of 52,000 for the Song Sparrow, and with a generation time of 1–2 yr, it might require 200,000 yr for evidence of a Wisconsinian split to be detected with our current mtDNA surveys. Thus the unstructured haplotype tree for the Song Sparrow could reflect an insufficient amount of time post-barrier. Alternatively, even if Song Sparrows were isolated in refugia during the Wisconsinian, subsequent gene exchange between refugial populations would negate the isolating effect of the glacier; Zink and Dittmann (1993) suggested that gene flow was high. Hence, although unstructured haplotype trees are consistent with post-glacial population expansion, they might also reflect polymorphism in pre-glaciation populations. Other analyses are required. Clearly, if isolation induced during the last glacial cycle has not yet yielded detectable genetic imprints within species, it did not cause speciation itself.

Few mismatch distributions have been computed for North American birds (Zink 1997). The mismatch distribution for the Song Sparrow (Fig. 4) exhibits the general shape of an exponentially expanding population but differs from the theoretical expectation. Possibly, populations of Song Sparrows have been increasing but not exponentially, which the predicted curve represents (Slatkin and Hudson 1991). Furthermore, some geographic heterogeneity (Fry and Zink 1998) might cause departure from expectation. For the California Gnatcatcher (*Polioptila californica*), found coastally from California to the southern extent of Baja California Sur (Mexico), the mismatch distribution is consistent with major recent population increases (Zink et al. unpubl. data). Hence, we conclude that mismatch distributions support the general expectation of recent post-glacial population expansions of North American birds. Clearly more geographically extensive studies are required.

The geographic pattern of nucleotide diversity in the Song Sparrow does not match the simple prediction of a linear south to north increase, as was found for the Red-winged Blackbird (Fig. 3). These are the only two studies in which a general continental pattern can be estimated. We can only conclude at this time that species recolonized deglaciated areas in idiosyncratic ways.

The distribution of the number of lineages vs time (Fig. 5) is also consistent with a growing population. No other analyses of this type are published although several are underway (G. F. Barrowclough, pers. comm.).

In summary, searching for the imprints of Late Pleistocene glaciations is hampered by lack of available data. Although the North American avifauna is probably the best studied genetically, few continent-wide surveys exist that permit tests of genetic expectations (Fig. 1). Available data support the general prediction of population increases associated with recolonization of recently deglaciated areas.

Acknowledgments

We thank A. Kessen for comments on the manuscript.

References

Arbogast BS, Slowinski JB (1998) Pleistocene speciation and the mitochondrial DNA clock. Science 282:1955a

Avise JC (1994) Molecular markers, natural history and evolution. Chapman and Hall, New York

Avise JC, Walker D (1998) Pleistocene phylogeographic effects on avian populations and the speciation process. Proc Roy Soc Lond B 265:457-463

Ball RM Jr., Avise JC (1992) Mitochondrial DNA phylogeographic differentiation among avian populations and the evolutionary significance of subspecies. Auk 109:626-636

Ball RM Jr., Freeman S, James FC, Bermingham E, Avise JC (1988) Phylogeographic population structure of Red-winged Blackbirds assessed by mitochondrial DNA. Proc Natl Acad Sci USA 85:1558-1562

Fleischer RC, Mcintosh CE, Tarr CL (1998) Evolution on a volcanic conveyor belt: using phylogeographic reconstructions and K-Ar-based ages of the Hawaiian Islands to estimate molecular evolutionary rates. Mol Ecol 7:533-545

Fry A, Zink RM (1998) Geographic analysis of nucleotide diversity and song sparrow (*Aves*: Emberizidae) population history. Mol Ecol 7:1303-1313

Goldman N (1993) Statistical tests of models of DNA substitution. J Mol Evol 36:182-198

Hewitt GM (1996) Some genetic consequences of ice ages, and their role in divergence and speciation. Biol J Linn Soc 58:247-276

Hillis DM, Mable BK, Moritz C (1996) Applications of molecular systematics: the state of the field and a look to the future. In: Hillis DM, Moritz C, Mable BK (Eds) Molecular systematics. Sinauer, Sunderland, MA, pp 515-543

Hubbard JP (1973) Avian evolution in the aridlands of North America. Living Bird 12:155-196

Huelsenbeck JP, Rannala B (1997) Phylogenetic methods come of age: testing hypotheses in an evolutionary context. Science 276:227-232

Kimura M (1980) A simple method for estimating evolutionary rates of base substitutions through comparative studies of nucleotide sequences. J Mol Ecol 16:111-120

Klicka J, Zink RM (1997) The importance of recent Ice Ages in speciation: a failed paradigm. Science 277:1666-1669

Klicka J, Zink RM (1998) Pleistocene speciation and the mitochondrial DNA clock: a response to Arbogast and Slowinski. Science 282:1955a

Klicka J, Zink RM (1999) Pleistocene effects on North American songbird evolution. Proc Roy Soc Lond B 266:695-700

Klicka J, Zink RM, Barlow JC, McGillivray WB, Doyle TJ (1999) Evidence supporting the recent origin and species status of the Timberline Sparrow. Condor (in press)

Mengel RM (1964) The probable history of species formation in some northern wood warblers (Parulidae). Living Bird 3:9-43

Mengel RM (1970) The North American Great Plains as an isolating agent in bird speciation. Univ Kans Dept Geol Spec Publ 3:279-340

Moore WS (1995) Inferring phylogenies from mtDNA variation: mitochondrial-gene trees versus nuclear-gene trees. Evolution 49:718-726

Nee S, Holmes EC, Rambaut A, Harvey PH (1995) Inferring population history from molecular phylogenies. Phil Trans Roy Soc Lond B 349:25-31

Pielou EC (1991) After the Ice Age, the return of life to glaciated North America. Univ Chicago Press, Chicago, IL, pp 1-366

Randi E (1996) A mitochondrial cytochrome *b* phylogeny of the *Alectoris* partridges. Mol Phyl Evol 6:214-247

Rising JD (1983) The Great Plains hybrid zones. In: Johnston RF (Ed) Current ornithology. Vol 1. Plenum Press, NY, pp 131-155

Rogers AR (1996) Genetic evidence for a Pleistocene population explosion. Evolution 49:608-615

Rogers AR, Harpending H (1992) Population growth makes waves in the distribution of pairwise genetic differences. Mol Biol Evol 9:552-569

Slatkin M, Hudson RR (1991) Pairwise comparisons of mitochondrial DNA sequences in stable and exponentially growing populations. Genetics 129:555-562

Sullivan, J, Holsinger KE, Simon C (1995) Among-site rate variation and phylogenetic analysis of 12S rRNA in sigmodontine rodents. Mol Biol Evol 12:988-1001

Swofford DL (1996) Paup*: Phylogenetic analysis using parsimony (and other methods) version 4.0. Sinauer, Sunderland, MA

Swofford DL, Olsen GJ, Waddell PJ, Hillis, DM (1996) Phylogenetic inference. In: Hillis DM, Moritz C, Mable BK (Eds) Molecular systematics. Sinauer, Sunderland, MA, pp 407-514

Webb T, Bartlein PJ (1992) Global changes during the last 3 million years: climatic controls and biotic responses. Annu Rev Ecol Syst 23:141-173

Zink RM (1994) The geography of mitochondrial DNA variation, population structure, hybridization, and species limits in the fox sparrow (*Passerella iliaca*). Evolution 48:96-111

Zink RM (1996) Comparative phylogeography in North American birds. Evolution 50:308-317

Zink RM (1997) Phylogeographic studies of North American birds. In: Mindell DP (Ed) Avian molecular evolution and systematics. Academic Press, San Diego, CA, pp 301-324

Zink RM, Dittmann DL (1993) Gene flow, refugia, and evolution of geographic variation in the song sparrow (*Melospiza melodia*). Evolution 47:717-729

Zink RM, Slowinski JB (1995) Evidence from molecular systematics for decreased avian diversification in the Pleistocene epoch. Proc Natl Acad Sci USA 92:5832-5835

19
Genetic Diversity of Human Populations in Eastern Asia

Keiichi Omoto

International Research Center for Japanese Studies, 3-2 Goryo Oeyama-cho, Nishikyo-ku, Kyoto 610-1192, Japan

Abstract

This chapter summarizes the results of my population genetic studies concerning various ethnic groups in eastern Asia and western Pacific with special reference to the origins of (1) the Japanese peoples, and (2) the so-called Mongoloid groups. On the first topic, I will show that Kazuro Hanihara's "dual structure model" is only partly supported by genetic evidence. As for the second, I will argue that the genetic diversity among eastern Asian, Pacific and American populations is so extensive that the classic racial concept of "Mongoloid" is no longer tenable. It is also indicated that contrary to the classic view of a single south-east Asian origin of these populations, there were at least two independent sources for human migrations in eastern Asia during the Upper Paleolithic Times from about 50 000 to 20 000 years BP.

1 Introduction

Since 1966, I have carried out a series of human population genetic studies in eastern Asia. My main objectives were twofold. On the one hand, I have tried to elucidate the genetic origins of some indigenous peoples such as the Ainu of Hokkaido, the northernmost island of Japan, and the Negritos of the Philippines, whose origins have been controversial among anthropologists. On the other hand, I have wished to understand the micro-evolutionary history of modern humans in

Key words. population genetics, molecular anthropology, blood groups, protein polymorphism, classic genetic marker, ethnic group, East Asia, Japanese, Ainu, Negrito, Mongoloids, genetic distance, dendrogram, Oceania, Sundaland

this area by analyzing geographical and genetic diversities. The methods I used were mainly a combination of field work on blood collections and laboratory work on "classic" genetic markers such as blood groups and protein types, followed by multivariate statistical analyses of genetic distance. Only after the 1980's was the DNA data, mostly of mtDNA, made available using the samples I have collected, largely resulting in similar findings of genetic relationship to those obtained by "classic" markers.

Here, I will first briefly introduce my earlier studies on two indigenous populations of eastern Asia, namely the Ainu of northern Japan and the Negritos of the Philippines. Then, I will focus on two topics which have arisen from my studies: (1) the origins of the Japanese populations, and (2) the genetic diversity of the so-called Mongoloid group.

1.1 The Genetic Origins of the Ainu

My study on the Ainu was carried out as part of the International Biology Program (IBP) and started in 1966, lasting for about 6 years. Altogether, I was able to obtain about 500 blood samples from the Ainu individuals living in the region of Hidaka, Hokkaido. At that time, studies on the protein polymorphisms detected by electrophoresis and genetic distance analysis with the construction of phylogenetic trees for human ethnic groups developed by L. L. Cavalli-Sforza and A. W. F. Edwards

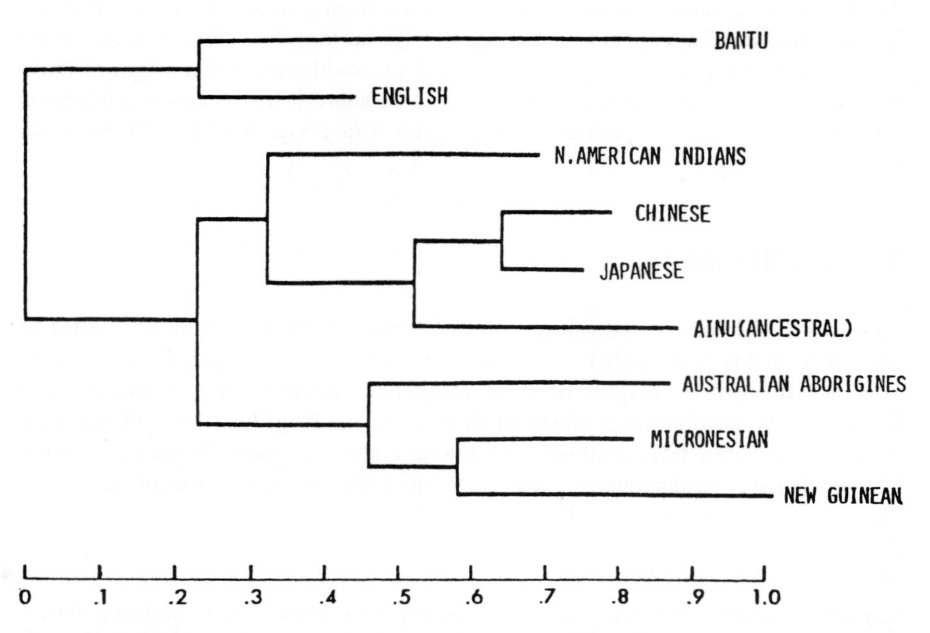

Fig. 1. Dendrogram based on the data from 16 classic genetic markers, indicating that the Ainu belong genetically to eastern Asian populations (Omoto 1972)

(1967) were at the frontier of human population genetic studies. Adopting these methods, I was able to negate the popular notion in classic anthropology that the Ainu are an old European race (Fig. 1). It was shown that they should belong to the "Mongoloid" or the truly Asian population group who may have migrated to Japan during the Upper Paleolithic Time (Omoto 1972, 1975).

1.2 The Genetic Origins of the Negritos

The study of the Philippine Negritos started in 1975 and lasted for about ten years. I was able to obtain nearly 1000 blood samples of different tribal groups widely scattered in remote areas of the Philippines. My main findings were twofold. Firstly, I was able to show that the Negritos have little genetic affinities to the Africans, contrary to the classic anthropological view based on morphological similarities, such as dark skin, frizzle hair and small stature. My interpretation was that the Negritos are derived from the inhabitants of Sundaland, a large landmass that existed during the last glacial maximum around 20000 years ago, covering the region

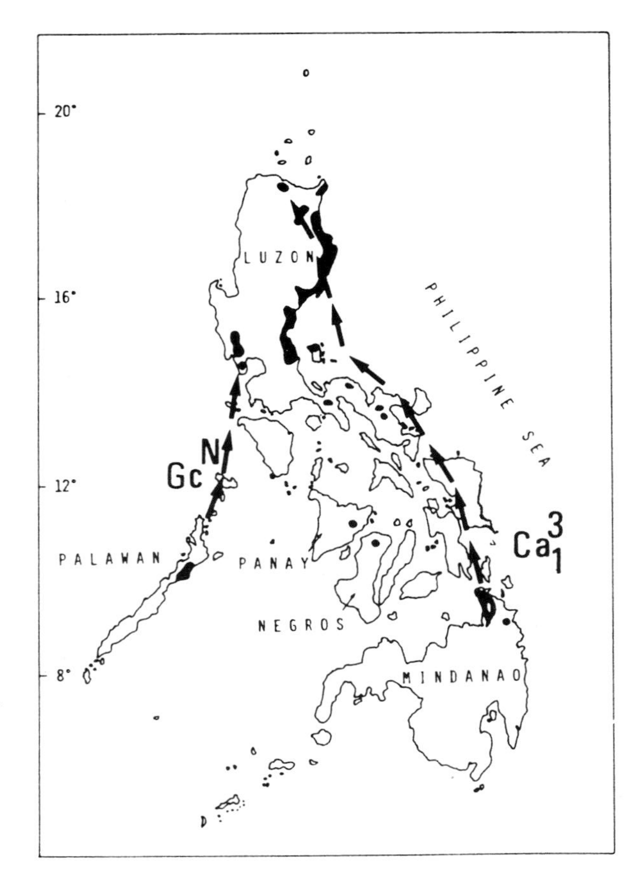

Fig. 2. Map indicating two possible migration routes of the Negritos of the Philippines inferred by the occurrence of rare variant alleles of red cell enzyme loci (Omoto 1984)

of what is now Indonesian islands. Their phenotypic resemblance to the African Pygmy is probably due to conversion resulting from adaptive evolution in tropical rain forests, since it was assumed that small size may have a selective advantage in that environment (Omoto 1984).

It was further found, based on the presence or absence of several unique alleles of red cell enzyme and serum protein types, that among the so-called Negrito populations of the Philippines, the Aeta of western-central Luzon and the Mamanwa of northern Mindanao are genetically quite different from each other. Thus, it was considered to be likely that during the Late Pleistocene Time, at least two different waves of migration should have given rise to the Negritos (Fig. 2). One wave came directly from Sundaland, perhaps via Borneo and Palawan to Luzon, and the other from the island area to the west of Sundaland referred to as Wallacea northward to Mindanao, and further north to the eastern coast of Luzon (Omoto 1984).

2 The Genetic Origins of the Japanese

The first main topic of this chapter is the origins of the Japanese peoples, that is, the modern human populations of the Japanese archipelago. Since the time, in the middle of the 19th Century, that Philip Franz von Siebold (1796-1866) considered in his magnificent book *Nippon* that the original inhabitants of the Japanese archipelago were the Ainu, there have been a long list of publications concerning this topic. Today, however, we can start discussions from the "dual structure model" of Kazuro Hanihara (1991). After examining most of the theories reported so far, including both morphological and genetic studies, he proposed a simple and testable hypothesis (Fig. 3).

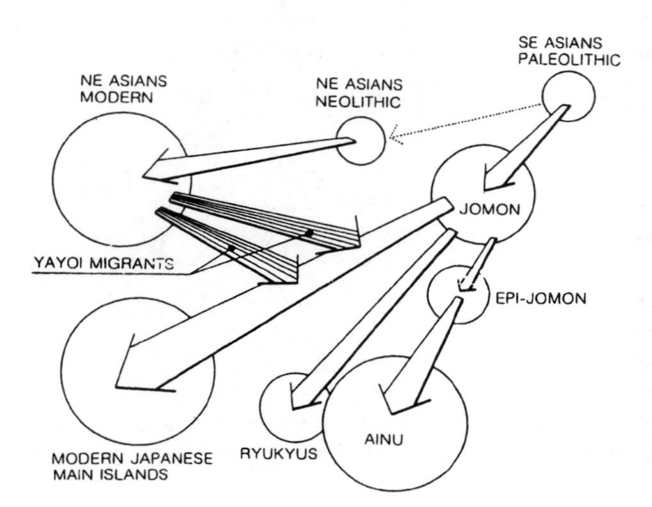

Fig. 3. The dual structure model of the formation of the modern Japanese peoples (Hanihara 1991)

His model can be divided into two sub-hypotheses as follows. Firstly, Hanihara considers that present Japanese populations were formed by a mixture of two essentially independent populations: the native Jomonese and the migrant Yayoi populations. The former were the native people of Japan during the Neolithic Jomon Period (ca. 12 000 - 2300 years BP (Before Present)) and the latter were peoples of north-eastern Asia who migrated to western Japan during the Eneolithic Yayoi Period (2300 - 1700 years BP). The ethnic minorities of present-day Japan, namely the Ainu of Hokkaido and the Ryukyuans of the Okinawa Islands are both relatively

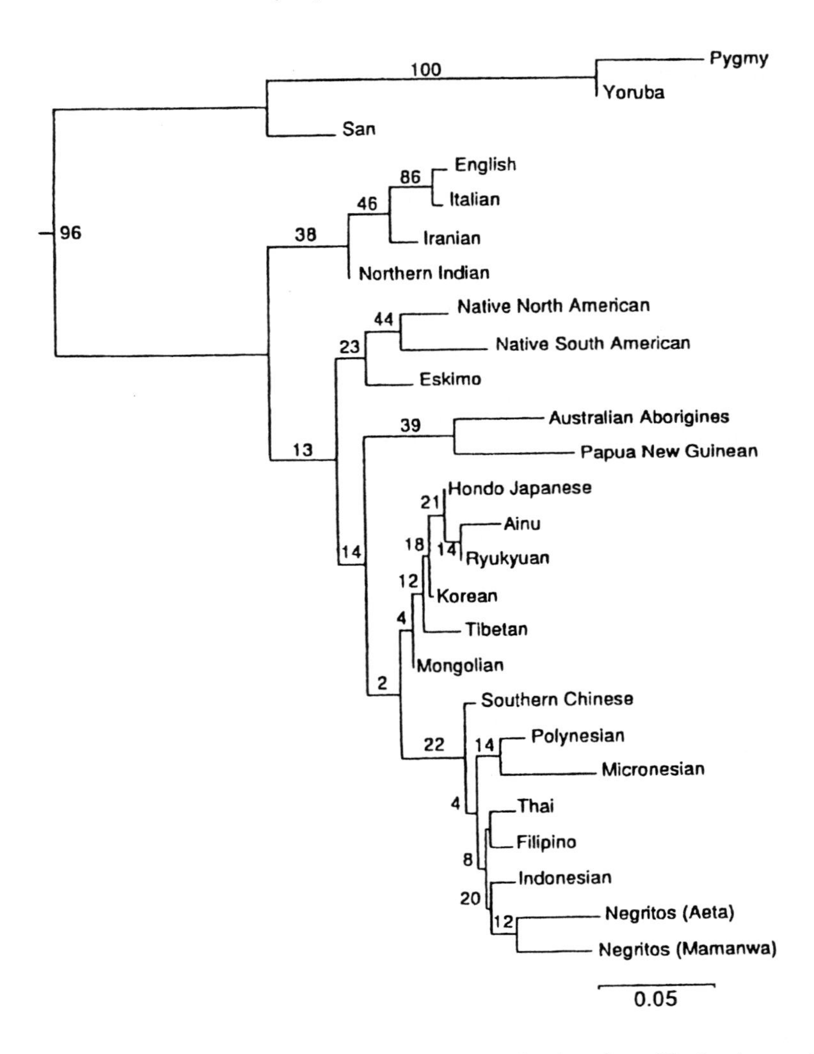

Fig. 4. A Neighbor-Joining (NJ) dendrogram based on the data from 23 classic genetic markers for 26 world populations. Note the presence of two separate cluster groups from eastern Asia, the northern and the southern (Omoto and Saitou 1997)

pure descendants of the Jomonese, while the majority of the Japanese, called Hondo-Japanese or Wajin, are a strongly mixed group. Secondly, it is assumed that the Jomonese were derived from the Upper Paleolithic population of Japan who originated in south-eastern Asia, while the Yayoi migrants were from north-eastern Asia.

Japanese archaeologists, however, find that cultural remains such as the stone tools of the Upper Paleolithic Times in Japan and also the pottery and pit type dwellings of the Jomon Period have strong affinities to those of north-eastern Asia, rather than to those of south-eastern Asia. Among geneticists, Nei (1995) criticized Hanihara's model on the basis of genetic distance analysis using the data of classic genetic markers, negating both the dual structure and the south-east Asian origin of the Jomonese.

Recently, we have tested Hanihara's two sub-models separately by genetic distance analyses, using data from up to 25 genetic loci of classic genetic markers (Omoto and Saitou 1997). In these analyses we used the "modified Cavalli-Sforza's" genetic distance (Nei et al. 1983) and the neighbor-joining (NJ) method for constructing phylogenetic trees (Saitou and Nei 1987). The results supported Hanihara's model only partly. While the "dual structure" is suggested to present genetically, the origin of the native inhabitants as indicated by the Ainu and the Ryukyuan genetic compositions does not seem to point to south-east Asia. Rather, they may have been derived from north-eastern Asian populations (Fig. 4). This finding in turn suggests that, contrary to the hitherto popular view of morphological anthropologists, there were at least two independent sources for human migrations in eastern Asia during Upper Paleolithic Times from about 50 000 to 20 000 years BP. I will come back to this point later.

Hanihara's model assumes a close relationship between the Ainu and the Ryukyuan, and considers that they have common descent from the Jomonese. In the study mentioned above, we found on the basis of 25 classic genetic markers a high bootstrap value between the Ainu/Ryukyuan cluster and the Hondo-Japanese/Korean cluster. This was interpreted to support Hanihara's assumption, although it was noted that the branch length separating the Ainu and the Ryukyuan populations is quite large (Fig. 5). More recently, however, Horai and Omoto (1998) failed to show evidence for a common descent of the two populations from the Jomonese based on the sequence data of a 482 bp (base pair) fragment of the mtDNA D-loop

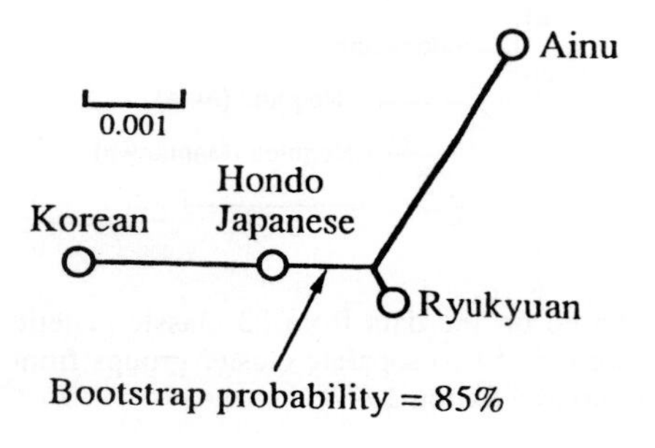

Fig. 5. A NJ genetic tree comparing three Japanese populations (Ainu, Ryukyuan, and Hondo-Japanese) with Koreans as a control, based on allele frequency data of 25 classic genetic markers. Note the bootstrap probability is much higher than that under the random expectation of 33% (Omoto and Saitou 1997)

region. Also, a recent morphological study using non-metrical traits of skulls indicates a remote relationship between the Ainu and the Ryukyuan (Dodo et al. 1998).

It is known that during the last glacial maximum, about 20 000 years BP, there were at least three possible routes of human migration to Japan: northern, western, and southern (Fig. 6). Only a few fossil specimens of *Homo sapiens sapiens* have been known in Japan prior to the Jomon Period. In mainland China, however, there are at least two well known sets of skull specimens: three specimens from Upper Cave of Zhoukoudian, north-eastern China, and a single specimen from Liujiang, southern China (Wu 1992).

The most complete Japanese fossil specimens are those excavated in the Minatogawa Cave of Okinawa Island (Suzuki 1982). They are dated to about 18 000 years BP, and are certainly of the Upper Paleolithic Times, but were without any accompanying cultural remains. Most anthropologists in Japan consider these specimens to be ancestral to the Jomonese, and also morphologically more similar to the Liujiang specimen than to those of Upper Cave. Thus, Hanihara's view that the Upper Paleolithic people of Japan who gave rise to the Jomonese originated in south-eastern Asia has been generally accepted by Japanese anthropologists.

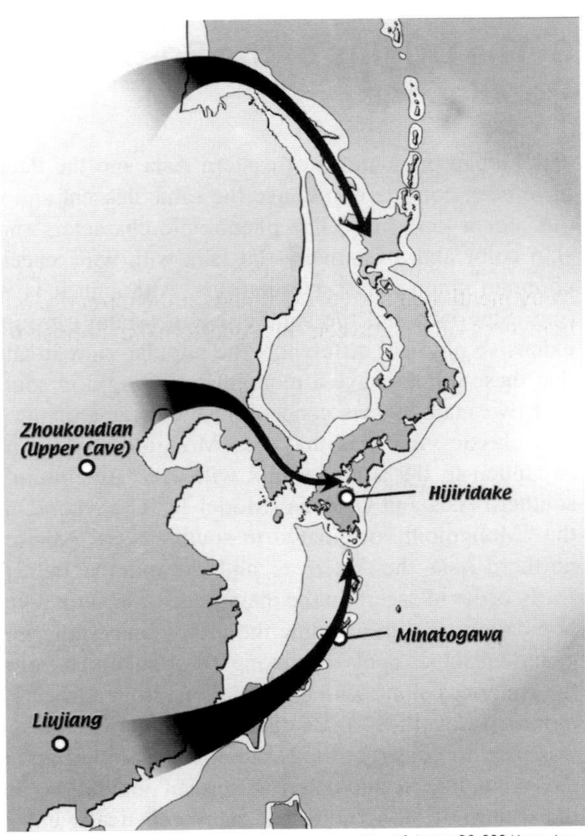

Fig. 6. Three possible routes of human migration to Japan during the last glacial maximum (ca.20 000 years BP). Place names indicate sites for Upper Paleolithic fossil specimens

However, there is an interesting fossil specimen, which may hint that an alternative migration could have existed. It is the occipital part of a skull found in 1962 in Hijiridake Cave of Oita Prefecture, Kyushu, accompanied with obsidian blade tools of the Upper Paleolithic type. According to the late Prof. T. Ogata, who investigated this specimen, it shows some morphological similarities to those of the Upper Cave (Ogata 1981). Although future studies are needed to confirm this, an alternative route other than that from south-eastern Asia to Japan, namely that from north-eastern China to western Japan, seems not to be excluded at present, as indicated in Fig. 6.

An important proponent of Hanihara's model is Christy Turner, who found a dichotomy of south-eastern and north-eastern populations on the basis of a complex of dental features. He distinguishes "sundadonty" for south-eastern Asia and "sinodonty" for north-eastern Asia. Moreover, he considers that the former is morphologically more generalized than the latter, suggesting that south-eastern Asia is the homeland of the entire "Mongoloid" populations (Turner 1987; 1995). As for the Japanese populations, the Hondo-Japanese show sinodonty, while the Jomonese and Ainu both show sundadonty, supporting Hanihara's model (Turner 1995).

3 The Origins of the So-Called Mongoloid Populations

The human populations of eastern Asia and the Pacific, as well as the Americas, have been considered to ..ave the same descent and were referred to as "Mongoloid" for a long time. The phenotypic characters such as straight hair, yellowish skin color and a relatively flat face with wide cheek bones are considered to be common among these populations. Although it is well known that the northern (e.g. Siberian) and the southern (e.g. Malay) groups of "Mongoloids" show an extensive physical difference, the popular view in anthropology until recently was that these groups have a monophyletic origin in south-eastern Asia. I have classified two major views concerning the origins of the "Mongoloid" (Omoto 1995). The classic view was that the "Mongoloid" emerged in north-eastern Asia and expanded to the south to mix with the "Australoid", a different racial group of southern Asia and Oceania (Model 1). The second, and more popular view is that the "Mongoloid" originated in south-eastern Asia and from there it dispersed in northern Asia, the Americas, and Oceania (Model 2).

In order to ascertain the magnitude of genetic diversity among the "Mongoloid" populations and to examine the views concerning their origins, I have carried out genetic distance analyses among 18 populations from eastern Asia, the Pacific and the Americas along with 6 populations from Africa and Europe (Omoto 1995). The modified Cavalli-Sforza's distance based on the data of 20 classic genetic markers was used to construct the NJ tree. It was found that the genetic diversities among the Asian, Pacific and native American populations are so extensive that the classic racial term of Mongoloid may seem genetically untenable.

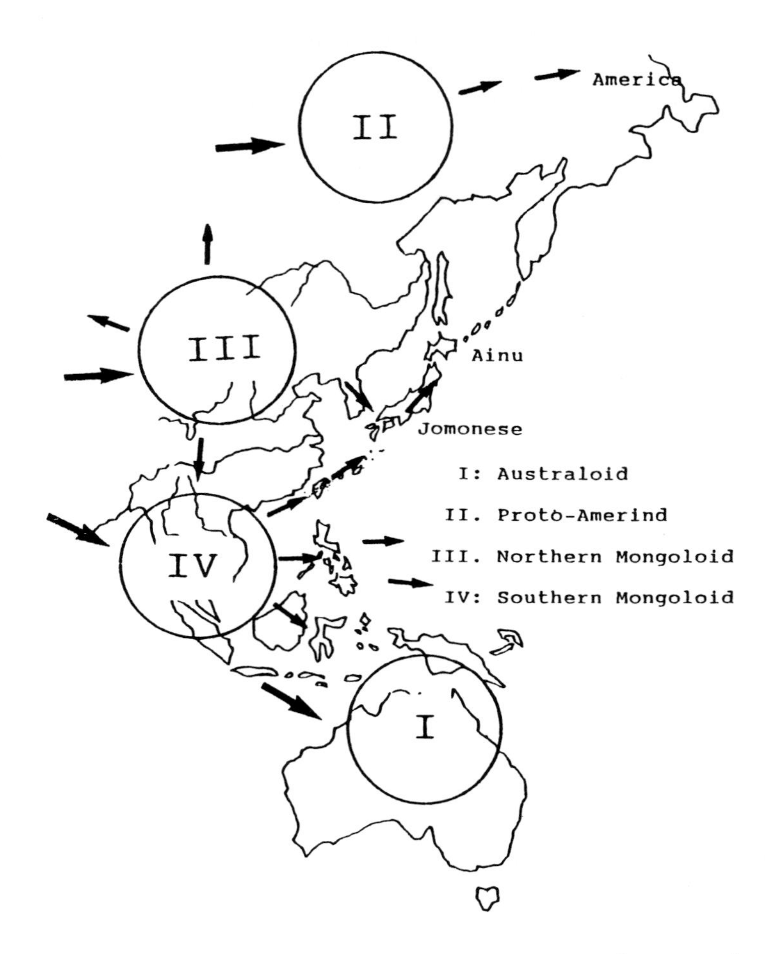

Fig. 7. A model of the origin and dispersal of eastern Asian populations, assuming at least two independent sources for migration since the Upper Paleolithic Time (Omoto 1995)

It is also noted that among the Asian-Pacific populations, there are essentially two genetically separate cluster groups, namely the northern and the southern groups. Although the "bootstrap" probability between the two clusters is not high and statistically insignificant, it may be hypothesized that this dichotomy was present in the past. It is known that there was a large-scale migration of the Austronesian speakers from the Asian continent to the Pacific about 6000 years ago (Bellwood 1979). They were pushed out from what is now China, probably by the population expansion of northern groups.

Based on these findings, I have put forward an alternative model of "Mongoloid" origin, which I refer to as Model 3 (Omoto 1995). This model assumes that there were at least two sources in eastern Asia for human diffusion since the Late Pleistocene Times, ca. 30 000-50 000 years BP, one in north-eastern Asia and the other in the south-east (Fig. 7). The model is consistent with the "Out-of Africa" model of the origin and dispersal of modern humans (Cann et al. 1987).

It is considered that when the ancestors of modern humans in eastern Asia migrated from western Asia since about 100 000 years BP, there have been at least two geographically possible routes: north or south of the Great Himalayas. It seems illogical to consider that only the southern route was possible, as the proponents of Model 2 do. The occurrence of the Upper Paleolithic stone tools in north-eastern Asia, showing affinities to western Siberia and Central Asia, seems to support Model 3 (Kimura 1993). The Upper Cave skeletal remains discussed above represent an important clue for the presence of the Upper Paleolithic peoples in north-eastern Asia prior to the population expansion of the Neolithic peoples who had typically flat faces, probably due to adaptation to extreme coldness.

As mentioned above, future studies including the "DNA palaeo-anthropology" will shed light on the whole question of the origins and dispersal of modern humans in eastern Asia, the Pacific and the Americas. Since the 1960's we have had two distinct methods for human evolutionary studies, namely morphological and molecular. Morphological studies of fossil material give us invaluable information about phenotypes, or the "individuals", such as size and physical appearance, ways of life, diseases, etc., in addition to the age and the ecological conditions of the habitat.

However, we cannot rely upon the phenotypes conclusively for delineating our origins and dispersal, since there are factors such as non-genetic individual variation of age and sex, effects of growth, nutrition and other habits, convergence or adaptive parallelism, and so on. On the other hand, molecular studies of DNA are far more decisive for phylogenetic information on origins and migrations, while they do not tell us about the individual phenotypes. We should use both of these methods, which are equally important and considered to be like the two wheels of anthropology, in order to understand the problems of the origins, adaptation and variation of humans.

Acknowledgements

During my studies I have had the invaluable cooperation of a number of colleagues. I wish to mention here just two, namely Professor Shogo Misawa, University of Tsukuba, who has joined most of my field work to obtain blood samples and provided me with his blood group data; and Dr. Naruya Saitou, National Institute of Genetics, who has helped me with recent statistical analyses. I thank also Ms. Takako Kimura and Mr. Tadahiko Fukumine for their assistance in preparing this manuscript. Most of my field and laboratory studies have been supported by Grants-in-Aid for Scientific Research from the Ministry of Education, Science, Sports and Culture.

References

Bellwood P (1979) Man's conquest of the Pacific. Oxford Univ Press, New York

Cann RL, Stoneking M, Wilson AC (1987) Mitochondrial DNA and human evolution. Nature 325:31-36

Cavalli-Sforza LL, Edwards AWF (1967) Phylogenetic analysis: models and estimation procedures. Am J Hum Genet 19:223-257

Dodo Y, Doi N, Ishida H (1998) Ainu and Ryukyuan cranial nonmetric variation: evidence which disputes the Ainu-Ryukyu common origin theory. Anthropol Sci 106:99-120

Hanihara K (1991) Dual structure model for the formation of the Japanese population. Japan Review 2:1-33

Horai S, Omoto K (1998) Peopling of Japan inferred from mitochondrial DNA polymorphisms in east Asians. In: Omoto K, Tobias PV (Eds) The origins and past of modern humans-towards reconciliation. World Scientific, Singapore, pp 54-73

Kimura H (Ed) (1993) The origin and dispersal of microblade industry in northern Eurasia. Proc Int Conference, Sapporo Univ Archeol Mus, Sapporo

Nei M (1995) The origins of human populations: genetic, linguistic, and archeological data. In: Brenner S, Hanihara K (Eds) The origin and past of modern humans as viewed from DNA. World Scientific, Singapore, pp 71-91

Nei M, Tajima F, Tateno Y (1983) Accuracy of estimated phylogenetic trees from molecular data. II. Gene frequency data. J Mol Evol 19:153-170

Ogata T (1981) Human remains of the Palaeolithic Period (Kyusekki-jidai jinkotsu). In: Ogata T (Ed) Jinruigaku-Koza (Anthropology). Vol 5. Yuzankaku, Tokyo

Omoto K (1972) Polymorphisms and genetic affinities of the Ainu of Hokkaido. Hum Biol Oceania 1:278-288

Omoto K (1975) Genetic affinities of the Ainu as assessed from data on polymorphic traits. In: Watanabe S, Kondo S, Matsunaga E (Eds) Anthropological and genetic studies on the Japanese. Univ Tokyo Press, Tokyo, pp 296-303

Omoto K (1984) The Negritos: genetic origins and microevolution. Acta Anthropogenet 8:137-147

Omoto K (1995) Genetic diversity and the origins of the "Mongoloids". In: Brenner S, Hanihara K (Eds) The origins and past of modern humans as viewed from DNA. World Scientific, Singapore, pp 92-109

Omoto K, Saitou N (1997) Genetic origins of the Japanese: a partial support for the dual structure hypothesis. Am J Phys Anthropol 102:437-446

Saitou N, Nei M (1987) The neighbor-joining method: a new method for constructing phylogenetic trees. Mol Biol Evol 4:406-425

Suzuki H (1982) Skulls of the Minatogawa Man. In: Suzuki H, Hanihara K (Eds) The Minatogawa Man: The Upper Pleistocene Man from the Island of Okinawa. Bull Univ Mus Univ Tokyo 19, pp 7-49

Turner CG II (1987) Late Pleistocene and Holocene population history of East Asia. Am J Phys Anthropol 73:305-321

Turner CG II (1995) Shifting continuity: modern human origin. In: Brenner S, Hanihara K (Eds) The origin and past of modern humans as viewed from DNA. World Scientific, Singapore, pp 216-243

Wu X (1992) The origin and dispersal of anatomically modern humans in East and Southeast Asia. In: Akazawa T, Aoki K, Kimura T (Eds) The evolution and dispersal of modern humans in Asia. Hokusen-sha, Tokyo, pp 373-378

20
Human Diversity and Its History

Henry C. Harpending and Elise Eller

Department of Anthropology, University of Utah, Salt Lake City, UT 84112, USA

Abstract

In the last decade a large amount of new genetic data from human populations has appeared. The most informative of the new loci are STR (short tandem repeat) polymorphisms, because they are not subject to the ascertainment biases that affect classical markers and SNPs (single nucleotide polymorphisms). These loci show a marked diversity cline away from Africa, as they should if a version of the SOM (single origin model) is correct for our species. But the new data have not given us many insights into ancient population history and movements: they generally show that neighboring populations are similar to each other and that similarity declines with geographic distance. Much interesting human history has been blurred and erased by recurrent local gene flow. Other genetic and non-genetic markers, like language and physical appearance, may have better "memories" and tell us more about ancient populations movements and relationships.

1 Introduction

New technology for ascertaining and typing genetic markers has given anthropologists a flood of data in the last decade. Today a single publication can present more and better data than the sum of everything available in the literature before 1985 or so. The new data have essentially confirmed the SOM model (single origin model) of human history, in which we are descended from a small founding population that was probably in Africa.

Many of us did not foresee that we would infer demographic history from genetic data. On the other hand we did foresee that more and better genetic data

Key words. modern human origins, Short Tandem Repeats (STRs), race, sexual selection

would let us read this history of population relationships, migrations, and the genesis of human genetic diversity. In this paper we suggest that the new data from neutral markers paint a rather dull picture of high levels of local gene flow everywhere and unremarkable correlations between genetic distances and geographic distances between populations. In a sense the new wealth in data has been a disappointment.

When populations exchange neutral genes there is essentially blending of gene frequencies, so red and white each become pink over time. Markers that do not blend in this way may give us better information about population history. Languages, for example, do not blend like gene frequencies. Instead there is "majority advantage" in which immigrants just learn the indigenous language (Renfrew 1987). Prominent visible "racial" traits may also have enjoyed such majority advantage, so that appearance can tell us more about ancient relationships than can gene frequencies.

2 Human Demographic History

2.1 Small Effective Size of Humans

Findings from many genetic systems suggest that the effective size of humanity is on the order of 10000 breeding individuals. Since this estimate is wrong by a factor of a million or so today, the implication is that the our ancestry is some specific small isolated population of archaic humans (SOM) rather than the whole array of *Homo erectus* relatives that occupied the temperate Old World for one to two million years (MRM, multiregional model).

Haigh and Maynard Smith (1972) suggested that there was a bottleneck in our ancestry on the basis of the spectrum of substitutions in hemoglobin. Subsequently the number 10000 bas become widely established as the summary effective breeding size of humans (Li and Sadler 1991). This could reflect either a population that was this small for a very long time else a severe transient bottleneck during which the number of our ancestors was much less than this. Estimates from nuclear genes (Harding et al. 1997; Hey 1997; Zietkiewicz et al. 1998), mtDNA (mitochondrial DNA) (Rogers and Jorde 1995), the HLA system (Takahata and Satta 1998, in press) and from human-specific *alu* insertions (Sherry et al. 1997) all converge on a similar figure.

2.2 Expansion from Small Size

The tree of human mitochondrial DNA is star-like as if it is recording a major population expansion in our history (Di Rienzo and Wilson 1991; Slatkin and Hudson 1991; Rogers and Harpending 1992; Harpending et al. 1993). The pattern in mtDNA is clear, but it could be the result of selection as well as population expansion. The expansion hypothesis has been in limbo for several years, since no such pattern is apparent in several nuclear genes (Harding et al. 1997; Hey 1997; Zietkiewicz et al. 1998). Recently, however, there have appeared several papers analyzing STR polymorphisms (Shriver et al. 1997; Di Rienzo et al. 1998; Kimmel et al. 1998; Reich and Goldstein 1998). These all find strong evidence of a major population expansion in our history. Given the new findings, it seems safe to go back to the mitochondrial estimates of the timing of this expansion: these vary from 120 000 to 30 000 years ago, with heterogeneity among populations. Uncertainty about mtDNA mutation rates is great, so these numbers should be treated with caution.

Can we relate the expansion visible in our DNA to anything in the fossil and archaeological records? There are at least four candidates for the correspondence. First, Mode 3 stone tool technologies appear about 250 000 years ago in Africa and Europe[1]. Second, human fossils that are equivocally modern appear in Africa and the Levant 100000 or more years ago associated with Mode 3 technologies. Third, the Toba super-eruption at 71 000 years ago may have caused widespread ecological devastation and extinction on earth, and the expansion may be recovery from that event (Ambrose 1998). Fourth, there is a dramatic invasion of Europe by bearers of Upper Paleolithic (Mode 4) technologies 40000 years ago. We do not believe that there is compelling evidence favoring any of these alternatives.

2.3 Dispersal of Modern Humans

In Europe and western Asia north of the Himalayas there is a clear record of the spread of modern humans from Africa. Most of Europe was occupied rapidly although pockets of Neanderthals apparently persisted for millennia, even adopting some of the new tool technologies.

South of the Himalayas the record is not so clear. If there was a separate southern exodus there is little record of it until Australia, which was occupied between 35000 and 40000 years ago (O'Connell and Allen 1998) by bearers of Mode 3 technology. Mode 4, thought by many to be diagnostic of modern humans, never appeared in Australia. There are at least two scenarios about the history of Australia that are plausible.

The first scenario is that there was a southern branch of the human expansion that did not bear Mode 4 technologies, implying that Mode 4 is not a marker of modern humanity but only a marker of a society where males were not continuously engaged in parental investment.

In a famous paper the Whitings (Whiting and Whiting 1975) point out that there are consistent differences between societies where males are familial and those where males are "aloof" from women and the family. In the latter males are more

[1] Modes refer to a regular sequence of complexity of stone tools made by humans and human ancestors during the Pliocene and Pleistocene (Clark 1977). Mode 1 is pebble tools, Mode 2 adds the Acheulian hand axe technology, Mode 3 contains flakes made from prepared cores, Mode 4 is the technology of the Upper Paleolithic with blades and worked bone, and Mode 5 is the microlithic technology associated with the Mesolithic.

likely to be belligerent and gaudy: it is in these among technologically primitive societies where fancy artistic expression occurs as either a manifestation or a side-effect of male competition. In the former kind of society males are likely to be occupied working to provision their offspring. Certainly there are many foraging societies where there is no production of fancy technology like that of Mode 4 cultures. We can compare the drab limited technology of Kalahari Bushmen, for example, with the rich diversity of weapons and art from central African societies.

The second scenario is that there was a single expansion of moderns out of Africa. Modern humans reached Australia from southeast Asia much later, and the old technologies and fossils from Australia, like Mungo and Kow Swamp, are those of another species of archaics, not of modern humans. This is not a popular scenario today, but it deserves consideration. Under this model the arrival of modern humans in Australia occurred not 40 000 years ago but perhaps 5000 with the appearance of a small tool technology, the dingo, and an apparent tenfold increase in population (Bellwood 1997).

2.4 Diversity Patterns

Figures 1 and 2 summarize world patterns in microsatellite diversity from two separate sources: Fig. 1 is based on results from Lynn Jorde's laboratory at the University of Utah (Jorde et al. 1995), while Fig. 2 is based on results from Kenneth Kidd's laboratory at Yale (Calafell et al. 1997).

The left panels in each figure show how genetic diversity declines away from Africa. The horizontal axes are genetic distances from the African mean. The vertical axes are average heterozygosity. The pattern is clear in each case: the more genetically different a population is from African populations the lower its genetic diversity. This pattern is in good agreement with a model of human expansion that includes series of bottlenecks during colonization episodes in which diversity was lost. By the time we reached interior South America, according to the evidence of the Surui of Amazonia, we had lost about one quarter of our neutral genetic diversity.

The right panels in each figure are plots of the leading two principal components of the normalized allele size covariance matrix among populations. This is the (least squares) best two dimensional picture of genetic distances among populations. In each case we see that there is a general concordance between geography and genetic distance: neighbors are genetically similar. In Fig. 1 there is apparent "clumping" into three traditional races: African, European, and Asian, with the Biaka and Mbuti as slight outliers. This could reflect real differences from other African populations or it could reflect the small number of individuals from each population in the Jorde data. In Fig. 2 the picture is dominated by the genetic distinctiveness of small and probably isolated populations that have undergone a lot of gene frequency change due to drift. Compare, for example, the isolated Surui with the large cosmopolitan Maya.

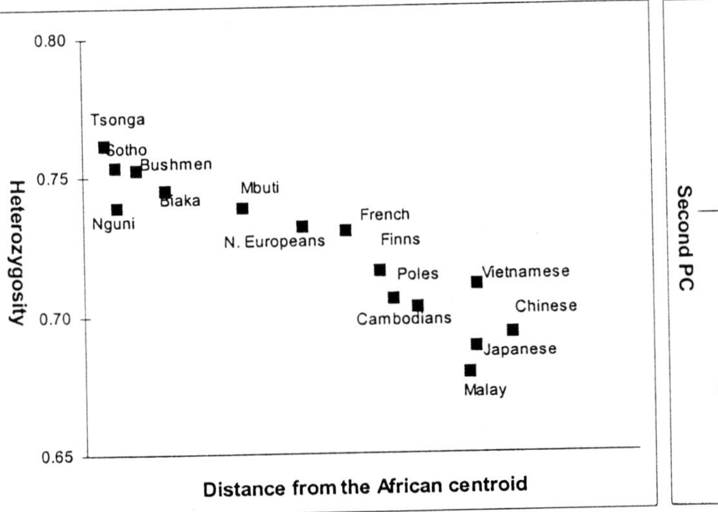

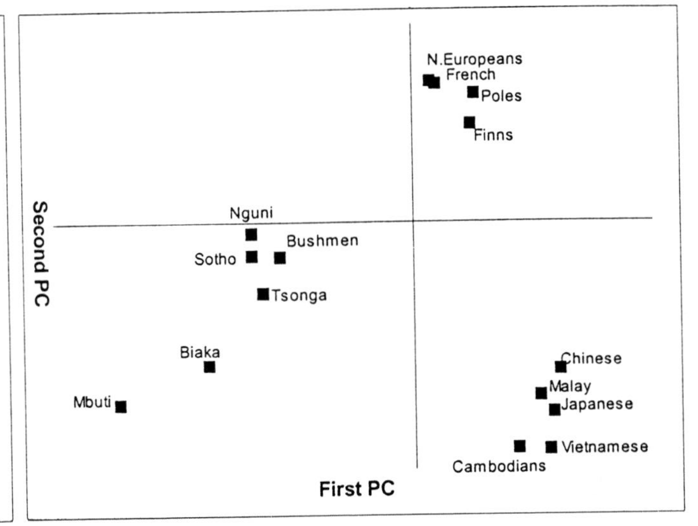

Fig.1. Jorde lab data. The *left panel* plots heterozygosity (y axis) against genetic distance from Africa (x axis) on the basis of 60 polymorphic microsatellite loci. The *right panel* shows principal coordinates of genetic distance among populations (Jorde et al. 1995)

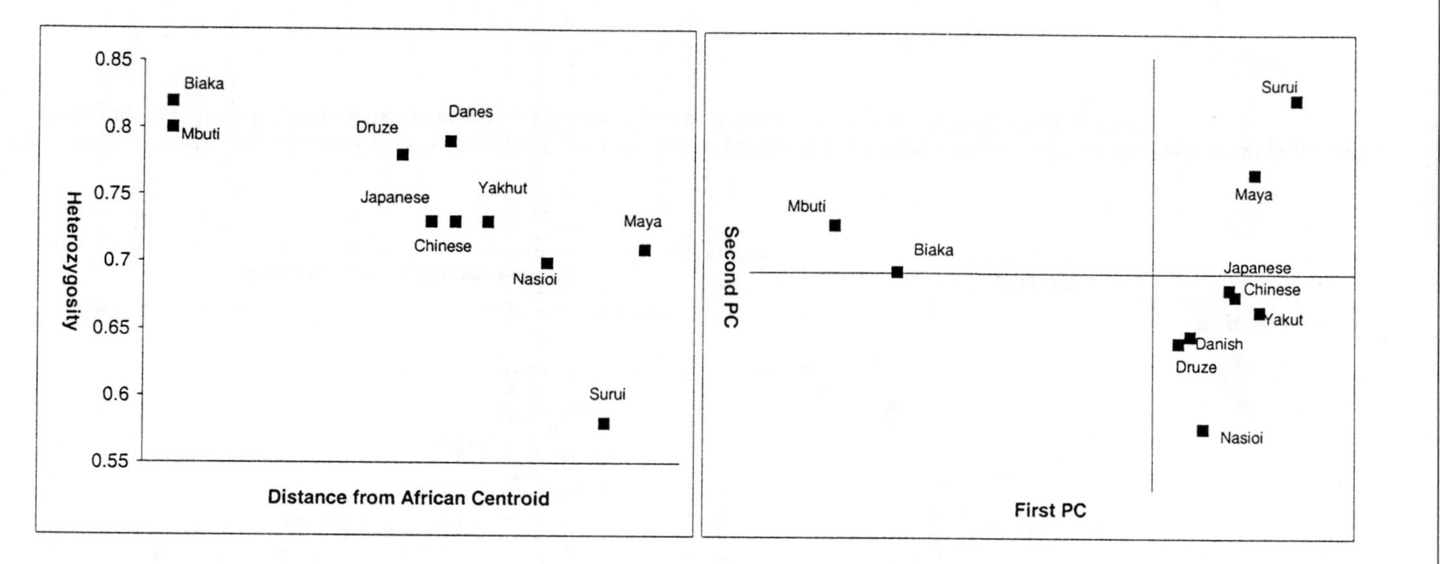

Fig. 2. Kidd lab data. The *left panel* plots heterozygosity (y axis) against genetic distance from Africa (x axis) on the basis of 92 polymorphic microsatellite loci. The *right panel* shows principal coordinates of genetic distance among populations (Calafell et al. 1997)

2.5 Problems

The rich new data that have become available support two general conclusions. First, F_{st}, a standard measure of heterogeneity among populations, is about ten percent among major continental groups. Essentially the same number has been known since 1972 or so. Second, there is a cline in neutral genetic diversity outward from Africa. This is a new finding that relies on microsatellite loci. With their large number of alleles they are not subject to the ascertainment bias that affects classical polymorphisms and SNPs. Older studies, summarized in Cavalli-Sforza et al. (1994), did not show any clear diversity gradient because most of them had been found in Europeans, thus selecting for markers most diverse in Europeans.

Any account of the expansion of modern humans must account for the ten percent difference among continental areas as well as the diversity cline away from Africa. A popular idea in anthropology is that these race differences developed *in situ* as a consequence of geographically restricted gene flow, but this does not seem possible. In a collection of completely isolated populations, each of size N, drawn from a common founding population, F_{st} should increase approximately as

$$F_{st}(t) \sim 1 - e^{\frac{-t}{2N}}$$

so that if $N \sim 10\,000$, for example after the expansion and dispersal of humans, it would require approximately 50 000 years for differences that we observe today to accumulate. We know that 50 000 years ago humans were about to begin their colonization of Europe and west Asia and may have already reached Australia, and there must have been many more than 10 000 in these populations spread over several continents. Since there is such clear geographic patterning in genetic distances between populations, there must also have been substantial gene flow among populations. Such gene flow would retard the accumulation of F_{st}. In populations with hundreds of thousands of members F_{st} is essentially frozen over time scales of interest to us.

How can we account simultaneously for an expansion from a small population of only several thousands of adults, a global F_{st} of ten percent, and the diversity cline away from Africa? Since F_{st} is essentially frozen in populations greater than several tens of thousands, these neutral differences must have accumulated in small populations. Two models that have been proposed are the "divided Eden" model in which the small ancestral population was itself subdivided into races, and the "Cain model" in which colonizing populations are small for a long time so that a series of founder effects occurs during colonization of new areas. We can generate neutral gene distributions in a computer simulation of human history that match our origin from a small population, contemporary F_{st}, and contemporary diversity clines by either mechanism, and so far we have not found a way to distinguish their effects.

3 A New Direction

The emerging picture from new and plentiful data about neutral markers is one of a smooth relationship between geographic and genetic distances with some perturbations due to recent history and to language. These findings suggest that there has been a lot of gene flow between neighbors and that isolation by distance, rather than history, dominates the distribution of neutral genes in human populations. We believe it is time to extend our domain of inquiry to marker systems where history is better preserved. In their pioneering study of human race differences, Nei and Roychoudhury (1974) hint at a new direction. They say that their finding of rather small race differences "does not apply to those genes which control morphological characters such as pigmentation and facial structure." Darwin (1871) thought that our race differences were driven by sexual selection, a view defended by Diamond (1992) who shows that environmental selection accounts of race differences in skin color and other traits do not bear close scrutiny. If skin color were a response to climate, Tasmanians should have turned white. If this view of differences in appearance is correct, even in part, then physical appearance, like language, could carry a signature of history much deeper than the signature in neutral genes.

When gene flow occurs between populations gene frequencies blend in a linear way, so that over time differences between the populations are erased as history yields to local migration-drift equilibrium. But both language and sexually selected appearance may respond very differently to contact and admixture. We think it is time for biological anthropology to turn its attention again to language and to external appearance, but we realize that both these topics are, putting it mildly, unfashionable. We will discuss a simple model showing how traits like this should preserve deep history and discuss several examples.

3.1 Language and Rh among Basques

The Basques of the Pyrenees speak a language that is perhaps distantly related to languages of the Caucasus mountains but is otherwise unique in the world. A common theory is that they are a relic of an earlier occupation of Europe that was overrun by later invaders, perhaps the Indo-Europeans. Their genes are mostly similar to those of their neighbors except that they have a frequency of Rhesus negative that is greater than one-half. Following Ruhlen (1994) imagine that there has been some small level of gene flow, one percent per generation, into the Basque population from their new neighbors for five thousand years, say 250 generations. Then of the neutral genes in the Basque population today only $.99^{250} \sim$ eight percent are descended from ancient Basque genes. The neutral parts of the genome have essentially been replaced.

But consider the language spoken by these immigrants. Since they arrive in small numbers they and their children learn Basque and, save for occasional loan words, have little effect on the language. The Basque language has persisted over millennia while the neutral genome has been replaced. Meanwhile, natural selection at the Rh locus is such that the common type is favored. If the original state was all or mostly Rh negative, then there would be ongoing selection against any Rh positive genes introduced by immigrants. In this way, both language and the Rh system preserve deeper history than neutral genes. Rh and language share the property that there is selection for the common type. Sexual selection for external appearance may follow similar dynamics.

3.2 A Model of Majority Advantage

Simple models of single diallelic loci can give us insights that are robust. Here we discuss the simplest model we can write of the process we envision, but the dynamics of a quantitative trait should be much the same.

Consider an allele A whose frequency in a population is p. The genic fitness of A is proportional to its frequency with selection intensity s , so that

$$W(A) = sp$$

$$W(a) = s(1-p).$$

Gene frequency change follows

$$\frac{dp}{dt} = sp(1-p)(2p-1)$$

If p is greater than one half selection will drive A to fixation while if p is less than one half the allele a will go to fixation.

Now put this population in an island model, so that it is one of many demes that receive M immigrants per generation from the whole array of islands. Assume that half the islands have $p>0.5$, half $p<0.5$, and that the overall mean is just $P=0.5$. The effective size of each island is G genes. Local frequency change is described by

$$\frac{dp}{dt} = sp(1-p)(2p-1) + \frac{M}{G}(\frac{1}{2} - p)$$

for which there are two interior stable points, one on either side of the grand mean $P=1/2$. Polymorphism persists if

$$\frac{2M}{sG} < 1$$

that is if twice the migration rate is less than the selection intensity. If the migration rate is high enough it overwhelms local selection and the whole system quickly goes to fixation, that is to monomorphism of either a or A.

At the equilibrium between migration and local selection for the common type diversity among islands due to selection is

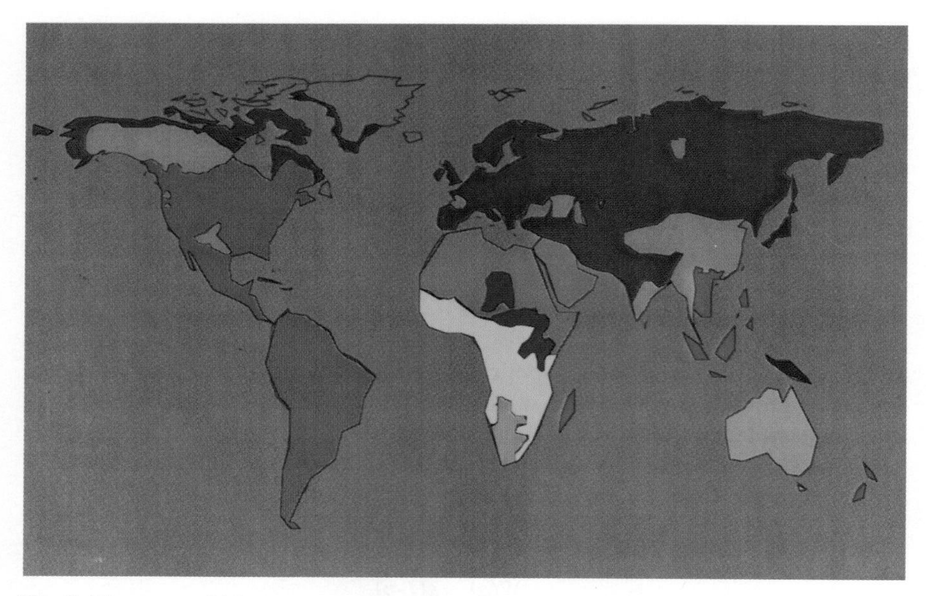

Fig. 3. The ten world language macro-families described in Ruhlen (1994)

$$F_{st} = 1 - \frac{2M}{sG} \qquad (1)$$

while neutral traits would follow migration-drift equilibrium for which

$$F_{st} = \frac{1}{2M + 1} \qquad (2)$$

These are abstract but we can obtain a feel for magnitudes involved by using some human data and plausible population parameters. F_{st} among continental groups is about ten percent, implying from equation (2) that $M \sim 4.5$ corresponding to 2 diploid migrants per generation. Relethford (Relethford 1994) estimates F_{st} for human skin color to be about 0.6, that is six times as great as that of neutral genes. If the effective size of our human islands is $G = 10000$, equation (1) implies s ~ 0.001. This is a very low intensity of selection, a level that is probably undetectable by epidemiological methods.

3.3 World Languages

Language ought in some cases to follow something like our model of common type advantage, as we proposed for Basque. Figure 3 is a map (redrawn from Ruhlen 1994) of world languages lumped into the largest possible groups by Ruhlen. This degree of lumping is very controversial, but it may suggest interesting patterns. There does seem to be support for a North–South, rather than East–West, division of Eurasia. In the North there are two language families: Dene-Caucasian in scattered patches including several in the New World, and Eurasiastic. The pattern is one in which the Eurasiatic languages underwent a later expansion at the expense of Dene-Caucasian leaving behind relics (including Chinese!). South of the Himalayas there are more groups each more restricted reflecting perhaps lower mobility or the absence of subsistence innovations that allowed the wide expansions seen in the North.

This pattern in world languages supports the southern-branch theory of the expansion of modern humans. There are two families north of the Himalayas, with Dene-Caucasian perhaps the more ancient and associated with a pre-agricultural expansion and Euro-Asiatic superimposed on it. In the tropical regions there are four families in Asia and Oceania. The Indo-European intrusion into India and the Sino-Tibetan intrusion into southeast Asia may be relatively recent events.

3.4 Kalahari Bushmen

An interesting case for the retention of old appearance in the face of gene flow is provided by Kalahari Bushmen. (They are called *San* in some recent literature, but this is a nasty word to use to someone in the central Kalahari, and we avoid it.) These are people who historically made their living by either foraging or keeping small stock in southern Africa. They resemble, to European eyes at least, east Asians. They have yellowish rather than black skin, epicanthic folds, shovel-shaped incisors, and many newborns have "Mongoloid spots" at the base of the spine. The Asian appearance is not just a perception of Europeans. In the !Kung language there are three kinds of mammals: !a is an edible animal like a warthog or a giraffe, !oma is an inedible animal like a jackal, hyena, black African, or European, and zhu is a person. Vietnamese in Botswana were immediately identifed as zhu by Bushmen. In other words, their perception of their similarity to Asians is the same as ours (i.e. Europeans').

Immediately to the north of the !Kung there are groups of so-called "Black Bushmen" across Africa. The most familiar of these are the Berg Dama of Namibia. These people speak Khoisan languages related to Nama Hottentot but in appearance they are completely like their Bantu-speaking neighbors. Our model suggests that intermediate appearance should be unstable and that selection should drive appearance rapidly toward one "type" or the other. The "Black Bushmen" may have received just enough gene flow from Bantu invaders to cause a switch to the other adaptive peak with rapid (in evolutionary time) loss of the earlier Bushman appearance.

Are there critical tests of this model? The model predicts retention of shared genes affecting external appearance between Bushmen and east Asians, but retention of nothing else. Hence, in agreement with the model, neutral Bushman genes seem completely African as shown in Fig. 2. The other more interesting prediction must wait for identification of the genetic basis of skin color and other aspects of physical appearance for a critical test.

If there is any validity to our hypothesis, other similarities in appearance that anthropologists in the early part of this century described might in fact be real signatures of ancient population expansions and movements that should be the basis for archaeological work. The idea of a Vedda-Australian-Ainu connection and the similarities of Negrito peoples around the Pacific rim and African Pygmies are examples of such hypotheses.

4 Conclusion

We have outlined some of the more interesting problems in the history of our species and considered how various categories of data may help resolve them.

It is widely accepted (but not necessarily correct!) that modern humans appeared in Africa and expanded from this homeland within the last 100 000 or so years. While the northern branch is readily apparent in the archaeological record of Europe and west Asia, the early appearance of humans in Australia at 40 000 to 50 000 years ago suggests that there was a southern branch of the human exodus that roughly followed the Indian ocean coast. The hypothesis of a southern branch receives support from the distribution of language macro-families in the Old World and from the presence of modern humans in Tasmania. This island had been separated from Australia for thousands of years before the other candidate for the peopling of Australia, the event 5000 years ago when dogs and microliths appeared and population increased by roughly a factor of ten. The suprising implication of the model of a separate southern branch is that Mode 4 technologies — fancy blade tools and worked bone — are not a distinctive signature of new human cognitive capacities but mere signs of male leisure in a new rather empty ecosystem.

Spectacular advances in typing neutral markers have provided large numbers of markers from various human populations. These markers should be free of the ascertainment bias that made comparisons of genetic diversity across populations unreliable. These markers show a diversity cline away from Africa and a pattern of population difference that looks like rather smooth isolation by distance but with marked clumping into groups that are rather like traditional "major races." More populations need to be sampled before this clumping can be carefully evaluated.

Populations that have seemed on other grounds to be distinct from their neighbors turn out not be distinct when we look at neutral markers. Bushmen of southern Africa, for example, appear as simply another African population. We suggest that other marker categories like language and external appearance might provide a deeper look at human history. These share with the Rh system the property that selection should favor the common type or, in the case of sexually selected appearance, some exaggeration of the common type. Under such dynamics they would "resist" the effects of gene immigration while neutral markers should "blend". We do not have a critical test now of this hypothesis but there are clear predictions about genes controlling external appearance that have not yet been identified.

The ultimate test of these hypotheses about human origins will be identifying traces of past movements in the archaeological record. We need to develop closer ties among linguistics, archaeology, and genetics in the study of human history.

References

Ambrose SH (1998) Late Pleistocene human population bottlenecks, volcanic winter, and differentiation of modern humans. J Human Evol 34:623-651

Bellwood PS (1997) Prehistory of the Indo-Malayasian Archipelago (revised edition). Univ Hawaii Press, Honolulu

Calafell F, Shuster A, Speed, WC, Kidd JR, Kidd KK (1997) Short tandem repeat polymorphism evolution in humans. European Journal of Human Genetics 6:38-49

Cavalli-Sforza LL, Menozzi P, Piazza A (1994) The history and geography of human genes. Princeton Univ Press, Princeton

Clark JGD (1977) World prehistory: a new perspective. Cambridge Univ Press, Cambridge

Darwin C (1871) The descent of man, and selection in relation to sex. John Murray, London

Di Rienzo A, Donnelly P, Toomajian C, Sisk B, Hill A, Petzl-Erler ML, Hanes GK, Barch DH (1998) Heterogeneity of microsatellite mutations within and between loci, and implications for human demographic histories. Genetics 148:1269-1284

Di Rienzo A, Wilson AC (1991) Branching pattern in the evolutionary tree for human mitochondrial DNA. Proc Natl Acad Sci USA 88:1597-1601

Diamond J (1992) The third chimpanzee. New York, HarperPerennial

Haigh J, Maynard Smith J (1972) Population size and protein variation in man. Genet Res 19:73-89

Harding R M, Fullerton SM, Griffith RC, Bond J, Cox MJ, Schneider, JA, Moulin DS, Clegg JB (1997) Archaic African and Asian lineages in the genetic ancestry of modern humans. Am J Hum Genet 60:772-789

Harpending HC, Sherry ST, Rogers AR, Stoneking M (1993) The genetic structure of ancient human populations. Curr Anthrop 34:483-496

Hey J (1997) Mitochondrial and nuclear genes present conflicting portraits of human origins. Mol. Biol Evol 14:166-172

Jorde LB, Bamshad MJ, Watkins WS, Zenger R, Fraley AE, Krakowiak P, Carpenter KD, Soodyall H, Jenkins T, Rogers AR (1995) Origins and affinities of modern humans: a comparison of mitochondrial and nuclear genetic data. Am J Hum Genet 57:523-538

Kimmel M, Chakraborty R, King JP, Bamshad M, Watkins WS, Jorde LB (1998) Signatures of population expansion in microsatellite repeat data. Genetics 148:1921-1930

Li W-H, Sadler LA (1991) Low nucleotide diversity in man. Genetics 129: 513-523

Nei M, Roychoudhury AK (1974) Genic variation within and between the three major races of man, Caucasoids, Negroids, and Mongoloids. Am J Hum Genet 26:421-443

O'Connell JF, Allen J (1998) When did humans first arrive in greater Australia and why is it important to know? Evol Anthropol 6:132-146

Reich DE, Goldstein DB (1998) Genetic evidence for a Paleolithic human population expansion in Africa. Proc Natl Acad Sci USA 95:8119-8123

Relethford JH (1994) Craniometric variation among modern human populations. Am J Phys Anthropol 95:53-62
Renfrew C (1987) Archaeology and language: the puzzle of Indo-European origins. Cambridge Univ Press, New York
Rogers AR, Harpending HC (1992) Population growth makes waves in the distribution of pairwise genetic differences. Mol Biol Evol 9: 552-569
Rogers AR, Jorde LB (1995) Genetic evidence on the origin of modern humans. Hum Biol 67:1-36
Ruhlen M (1994) The origin of language. John Wiley and Sons, New York
Sherry ST, Harpending HC, Batzer MA, Stoneking M (1997) *Alu* evolution in human populations: using the coalescent to estimate effective population size. Genetics 147:1977-1982
Shriver MD, Jin L, Ferrell RE, Deka R (1997) Microsatellite data support an early population expansion in Africa. Genome Res 7:586-591
Slatkin M, Hudson RR (1991) Pairwise comparisons of mitochondrial DNA sequences in stable and exponentially growing populations. Genetics 129:555-562
Takahata N, Satta Y (1998, in press) Footprints of intragenic recombination at HLA loci and human evolution. Immunogenetics
Whiting JWM, Whiting BB (1975) Aloofness and intimacy of husbands and wives: a cross-cultural study. Ethos 3:183-207
Zietkiewicz E, Yotova V, Jarnik M, Korab-Lakowska M, Kidd KK, Modiano D, Scozzari R, Stoneking M, Tishkoff S, Batzer M, Labuda D (1998) Genetic structure of the ancestral population of modern humans. J Mol Evol 47:146-155

Subject Index

A

B

E

F

G

H

I

J

K

L

O

P

Q

R

S

T